Technology Policy and Its Impact on Green Governance and Sustainability

Penpim Phuangsuwan
https://orcid.org/0000-0003-2371-7998
University of Phayao, Thailand

Parichat Jaipong
https://orcid.org/0000-0002-9249-3169
Manipal GlobalNxt University, Malaysia

IGI Global
Scientific Publishing
Publishing Tomorrow's Research Today

Vice President of Editorial Melissa Wagner
Director of Acquisitions Mikaela Felty
Director of Book Development Jocelynn Hessler
Production Manager Mike Brehm
Cover Design Jose Rosado

Published in the United States of America by
IGI Global Scientific Publishing
701 East Chocolate Avenue
Hershey, PA, 17033, USA
Tel: 717-533-8845 | Fax: 717-533-7115
Website: https://www.igi-global.com E-mail: cust@igi-global.com

Library of Congress Cataloging-in-Publication Data

Names: Phuangsuwan, Penpim, 1972- editor | Jaipong, Parichat, 1992- editor
Title: Technology policy and its impact on green governance and sustainability / edited by: Penpim Phuangsuwan, Parichat Jaipong.
Description: Hershey PA : IGI Global Scientific Publishing, [2026] | Includes bibliographical references. | Summary: "This book examines how technology policy drives green governance and sustainable development. It explores the role of digital innovation, regulatory frameworks, and eco-friendly solutions in advancing low-carbon economies, resource efficiency, and environmental protection, offering strategies for building resilient and sustainable systems worldwide"-- Provided by publisher.
Identifiers: LCCN 2025041865 (print) | LCCN 2025041866 (ebook) | ISBN 9798337368214 hardcover | ISBN 9798337368221 paperback | ISBN 9798337368238 ebook
Subjects: LCSH: Technology and state | Sustainable development--Government policy
Classification: LCC T49.5 .T447 2026 (print) | LCC T49.5 (ebook)
LC record available at https://lccn.loc.gov/2025041865
LC ebook record available at https://lccn.loc.gov/2025041866

British Cataloguing in Publication Data
A Cataloguing in Publication record for this book is available from the British Library.

Table of Contents

Detailed Table of Contents

Community enterprises make significant contributions to the social economy by providing innovative services, promoting economic growth, and creating future opportunities. This chapter examines the determinants affecting the sustainability of community enterprises and the social economy in Thailand. Using a quantitative approach, data were collected from 1,240 respondents via an online questionnaire and analyzed through binary regression using convenience sampling. Results reveal that age, gender, indebtedness, use of digital communication platforms, and overall readiness influence sustainability. Strengthened governmental and administrative support, along with corporate commitment to sustainable leadership and corporate social responsibility, are vital for long-term sustainability. This study extends the existing literature on factors influencing community enterprises and provides a framework for future research.

This chapter presents a systematic review that shows a technology-oriented e-waste governance framework emphasizes the use of technology to oversee the entire lifecycle of electronics—from manufacturing to disposal—through policies such as Extended Producer Responsibility (EPR), fostering a circular economy, and employing technology for collection and recycling. Essential elements encompass implementing Extended Producer Responsibility (EPR) to ensure manufacturer accountability, promoting a circular economy that emphasizes repairability and recyclability, and utilizing modern technologies, such as artificial intelligence, for electronic waste processing. This strategy integrates legislative measures, technological advancements, and economic incentives to reduce environmental damage and reclaim valuable resources.

This chapter covers the evolution, mediation, moderation, and research directions of green HRM. PRISMA-based systematic literature reviews excluded 549 papers, leaving 140. Analysis of themes involved comparisons. GHRM fosters financial, operational, and innovative performance and well-being (mental health, contentment, engagement, retention) through employee involvement, green behavior, green innovation, and organizational culture. AI, context, and digital technology improve these encounters. This unique analysis presents GHRM research from 2006 to 2025, integrates ecology-inclusivity (DEI) into Ability, Motivation, Opportunity, and Social Exchange Theory, and prioritizes research. The theoretical implications support GHRM as a social-exchange-based mechanism that uses the Ability-Motivation-Opportunity pathway to turn HR skills into sustainable advantages and requires a model that balances economic, social, and environmental aims.

Technology policy is essential for promoting sustainability and green governance in various sectors and countries. Green practices, renewable energy solutions, and resource-efficient systems are adopted through technology policy, which creates legislative frameworks, incentives, and innovation pathways. This study investigates how strategic technology policies support circular economy models, impact environmental management, and facilitate digital breakthroughs such as blockchain, AI, and IoT to improve sustainability outcomes. Additionally, it examines how well policy frameworks align with international environmental objectives, such as the Paris Agreement and the Sustainable Development Goals (SDGs) of the UN. The results show that well-designed technology policies increase institutional capacity for green governance and hasten the shift to low-carbon economies, guaranteeing long-term ecological balance and socioeconomic resilience.

Governance models are needed in this age when climate change, resource depletion, and environmental degradation are now accelerating the pace of life. This chapter is a systematic review method using thematic-content analysis, in reality, critically analyzes public-private collaborative ventures as public-private partnership (PPP) matters, positioning them as the device to connect government policy frameworks with private sector technological expertise and investment capacity. With renewable energy and waste management, among other low-carbon technology applications, coupled with much public ground governance, one finds PPPs at the interface between technology policy and a green governance agenda. The chapter delves into the dynamics of government factors, barriers, and replicable strategies. It applies them here using the case-study methodology of the Rewa Ultra Mega Solar Project and the Indore Smart City waste management initiative.

This chapter discusses the disruptive role of AI in governance and sustainable policymaking regarding climate issues. It examines the role of AI in both defending and making evidence-based decisions, carbon accounting, optimization of renewable energy, decreasing and decarbonizing transportation, and strengthening urban resilience, as well as informing international and regional collaboration. This chapter employed a systematic review. The findings show examples of AI implementation across different countries reveal its potential to enhance emissions reporting, policy

Foreword

To steer green governance and sustainability efforts, technology policy is essential. It does this by establishing the legal frameworks and incentives required for innovation and the broad adoption of environmentally friendly practices. Policy could be used to hasten the transition to a low-carbon economy by giving research and development in areas such as sustainable agriculture, smart grids, and renewable energy a higher priority.

Important policy instruments, such as carbon pricing, green subsidies, and energy-efficiency requirements, directly impact market dynamics, making environmentally friendly products more economically viable and competitive. Furthermore, good governance is primarily dependent on technology for thorough environmental monitoring, data gathering, and transparent reporting.

This helps ensure compliance and enables decision-making based on evidence. The widespread influence extends to urban planning through the implementation of smart city programs that maximize resource utilization and minimize waste. For the purpose of creating resilient systems and achieving a harmonious balance between economic development and ecological preservation, it is ultimately vital to have a technology strategy that is forward-thinking and integrated with environmental aims.

Thus, this book, "Technology Policy and Its Impact on Green Governance and Sustainability," may assist academicians, industry, policymakers, researchers, and stakeholders in improving Green Governance based on its policies to achieve sustainability across educational, industry, and related sectors.

Supaprawat Siripipatthanakul

Faculty of Public Health at Bangkokthonburi University, Thailand & School of Business Administration, GlobalNxt University, Malaysia & Faculty of Economics and Business, Universitas Terbuka, Indonesia

Preface

AN OVERVIEW OF THE SUBJECT MATTER

By guiding the development and adoption of environmentally friendly innovations, fostering resource efficiency, reducing emissions, and improving environmental monitoring through digital tools, technology policy significantly impacts green governance and sustainability. However, challenges such as affordability and potential job impacts require careful policy design in order to strike a balance between economic growth and ecological preservation. To achieve sustainability objectives, policies that create effective frameworks for green technology, accelerate transitions to low-carbon technologies, and strengthen accountability within institutions and enterprises are essential.

A description of the topic fits in the world today

Today, technology policy is essential for guiding green governance and sustainability. It does this by providing incentives for green technology, establishing environmental standards (such as carbon pricing), and encouraging innovation for low-carbon economies. However, to meet global goals such as the Sustainable Development Goals (SDGs), it must strike a balance between the benefits of technology (clean energy, efficiency) and the footprint it leaves (e-waste, resource use). Impacting everything from energy grids to waste management and employment markets, effective policies integrate legislation, investment, and international collaboration to guide transitions towards resource efficiency, circularity, and sustainable development. These changes affect everything from job markets to energy grids.

Target Audience

* Local, national, and international legislative and executive branch officials (e.g., environmental protection agencies, energy departments, city planning offices) who design, implement, and regulate technology policies and green initiatives. Tech company executives, R&D teams, and strategists who develop green technologies (e.g., solar, wind, smart grids) and integrate sustainability into their business models and product development.
* Environmental scientists:
* University and research institution academics and researchers who investigate climate change, technology's ecological implications, and the efficacy of green policies.
* NGOs and environmental advocates, such as Greenpeace, campaign for stronger environmental regulations, evaluate environmental performance, and educate the public.
* Venture capitalists, asset managers, and banks are interested in sustainable finance, green bonds, and ESG considerations in their portfolios
* Urban planners and architects: Specializing in green building, sustainable city design, and smart city technology for resource efficiency
* The Public and Community Leaders: Informed citizens, local activists, and community leaders interested in local green initiatives, policy advocacy, or adopting sustainable technologies

Description

As the world faces escalating environmental challenges, the intersection of technology policy and green governance has become increasingly critical to achieving sustainability goals. Technology policy plays a pivotal role in directing the development and deployment of green technologies. These policies influence how societies transition to low-carbon economies, manage natural resources, and promote environmental accountability. Technology policy affects green governance structures and sustainability outcomes.

Technology Policy and Its Impact on Green Governance and Sustainability explores both governmental institutions and business entities shifting towards low-carbon and environmentally sustainable practices. This book promotes sustainable environmental practices within enterprises. Covering topics such as sustainability, green governance, and technology policy, this book is an excellent resource for policymakers, organizational leaders, technology developers, scholars, and educators.

Coverage

The many academic areas covered in this publication include, but are not limited to:

- AI Frameworks
- AI Surveillance
- Blockchain
- Climate Culture
- Community Enterprise
- Cyber Security
- Energy Policy
- Environmental Factors
- Financial Literacy
- Green Banking
- Green Energy
- Green Finance
- Renewable Energy
- Sustainability

The book includes 11 chapters that could empower academic leaders and strategic planners in strategic planning for *Technology Policy and Its Impact on Green Governance and Sustainability*, encouraging researchers to adopt collaboration, creativity, communication, and critical thinking among researchers, organizations, and industries.

Chapter 1. Community Enterprise and Social Economy for Sustainability in a Developing Country

Yarnaphat Shaengchart, Pongsakorn Limna, Tanpat Kraiwanit, Supaprawat Siripipatthanakul, and Kris Jangjarat

This chapter discusses the social economy benefits from community entrepreneurs' innovative services, economic growth, and future potential. This chapter analyzes the sustainability factors of Thai community enterprises and the social economy. Quantitative data from 1,240 respondents to an online questionnaire were examined using binary regression. Results show that age, gender, indebtedness, digital communication, and readiness affect sustainability. Sustainability requires government and administrative assistance, sustainable leadership, and corporate social responsibility. It expands the knowledge on community enterprise factors and provides a platform for future research.

Chapter 2. Technology-Driven Organization Policies: E-waste Governance Framework

Astri Dwi Andriani and Supaprawat Siripipatthanakul

This chapter discusses e-waste governance, emphasizing the use of technology to manage the entire lifecycle of electronics, from manufacturing to disposal, through Extended Producer Responsibility (EPR), a circular economy, and collection and recycling technologies. Extended Producer Responsibility (EPR) to hold manufacturers accountable, a circular economy that prioritizes repairability and recyclability, and artificial intelligence for electronic waste processing are essential. This technique reduces environmental damage and recovers resources through legislation, technology, and economic incentives.

Chapter 3. Integrating Green HRM Policies with AMO & SCT Theory to Promote Sustainable Performance and Employee Well-being

Syamsul Hadi, Kusuma Chandra Kirana, Ambar Lukitaningsih, Utik Bidayati, and Dasmadi Dasmadi

This chapter discusses the development, mediation, moderation, and research of Green HRM. Topic analysis involved comparisons. GHRM promotes financial, operational, and innovative performance and well-being (mental health, contentment, engagement, retention) through employee involvement, green behavior, green innovation, and organizational culture. AI, context, and digital technologies enhance these interactions. This unique approach incorporates ecology-inclusivity (DEI) into Ability, Motivation, Opportunity, and Social Exchange Theory, prioritizing research. The theoretical implications support GHRM as a social-exchange-based mechanism that employs the Ability-Motivation-Opportunity pathway to turn HR skills into sustainable advantages and requires a model that balances economic, social, and environmental goals.

Chapter 4. Digital Infrastructure Transforming Healthcare, Cities, and Supply Chain: Green Governance Policies for Sustainable Resource Management

Sharmila A, Raja Shree K. C, and Gigi G. S

This chapter discusses sustainability and green governance in numerous industries and governments that require technology policy. Technology policy creates legal frameworks, incentives, and pathways for innovation in green practices, renewable energy, and resource-efficient systems. This paper examines how strategic technology policies strengthen circular economy models, environmental management, and digital innovations like blockchain, AI, and IoT to increase sustainability. It also assesses the alignment of policy frameworks with international environmental goals, such as the Paris Agreement and the UN SDGs. Well-designed technology policies boost institutional capacity for green governance and accelerate the transition to low-carbon economies, ensuring long-term ecological balance and socioeconomic resilience.

Chapter 5. Collaborative Ventures between Public and Private Sectors in Technology and Sustainability

Ashka Prince, Babita Chaudhary, and Charu Sehgal

This chapter discusses rapid climate change, resource depletion, and environmental degradation, and the need for governance models. In actuality, this chapter critically studies public-private collaboration ventures as public-private partnership (PPP) matters, positioning them to combine government policy frameworks with private sector technology competence and investment capacity. PPPs sit at the intersection of technological policy and green governance, with renewable energy, waste management, and other low-carbon technology applications and public ground governance. The chapter examines government factors, challenges, and replicable techniques using the Rewa Ultra Mega Solar Project and Indore Smart City waste management effort as case studies.

Chapter 6. AI-Driven Policy Frameworks for Net-Zero Economies: Digital Pathways to Decarbonization

Sitesh Kumar Singh and Jaishri Gothania.

This chapter discusses how AI disrupts climate governance and sustainable policymaking. AI is used to defend, make evidence-based judgments, perform carbon accounting, optimize renewable energy, reduce and decarbonize transportation, increase urban resilience, and guide international and regional collaboration. AI can improve emissions reporting, policy modeling, and net-zero transitions, as shown by international examples. They highlight ethical governance, equitable access, and data sovereignty as potential concerns. The chapter concludes that AI can alter finance, energy, and cities in second-generation climate solutions. To accelerate resilient and net-zero futures, it advocates inclusive governance structures that can implement AI-driven policies that are transparent, participative, and internationally equitable.

Chapter 7. Revisiting Environmental Security and Climate Justice: A Human Security Perspective

Jipson Joseph, Jobin Joseph, and S Srinivasan

This chapter discusses how life depends on a safe and healthy environment. Human interactions—individual, social, and economic—depend on the ecosystem. However, humanity cruelly exploited the biosphere beyond its regenerative potential, damaging environmental sustainability. Profit-seeking and egoism triggered unprecedented environmental destruction, including global warming, biodiversity loss, and anthropogenic climate change. This precarious position forced world leaders to prioritize environmental security under the 1987 sustainable development concept. In 1994, the UNDP proposed the human security paradigm, which includes environmental security. The Millennium and Sustainable Development Goals reflect a people-centered approach to human security. This chapter critically examines environmental security and climate justice through the human security paradigm.

Chapter 8. The Legal Architecture of Energy Diplomacy: India-China Pathways Toward Green Governance and Sustainable Transitions

Abhinash Mohanta, Shaikh Sahanwaz Islam, and Nikita Biswal

This chapter compares the legal architecture of energy diplomacy in India and China, focusing on how both countries balance foreign policy, sustainability, and clean technology governance. It contends that energy diplomacy has shifted from a geopolitical to a legal and technological focus, shaping national and global green transitions. The analysis compares treaties, national laws, and case law to show gaps and convergence in climate accountability, technology transfer, and environmental justice. It advocates "Green Law Diplomacy" as a normative foundation of international governance, favoring enforceable legal instruments over voluntary climate promises for equitable and sustainable transitions.

Chapter 9. Shaping Sustainable Growth with AI in Technology-Driven Organizations by

Pongsakorn Limna

This chapter discusses how AI revolutionizes green and sustainable growth in technology-driven companies. The study finds that AI improves operational efficiency, resource utilization, and the transition to a circular economy. The findings show that AI applications dramatically minimize environmental impacts and boost economic competitiveness. Governance and policy frameworks ensure transparency, accountability, and ESG compliance, enabling responsible AI use. AI, with the development of technologies, has also opened new doors for predictive governance and sustainability innovation. The study highlights AI's potential as a foundation for organizational transformation toward long-term ecological resilience and sustainable development, despite regulatory fragmentation, high implementation costs, and ethical concerns.

Chapter 10. The Net Zero Transition for Green HRM and Governance Policies

Syamsul Hadi, Penpim Phuangsuwan, Lin Dar Ong, Saizal Pinjaman, Auliya Rosiana, and Ambar Lukitaningsih

This chapter discusses the consolidated conceptual advancements in Net Zero Human Resource Management (HRM). It analyzes the integration of green employment, training, and performance evaluation with senior-level green governance policies to promote low-carbon organizational management strategies. This review demonstrates a favorable correlation between green human capital and green recruiting and training. In contrast, the green performance appraisal process is significantly more troublesome for various reasons, particularly due to persistent measurement challenges.

Chapter 11. Corporate Sustainability and Green Governance Environmental Policy Effectiveness

Syamsul Hadi, Parichat Jaipong, Saizal Pinjaman, Lin Dar Ong, Kusuma Chandra Kiran, and Ambar Lukitaningsih

This chapter discusses corporate sustainability, impacts, emissions, and waste policy. Coded governance, regulation, and environment are discussed. Analysis demonstrates widespread "institutional decoupling," in which governance rarely influences operations and compliance is symbolic. Circular economy and markets may cut industry emissions. Generic regulations that overlook high-emission industries may lead to greenwashing. Epistemological Paradox: Self-reported data hides greenwashing risks despite methodological advances. Business aims and conduct differ in stakeholder and Neo-Institutional frameworks. Future models should include internal governance and external constraints to forecast performance. Policymakers must advise sectors, not horizontally.

Through thoughtful collaboration among researchers and professors on these diverse themes, "*Technology Policy and Its Impact on Green Governance and Sustainability*" provides academic leaders and strategic planners with ways to strengthen organizational performance for green governance and sustainability.

Penpim Phuangsuwan
University of Phayao, Thailand

Parichat Jaipong
Manipal GlobalNxt University, Malaysia

Chapter 1
Community Enterprise and Social Economy for Sustainability in a Developing Country

Yarnaphat Shaengchart
https://orcid.org/0000-0002-0157-3682
Pathumthani University, Thailand

Pongsakorn Limna
https://orcid.org/0000-0002-7448-5261
Pathumthani University, Thailand

Tanpat Kraiwanit
Pathumthani University, Thailand

Supaprawat Siripipatthanakul
https://orcid.org/0000-0001-6671-2682
Bangkokthonburi University, Thailand

Kris Jangjarat
Pathumthani University, Thailand

ABSTRACT

Community enterprises make significant contributions to the social economy by providing innovative services, promoting economic growth, and creating future opportunities. This chapter examines the determinants affecting the sustainability of community enterprises and the social economy in Thailand. Using a quantitative approach, data were collected from 1,240 respondents via an online questionnaire

DOI: 10.4018/979-8-3373-6821-4.ch001

and analyzed through binary regression using convenience sampling. Results reveal that age, gender, indebtedness, use of digital communication platforms, and overall readiness influence sustainability. Strengthened governmental and administrative support, along with corporate commitment to sustainable leadership and corporate social responsibility, are vital for long-term sustainability. This study extends the existing literature on factors influencing community enterprises and provides a framework for future research.

1. INTRODUCTION

Community enterprise has emerged as a transformative driver for local economic systems, heralding new pathways for sustained growth while epitomizing the profound influence of information technology (IT) on the digital economy. By integrating traditional community values with modern technological advancements, community enterprises not only stimulate local markets but also build resilience against global economic fluctuations. For the realization of enduring economic targets, it is imperative that the aspirations of community leaders harmonize with digital marketing strategies that are bolstered by contemporary IT solutions, enabling broader outreach, efficient resource allocation, and enhanced competitiveness in regional and global markets (Kim & Lim, 2017; Romprasert & Trivedi, 2021; Tariq, 2025). Such enterprises, characterized by collective community efforts to achieve mutual objectives, are deeply rooted in local entrepreneurial ventures that capitalize on social assets, frameworks, and networks. These ventures embody a participatory model of economic development, where governance is community-driven and benefits are distributed to ensure long-term communal and individual welfare (Petcho et al., 2019; Wongadisai et al., 2020; Yu et al., 2025).

Within this framework, the social economy emerges as an innovative sector that harmonizes social innovation with community-centered problem-solving, giving rise to hybrid business models that prioritize social objectives over financial gains while still maintaining economic viability (Chaves & Monzón, 2012; Pongracz, 2020). This paradigm shift underscores the capacity of community enterprises to serve as engines of inclusive development, where profit is balanced with equity, sustainability, and social justice. Concurrently, the academic sphere has witnessed a surge in sustainable development studies, with a particular emphasis on the triadic pillars of sustainability—social, economic, and environmental. The Triple Bottom Line (TBL) framework offers a critical lens for balancing resource demands across generations, ensuring that growth is achieved without depleting natural resources or exacerbating social and economic inequalities. In this context, entrepreneurial acumen and innovation play pivotal roles in identifying, designing, and exploiting opportunities conducive to sustainable progress, where business growth is aligned with ecological responsibility and social advancement (Atiku et al., 2024; Bansal et al., 2019; Khan et al., 2021; Khiawnoi et al., 2025; Nogueira et al., 2025).

According to Montiel (2008) and Tien et al. (2019), the distinction between corporate sustainability and traditional corporate social responsibility (CSR) requires emphasis. While CSR often represents a set of voluntary, philanthropic, or compliance-oriented initiatives, corporate sustainability embodies a holistic, long-term strategic approach that integrates sustainability into the core of organizational operations. It not only balances economic performance with societal contributions but also embeds environmental stewardship, social equity, and community welfare as central components of corporate decision-making and governance. This shift reflects a growing recognition that sustainable success cannot be achieved solely through financial outcomes; rather, it requires creating shared value that addresses the needs of multiple stakeholders, including employees, communities, and the natural environment (Pfajfar et al., 2022; Yusif & Hafeez-Baig, 2024).

The advent of Industry 4.0 (I4.0), characterized by its IT-driven advancements, presents unprecedented opportunities for sustainable industrial practices. By harnessing the integration of cyber-physical systems, artificial intelligence (AI), the Internet of Things (IoT), and advanced analytical tools, enterprises can revolutionize production processes, reduce resource consumption, and minimize environmental impacts. For instance, smart manufacturing enabled by IoT can optimize energy use, predictive analytics can enhance supply chain efficiency, and AI-driven insights can support sustainable product design. Collectively, these innovations foster a paradigm shift toward data-driven, adaptive, and resource-efficient practices, strengthening both organizational competitiveness and ecological responsibility (Bouyahrouzi et al., 2025; Hadi et al., 2025; Khan et al., 2021; Khan et al., 2025; Pivoto et al., 2021; Siripipatthanakul et al., 2025; Srivastava et al., 2025; Zong & Guan, 2025).

Despite these advancements, a paucity of research remains on individual perceptions and behavioral factors that impact the sustainability of community enterprises and the social economy. These entities are vital for fostering job creation, social innovation, and economic resilience, particularly in developing countries where local enterprises form the backbone of livelihoods. This study, therefore, seeks to elucidate the determinants—including demographic, financial, technological, and readiness-related variables—that shape the sustainability of community enterprises within the digital economic framework. The insights generated are expected to serve as instrumental benchmarks for future social economy projects, guiding both policy development and practical interventions. Moreover, enhanced governmental and administrative backing for community enterprises and the social economy is essential to catalyze economic improvements, reinforce resilience, and bolster long-term sustainability.

1.1 Research Objective

This study aims to investigate the key determinants influencing the sustainability of community enterprises and the social economy in Thailand.

1.2 Research Question

What factors significantly affect the sustainability of community enterprises and the social economy in Thailand?

In pursuit of the study's objective, this manuscript is structured into six segments to ensure a comprehensive exploration of the research problem. The first segment provides an introductory overview, situating the study within contemporary debates and highlighting its relevance and significance. The second segment presents the theoretical underpinnings, critically examining existing frameworks and concepts that inform the research. The third segment offers a detailed methodological exposition, elaborating on the positivist, quantitative approach, including the research design, sampling strategy, data collection instruments, and analytical procedures employed. The fourth segment presents the results and analysis, systematically interpreting the empirical findings and identifying key patterns and trends. The fifth segment presents a thorough discussion, comparing the findings with those of previous research. Finally, the sixth segment concludes the manuscript by summarizing the study's contributions, acknowledging limitations, and providing strategic recommendations for scholars, practitioners, and policymakers to inform future research and practice.

2. LITERATURE REVIEW

Over recent decades, scholarship has increasingly affirmed that environmental contexts exert a profound influence on entrepreneurial activities, with recent emphasis—particularly over the last five years—on delineating the elements that constitute entrepreneurship ecosystems and their efficacy in fostering high-growth enterprises. Although prior research has acknowledged the heterogeneity of environmental factors such as policies, financing, cultural attributes, labor market structures, research and development (R&D), and market dynamics across various regions and countries, there remains an exigent need to explore how these factors differentially influence the stages of entrepreneurship within specific regional and national contexts, especially in relation to the technological dimensions of entrepreneurial ventures. In their empirical investigation, Guerrero et al. (2021) examined the environmental determinants that facilitate or hinder the evolution of individuals from potential entrepreneurs to nascent, new, and eventually established and consolidated business figures in both developed and developing nations. This analysis highlights the importance of contextual and demographic variability in entrepreneurship, identifying key positive influencers, including professional support frameworks, incubators, accelerators, diverse networking opportunities, and robust R&D funding. Conversely, identified detriments encompass insufficient financing avenues, restrictive labor market conditions, and adverse social norms. Complementarily, Fatimah et al. (2020) endeavored to develop a smart and sustainable waste management framework at the national level by applying Industry 4.0 technologies, addressing pressing environmental and operational challenges. Their comprehensive approach encompassed a systematic review of existing literature, focus groups, and empirical observations within Indonesian urban settings, culminating in the formulation of a novel strategy for waste management. This strategy not only assessed the technical maturity of existing waste systems but also proposed innovative business processes and systems designs for waste management. The processes they outlined include waste collection, segregation, transport, treatment, and disposal, all framed within the context of a circular economy. Notably, their proposed system design utilizes IoT technology for the separation and characterization of municipal waste, while also identifying viable technologies for sustainable waste treatment. This study illustrates a multifaceted approach towards integrating technological advancements with environmental sustainability in addressing complex waste management issues.

Furthermore, Khiawnoi et al. (2025) highlight the pivotal role of community enterprises in promoting sustainable development, noting that the lack of comprehensive strategic frameworks often limits their capacity to integrate economic, social, and environmental objectives effectively. To address this gap, the study examines the strategic roles undertaken by community enterprise entrepreneurs through a mixed-methods approach, combining qualitative insights from in-depth interviews and participatory observations with quantitative analysis of 215 standardized survey responses. The findings identify five critical strategic roles—vision and goal-setting, resource management, networking and collaboration, innovation leadership, and strategic leadership—each receiving high mean ratings between 4.45 and 4.54 on a five-point Likert scale. Qualitative evidence further demonstrates how entrepreneurs employ these strategies to optimize resources, foster cross-sector partnerships, drive innovation, and enhance organizational resilience. Based on these insights, the study recommends strengthening infrastructure, developing entrepreneurial capacities, and promoting multi-sectoral ecosystems to support community enterprises. Importantly, the research contributes to theoretical discourse by proposing a five-dimensional strategic entrepreneurship model for sustainable community development, providing practical guidance for policymakers and entrepreneurs seeking to cultivate resilient, inclusive, and sustainable local economies.

The expansion of China's resource-based cities has accentuated the tensions between the social economy and the ecological environment, spotlighting the necessity for their concurrent development to ensure sustainable urban progression. In addressing these issues, Li et al. (2020) implemented a coupling coordination degree model to evaluate the simultaneous growth of the social economy and the ecological backdrop in 19 resource-based cities within Northeastern China. Their findings reveal a less-than-optimal state of harmonized development, primarily attributed to the lagging progression of the social economy in relation to the ecological environment. This disparity highlights the urgent need for balanced development strategies that address both economic and environmental concerns equally. Further enriching the discourse, Tien et al. (2019) explored the dynamics of social enterprise and social entrepreneurship, highlighting their potential to forge and leverage social capital for the nurturing of entrepreneurial ventures. By delineating the prevalent forms of entrepreneurship across different regions, their research offers insightful strategies for intertwining social and commercial entrepreneurial efforts, particularly in rural settings. This approach proposes a model for fostering symbiosis between economic ventures and social well-being.

Additionally, Nguyen et al. (2021) delved into the evolving landscape of CSR and its significance in modern business practices, particularly with regard to sustainability. Their study illuminates the intricate interplay between human behavior in commercial contexts and the broader societal obligations, advocating for a nuanced approach to CSR that aligns with contemporary economic and social shifts. This research highlights the increasing importance for businesses to integrate social responsibility into core management practices, thereby fostering a more sustainable and ethically conscious business model. Collectively, these studies contribute to the complex tapestry of sustainable development, highlighting the multifaceted challenges and opportunities at the intersection of economic growth, social equity, and environmental stewardship. They underscore the urgent need for integrated approaches that prioritize balanced development and the ethical dimensions of entrepreneurship and corporate conduct.

Terdpaopong and Kraiwanit (2021) conducted an evaluative study to identify the elements that shape digital technology readiness within Thailand's Office of the Welfare Promotion Commission for Teachers and Educational Personnel (OTEP), a pivotal governmental agency. The research identified several independent variables, including age, technological literacy, experience with technology, attitudes towards technology, organizational culture, leadership quality, and availability of learning facilities, setting digital technology readiness as the dependent outcome. The analysis revealed that OTEP exhibits a moderate level of readiness towards digital technology, pinpointing learning environments, technological literacy, leadership, and organizational culture as significant determinants. This diagnosis indicates existing hurdles and constraints in advancing human resource capabilities, especially in digital technology realms, within substantial government-controlled organizations.

Concurrently, Romprasert and Trivedi (2021) explored the use of IT for promoting digital marketing within community enterprise frameworks, aiming to integrate knowledge for effective digital marketing strategies that support a sustainable economy. Their investigation defines community enterprise as a structured form of collective action distinct from general community endeavors, highlighting the crucial role of effective management in its success. The study asserts that the management of community resources, driven by social and cultural dynamics, is integral to achieving sustainable development objectives. Community enterprise initiatives, rooted in local wisdom and civic involvement, engender substantial communal advancements. Nonetheless, the impediments, such as lackluster regulatory mechanisms, can stifle the growth of community enterprises. The researchers suggest that overcoming these barriers requires active engagement and leadership from community figureheads, who should foster awareness and appreciation for local ecological and resource values. Recommendations include pursuing community enterprises with a deep-seated spiritual dedication and adherence to socio-cultural principles, while also navigating governmental statutes and regulations to cultivate a synergistic alliance between community enterprises and external bodies, thereby facilitating a fruitful exchange of knowledge within the digital economic landscape.

In a pioneering study conducted by Younis (2023), the significance of sustainability reports and their subsequent effect on a firm's value were scrutinized. To evaluate the firm's value, the research employed four distinct metrics: Tobin's Q, market value per share, the price/book value ratio per share, and security return. The study introduced two innovative indicators designed to assess both the level and quality of sustainability report disclosures, utilizing content analysis for this purpose. Results from regression analysis illustrated a positive association between the extent and quality of sustainability disclosures and the overall value of the firm. Younis's findings advocate for the establishment of a dedicated accounting standard specifically for sustainability report disclosures. This suggests a systematic approach to enhancing transparency and accountability in business practices related to sustainability. Furthermore, the study recommends integrating sustainable development principles into the academic curricula of Saudi universities, emphasizing the importance of educating future business leaders about sustainability. The research also underscores the necessity of convening dialogues among accountants, members of the Saudi capital market, and firms. These meetings aim to address and overcome the challenges encountered in preparing comprehensive sustainability reports, thereby fostering a deeper understanding and broader implementation of sustainable business practices. This study makes a significant contribution to the discourse on sustainability in business, highlighting the tangible benefits of sustainability reporting on firm valuation and advocating for systemic changes to promote sustainable development.

Wang et al. (2024) examine the potential of government-enterprise green collaborative governance as a long-term mechanism for addressing the climate crisis and promoting sustainable development. Recognizing that reliance solely on governments or enterprises is influenced by economic interests and regulatory efficiency, the study investigates whether private capital involvement in government-led environmental initiatives can enhance collaborative governance and reduce urban carbon emissions. Utilizing data on public-private partnerships (PPPs), ChatGPT-assisted identification of green PPP projects, and a generalized difference-in-differences framework based on China's 2016 Guidance on Building a Green Financial System, the study demonstrates that government-enterprise collaborative governance significantly mitigates urban carbon emissions. The results remain robust across multiple tests, including adjustments for climate policies, variable settings, and self-selection issues. Heterogeneity analyses reveal that the effects of carbon reduction vary with both project characteristics—such as return mechanisms, investment volume, and cooperation term—and urban attributes, including geographical location, city type, and size. Mechanism tests further indicate that collaborative governance influences emissions primarily through structural, technological, and co-investment effects. Overall, the study offers a valuable framework for fostering effective co-governance between government and enterprises, leveraging market mechanisms to optimize resource allocation and mitigate climate risks.

Ramadani et al. (2022) conducted a comprehensive investigation to identify the sustainable competencies essential for the success of social entrepreneurs in diverse locations of Uttar Pradesh, India. Employing methodologies such as in-depth interviews, exploratory factor analysis, and questionnaires, the study successfully identified eight essential sustainable competencies that are instrumental for social entrepreneurs aiming to foster sustainable development. These competencies include entrepreneurial resilience, empathy, emotional intelligence, and strategic thinking, each of which is underscored as a foundational trait for the effective and impactful practice of social entrepreneurship. In a parallel scholarly endeavor, Suriyankietkaew et al. (2022) delved into the domain of sustainability leadership within community-based social enterprises, specifically targeting the underprivileged sectors at the bottom of the pyramid, with a focused lens on the challenges presented by the COVID-19 pandemic. Their research uncovered the core principles, values, and behaviors essential for sustainability leaders operating within the challenging environments of underdeveloped small and medium-sized enterprises (SMEs). The findings advocate for a leadership paradigm that encompasses long-term visioning, profound care for people and stakeholders, ethical and innovative business management, and the harmonious balancing of social, environmental, and economic imperatives to steer towards sustainable growth. Moreover, the authors accentuated the critical necessity for the development of additional competencies aimed at enhancing the skills and abilities associated with sustainability leadership. This aspect highlights the ongoing and evolving nature of leadership in the sustainability context, emphasizing that continuous learning and adaptation are crucial for navigating the complex landscape of community-based social entrepreneurship, particularly during and post-crisis scenarios, such as the COVID-19 pandemic. Collectively, these studies significantly contribute to the expanding body of knowledge on sustainable competencies and leadership, offering valuable insights for practitioners and policymakers seeking to advance sustainable development through social entrepreneurship.

Wang and Wang (2023) examine the vertical management of environmental policy in China, where the central government formulates policies that are implemented by local governments, positioning provincial green governance as a key driver of enterprise green investment. The study examines the impact of provincial green governance on enterprise green investment and investigates the mediating role of government subsidies. By focusing on enterprise green investment rather than green innovation, this research complements and extends existing studies. Using data from corporate social responsibility reports of high-polluting enterprises listed on stock exchanges from 2012 to 2017, the study assesses the impact of provincial green governance, measured through three specific indicators, on enterprise green investment and the role of government subsidies. Findings indicate that provincial green governance significantly promotes green investment in enterprises, and government subsidies further strengthen this effect. The study provides valuable insights for fostering collaborative governance among government, enterprises, and the public, thereby supporting the balanced development of economic and environmental objectives.

3. METHODOLOGY

This study adopted a quantitative research methodology, grounded in the positivist paradigm, which emphasizes empirical evidence and observable phenomena as the foundation of valid knowledge, while deliberately eschewing metaphysical or theological explanations (Ahmadin, 2022). The positivist stance assumes that social phenomena can be studied with the same rigor as those in the natural sciences, enabling the use of standardized instruments and statistical techniques to derive generalizable insights. Within this framework, the study applied a structured survey design, ensuring consistency in responses and facilitating the collection of large-scale data that could be objectively measured, compared, and analyzed. This paradigm enables the objective measurement and analysis of variables, allowing for the testing of hypotheses and identification of causal relationships between determinants—such as age, gender, indebtedness, digital communication usage, and readiness—and outcomes related to the sustainability of community enterprises and the social economy. By employing statistical tools, particularly binary logistic regression, the study was able to quantify the extent to which these factors influence sustainability outcomes, while controlling for potential confounding variables. Such an approach enhances both the validity and reliability of the findings, providing robust evidence that can inform policy formulation, strategic interventions, and academic discourse. Furthermore, the adoption of a quantitative, positivist approach ensures replicability and transparency, as the procedures and measures can be systematically applied in future studies or across different contexts. This not only strengthens the study's contribution to the empirical literature on community enterprise sustainability but also offers a methodological template for subsequent research in similar socio-economic environments. By situating the investigation within the positivist tradition, the study underscores the value of measurable, evidence-based insights in shaping strategies for sustainable development in community enterprises.

3.1 Research Design

A structured survey design was employed to systematically gather quantitative data on the determinants of sustainability in community enterprises and the social economy. The questionnaire was composed primarily of closed-ended questions, which allowed for standardized responses across participants and facilitated statistical comparison and analysis. This design ensured both consistency and objectivity, reducing interviewer bias and enhancing reliability. Prior to full-scale deployment, a pilot test was conducted with 30 respondents to assess the clarity, reliability, and validity of the items. Feedback from this pilot phase informed revisions to eliminate ambiguities, refine wording, and ensure cultural appropriateness. Both content validity and construct validity were evaluated using established research metrics, with items reviewed by subject matter experts to confirm alignment with the intended constructs related to demographic factors, financial behaviors, social media usage, and organizational readiness. By using this structured approach, the research design enhanced internal validity and provided a solid foundation for analyzing causal relationships through regression techniques.

3.2 Population and Sampling

The target population consisted of Thai residents aged 18 years and above, encompassing individuals who were either directly engaged in or indirectly affected by community enterprises and the broader social economy. This inclusive criterion ensured that perspectives were gathered from a diverse pool of participants, capturing a wide range of demographic, social, and financial backgrounds. To determine an adequate sample size, Cochran's formula was applied at a 95% confidence level and a 0.05 precision level, yielding a minimum requirement of 384 respondents (Uakarn et al., 2021). To enhance robustness, the initial target was expanded to 1,240 participants, thereby reducing the risk of sampling error and accommodating possible non-responses. Convenience sampling, a non-probability technique, was employed to select participants due to the practical constraints of time and accessibility, particularly in reaching respondents across diverse geographic regions through online platforms. While this approach facilitated efficient data collection and ensured broad participation, it is acknowledged that it introduces limitations regarding representativeness and generalizability (Obilor, 2023; Shamsudin et al., 2024). Nevertheless, the achieved sample size aligns with recommendations for large-scale survey studies, thereby strengthening the reliability of the findings.

3.3 Variables and Measures

The study investigated multiple independent variables hypothesized to influence the sustainability of community enterprises and the social economy in Thailand. Demographic variables, including age, gender, and education, enabled the analysis of individual characteristics that may shape perceptions and practices. Financial factors were operationalized through the level of indebtedness, which reflected the economic stability and constraints faced by participants. Technological engagement was captured through frequency and type of social media usage (Instagram, Line, Pinterest, WhatsApp, YouTube, and Facebook), given the increasing role of digital platforms in entrepreneurial communication, networking, and marketing. Readiness was measured as participants' willingness and preparedness to adopt new practices, a key factor in determining adaptive capacity. The dependent variable was the sustainability of community enterprises and the social economy, operationalized as a binary outcome. This measurement approach enabled robust logistic regression modeling, allowing for the identification of significant predictors and their relative influence on sustainability outcomes.

3.4 Data Collection Procedure

Data were collected over a two-month period, from March to April 2025, using an online questionnaire distributed through multiple digital channels, including email invitations, community enterprise networks, and widely used social media platforms. This distribution strategy ensured broad geographic reach and access to diverse respondents across different age groups and socioeconomic categories. Participation was entirely voluntary, and respondents were provided with informed consent forms detailing the purpose of the study, data confidentiality measures, and their right to withdraw at any stage without penalty. To encourage honest and unbiased responses, anonymity was strictly maintained, and no personally identifiable information was collected. The online format facilitated real-time data collection, minimized logistical constraints, and aligned with the increasing digital integration of Thai communities. This method not only adhered to ethical research standards but also enhanced the reliability and efficiency of the data-gathering process.

3.5 Ethical Considerations

This study adhered to strict ethical protocols in conducting questionnaire-based research through an online survey. Before participating, all respondents were fully informed about the study's aims and assured that their responses would be used solely for academic purposes. Only individuals aged 18 years or older were included, and no vulnerable populations were involved, as the study did not involve medical procedures or interventions. Participation was entirely voluntary, and incomplete or withdrawn responses were excluded to preserve data integrity and voluntariness. Anonymity and confidentiality were rigorously maintained, with no personally identifiable information collected or disclosed. Participants provided informed consent after receiving clear explanations of the study's objectives, procedures, and any potential risks or benefits. Data were analyzed in aggregate form to prevent the identification of individual respondents. According to Thailand Science Research and Innovation (TSRI) guidelines, research in behavioral, social sciences, and humanities that does not impact the body, mind, cells, or genetic material—and which uses anonymous instruments such as questionnaires—is exempt from formal ethics committee approval (Guidance No. 3(3)) (Phuangsuwan et al., 2024). This study qualified for such an exemption while nonetheless fully adhering to ethical standards, ensuring the protection of participants' rights, dignity, and well-being throughout the research process.

3.6 Data Analysis

The collected data were analyzed using binary logistic regression, a statistical technique suitable for examining relationships between multiple independent variables and a binary dependent outcome. This method was selected because it allows for the estimation of the probability of sustainability (favorable vs. unfavorable) while controlling for the influence of confounding variables. Regression coefficients (B values), odds ratios (Exp(B)), and significance values were calculated to determine the relative contribution of each predictor. The robustness of the model was assessed through omnibus tests of model fit, R^2 statistics (Cox & Snell, Nagelkerke), and classification accuracy rates, ensuring that the results met statistical validity requirements (Boateng & Abaye, 2019; Gomila, 2021). This analytical approach enabled a nuanced interpretation of how demographic characteristics, financial behaviors, digital communication practices, and readiness levels shape sustainability outcomes. The findings provide not only theoretical contributions to the literature on social economy sustainability but also evidence-based insights for policymakers, community leaders, and corporate stakeholders aiming to design interventions that strengthen community enterprises.

4. RESULTS

The demographic profile of the respondents revealed that the majority were male and predominantly fell within the 31–40 years age bracket. A substantial portion of participants had attained a bachelor's degree as their highest level of education. In terms of financial status, a considerable number reported monthly incomes exceeding 30,000 baht, although many also indicated some level of indebtedness. Regarding digital engagement, the majority of respondents reported using online platforms on a daily basis, and a significant proportion actively engaged with multiple platforms simultaneously, reflecting high levels of technological adoption and integration into everyday life.

Table 1. Omnibus test of the model's performance using all independent variables

		Chi-square	Degrees of freedom (df)	Sig.
Step 1	Step	480.253	11	0.000
	Block	480.253	11	0.000
	Model	480.253	11	0.000

Based on Table 1, the chi-square value is 480.253, and the degrees of freedom (df) is 11. This suggests that all the independent variables in the study can explain the dependent variable at a significance level of 0.05.

Table 2. Model summary using all independent variables

Step	-2 log likelihood	Cox & Snell R-squared	Nagelkerke R square
1	1196.664[a]	0.322	0.434

a. Estimation terminated at iteration number 3 because parameter estimates changed by less than .001.

According to the data presented in Table 2, the regression model accounts for approximately 43.4% of the variance in the outcome, which, in this context, relates to the sustainability of community enterprises and the social economy in Thailand. This percentage indicates the model's ability to explain a substantial portion of the variability in sustainability outcomes based on the independent variables included.

Table 3. Classification table for back-testing using all independent variables

			Predicted		
	Observed		**Sustainability**		**Percentage correct**
			No	**Yes**	
Step 1	Sustainability	No	575	149	79.4%
		Yes	156	356	69.5%
	Overall percentage				75.3%

Note: The cut-off value is .500.

The classification results presented in Table 3 reveal that the predictive model, which encompasses all independent variables, accurately forecasts the sustainability of community enterprises and social economies in Thailand in 75.3% of instances. This accuracy rate is determined using a cut-off value of 0.500 or 50%, meaning that the model is deemed successful in its predictions if it assigns a probability of at least 50% for a case to be classified as contributing to sustainability.

Table 4. Variables in the model using all independent variables

		B	**S.E.**	**Wald**	**df**	**Sig.**	**Exp(B)**
Step 1[a]	Age	0.077	0.010	57.367	1	0.000	1.080
	Education	0.094	0.034	7.488	1	0.006	1.098
	Gender	0.406	0.148	7.460	1	0.006	1.500
	Debt	0.000	0.000	172.190	1	0.000	1.000
	Instagram	- 23.693	3414.486	0.000	1	0.994	0.000
	Line	1.418	0.223	40.287	1	0.000	4.131
	Pinterest	0.725	0.169	18.515	1	0.000	2.066
	WhatsApp	0.460	0.234	3.871	1	0.049	1.584
	YouTube	0.352	0.244	2.076	1	0.150	1.421
	Facebook	0.329	0.268	1.511	1	0.219	1.390
	Readiness	0.117	0.047	6.237	1	0.013	1.125
	Constant	- 9.055	0.852	112.931	1	0.000	0.000

The predictive regression equation of Model 1 using the coefficients from Table 4 can be described by the following equation:

$P = \frac{1}{1+e^{-z}}$ ------------ Model 1

where P is the community enterprise and social economy for sustainability in Thailand,

and $Z = -9.055 + 0.077(\text{Age}) + 0.940(\text{Education}) + 0.406(\text{Gender}) + 0.000(\text{Debt}) + 1.418(\text{Line}) + 0.725(\text{Pinterest}) + 0.460(\text{WhatsApp}) + 0.117(\text{Readiness})$.

The data presented in Table 4 illustrates how various independent variables significantly impact the sustainability of community enterprises and social economies in Thailand. According to the regression analysis results, significant predictors include age, education, gender, debt, Line, Pinterest, WhatsApp, and readiness, while Instagram, YouTube, and Facebook do not show significant effects. In detail, the analysis indicates that an increase in age correlates with a rise in sustainability, with a coefficient of 1.080. This suggests that older individuals may be more inclined or able to contribute to sustainable practices within community enterprises and social economies. Similarly, an increase in educational levels leads to a 1.098-fold increase in sustainability. This reflects the critical role that education plays in fostering an understanding of and implementing sustainable practices. Gender, specifically being male, significantly influences sustainability, with a factor of 1.500 increase. This indicates a notable gender-based difference in contributing to sustainability, although the direction (positive or negative) depends on the context in which this coefficient is interpreted. The model also highlights that an increment in debt levels marginally affects sustainability, with a coefficient increase of 1.000. This minimal impact suggests that the debt level is not a major determinant of sustainability practices in this context. The use of social media platforms shows varied effects; Line users contribute to a 4.131% increase in sustainability, Pinterest by 2.066%, and WhatsApp by 1.584%. These coefficients suggest that engagement with these specific platforms could enhance the communication, collaboration, and dissemination of information related to sustainable practices. Finally, an increase in readiness, or the degree of preparedness to engage in sustainable activities, correlates with an increase in sustainability, as indicated by a coefficient of 1.125. This highlights the importance of individual and collective willingness to adopt sustainable practices within the community and social enterprises.

Following the initial data analysis, only statistically significant independent variables were retained for further examination. This approach focused on variables that meaningfully influence the sustainability of community enterprises and the social economy in Thailand. By concentrating on these key determinants, the analysis reduced unnecessary complexity and minimized potential distortion from irrelevant factors. This streamlined process enhanced the reliability of the findings, allowed for clearer interpretation, and supported the development of targeted recommendations to strengthen community enterprises and promote long-term social and economic sustainability.

Table 5. Omnibus test of the model's performance using only significant variables

		Chi-square	Degrees of freedom (df)	Sig.
Step 1	Step	185.753	8	0.000
	Block	185.753	8	0.000
	Model	185.753	8	0.000

Table 5 shows that the chi-square value is 185.753, with degrees of freedom (df) equal to 8. Thus, a dependent variable can be explained by all of the independent variables at a significance level of 0.05.

Table 6. Model summary using only significant variables

Step	-2 log likelihood	Cox & Snell R-squared	Nagelkerke R square
1	1491.164[a]	0.140	0.188

a. Estimation terminated at iteration number 3 because parameter estimates changed by less than .001.

According to Table 6, the model explained approximately 18.8% of the variation in the result with a significance value of 0.05.

Table 7. Classification table for back-testing only significant variables

			Predicted		
	Observed		Sustainability.		Percentage correct
			No	Yes	
Step 1	Sustainability	No	574	150	79.3%
		Yes	244	268	52.3%
	Overall percentage				68.1%

Note: The cut-off value is .500.

Based on Table 7's classification, the model that included all of the independent variables correctly predicted community enterprise and social economy for sustainability in Thailand in 68.1% of cases when using a cut-off value of 0.500 or 50%.

Table 8. Variables in the model using only significant variables

		B	S.E.	Wald	df	Sig.	Exp(B)
Step 1[a]	Age	0.022	0.008	7.094	1	0.008	1.022
	Education	- 0.003	0.029	0.009	1	0.924	0.997
	Gender	- 0.264	0.132	4.005	1	0.045	0.768
	Debt	0.000	0.000	108.749	1	0.000	1.000
	Line	0.494	0.188	6.924	1	0.009	1.640
	Pinterest	0.047	0.149	0.101	1	0.750	1.049
	WhatsApp	0.418	0.173	5.849	1	0.016	1.518
	Readiness	0.244	0.042	33.233	1	0.000	1.276
	Constant	- 4.439	0.632	49.290	1	0.000	0.012

The predictive regression equation of Model 2 using the coefficients from Table 8 can be described by the following equation:

$$P = \frac{1}{1 + e^{-z}} \text{ ------------ Model 2}$$

where P is the community enterprise and social economy for sustainability in Thailand,

and Z = – 4.439 + 0.022(Age) - 0.264(Gender) + 0.000(Debt) + 0.494(Line) + 0.418(WhatsApp) + 0.244(Readiness).

The statistical outcomes presented in Table 8 highlight the relationship between various independent variables and the dependent variable, which in this study is the sustainability of community enterprises and social economies in Thailand. The analysis identifies age, gender, debt, Line, WhatsApp, and readiness as significant predictors, while education and Pinterest do not exhibit significant impacts on the dependent variable. Specifically, the analysis reveals that with every additional year in age, there is a 1.022 increase in sustainability, indicating that older individuals may contribute positively to the sustainability of community enterprises and the social economy. Gender also plays a discernible role, with male respondents showing a slight decrease in sustainability, by a margin of 0.003 (from 1 to 0.997), suggesting a nuanced influence of gender on sustainability efforts. Furthermore, an increment by one unit in debt marginally elevates sustainability by a factor of 1.000, suggesting that financial indebtedness has a very subtle, albeit positive, effect on the sustainability of such enterprises and economies. The usage of social media platforms like Line and WhatsApp is significantly associated with higher sustainability, with coefficients of 1.640 and 1.518, respectively. This implies that engagement with these platforms is correlated with an increase in the sustainability of community enterprises and social economies, potentially due to enhanced communication and networking capabilities. Lastly, an increase in readiness, or preparedness to engage with sustainability practices, correlates with a notable increase in sustainability, with a coefficient of 1.276. This underscores the importance of readiness and willingness among individuals to adopt sustainable practices within the context of community enterprises and social economies. These findings suggest that demographic factors (such as age and gender), financial conditions, technological engagement, and the general readiness to embrace sustainability significantly influence the success of sustainable practices within community enterprises and social economies in Thailand.

5. DISCUSSION

The current research was conducted to examine the determinants that impact the sustainability of community enterprises and the social economy in Thailand. The investigation identified variables such as age, gender, indebtedness, social media platform usage (Line, WhatsApp), and level of preparedness as significant contributors to the sustainability of community enterprises and the social economy within the Thai context. These outcomes are consistent with the study by Marín et al. (2019), which suggested that individuals who are older, possess higher educational levels, and are female tend to display a more profound social orientation. Additionally, the research by Ross et al. (2019) highlighted the considerable influence of social media networks on public opinion formation. The relevance of the social economy, particularly in upholding people-centric values, has been increasingly recognized for its capacity to achieve sustainable development objectives, as noted by Núñez et al. (2020). In line with this, Del-Aguila-Arcentales et al. (2022) discovered a positive correlation between social and environmental sustainability objectives and economic sustainability goals, thereby enhancing entrepreneurial spirit and competitive advantage. Furthermore, Jiatong et al. (2021) found a positive association between a social mission and social innovation, leading to sustainable enterprise growth. In this context, social networks served as a mediating function, while governmental regulations contributed a positive regulatory and moderating effect. The findings by Jangjarat et al. (2023) indicated that demographic aspects, including age and place of residence, as well as technological preparedness and the use of online communication tools such as Instagram, Pinterest, Facebook, and blogs, exert an influence on the social economy.

Hence, promoting career opportunities by fostering business initiatives that support community and social endeavors across various age groups can enhance community engagement, strengthen societal resilience, and contribute to broader sustainable development goals. Within this context, effective governance plays a crucial role in ensuring that community enterprises operate within frameworks that strike a balance between economic growth, social inclusion, and environmental stewardship. Strengthened governmental and administrative support is therefore essential, not only to sustain economic pursuits but also to guide enterprises toward environmentally responsible practices through regulatory standards, incentives, and transparent oversight. As industries evolve, the risk of economic volatility underscores the importance of embedding governance mechanisms that promote adaptive capacity, accountability, and long-term resilience. Equally, businesses and corporations must embrace sustainable leadership by integrating green governance principles into their corporate social responsibility strategies, thereby fostering ethical practices that safeguard both society and the environment.

Building on the discussion, this study proposes an extended model that integrates technological policy with sustainable green governance as a framework for advancing community enterprises and the social economy. The model emphasizes that sustainability requires more than demographic and financial readiness; it demands a coordinated approach that aligns technological innovation with ecological responsibility and governance mechanisms. First, technological policy serves as the foundation for enabling digital transformation in community enterprises. By providing regulatory clarity, ensuring digital infrastructure, and promoting equitable access to technological resources, policymakers can reduce barriers that hinder participation in the digital economy. Technology policies should also prioritize digital literacy, cybersecurity, and responsible data governance to ensure that community enterprises adopt innovations in a secure and inclusive manner. Second, sustainable green governance complements technological policy by embedding environmental stewardship into economic decision-making. This governance model involves multi-level collaboration among governments, enterprises, and communities to establish accountability systems, environmental standards, and incentive structures that encourage green practices. Such practices include resource-efficient production, renewable energy adoption, and circular economy initiatives that reduce waste and carbon emissions. Third, the integration of technological policy and green governance creates a synergistic effect. Technology accelerates the adoption of eco-friendly practices through innovations such as artificial intelligence for predictive resource management, IoT applications for real-time monitoring, and blockchain for transparent supply chains. At the same time, governance frameworks ensure that these technologies are deployed ethically, equitably, and sustainably, preventing potential misuse or ecological harm. The extended model therefore highlights three interconnected pillars: (1) technological policy for digital readiness, (2) green governance for environmental accountability, and (3) community engagement as the mediating force. By situating community enterprises at the center of this triad, the model underscores their role as local engines of sustainability that bridge top-down policy initiatives with grassroots practices. Ultimately, this framework offers policymakers, practitioners, scholars, and other stakeholders, a holistic pathway for strengthening resilience in community enterprises. It positions technological innovation not merely as a driver of competitiveness, but as a tool for advancing environmental sustainability and social equity, ensuring that economic progress aligns with long-term ecological and societal well-being in the digital age.

6. CONCLUSION

This study has provided empirical evidence on the determinants influencing the sustainability of community enterprises and the social economy in Thailand, offering insights that extend both theoretical and practical understanding. The findings revealed that demographic characteristics, such as age and gender, financial conditions including indebtedness, and the utilization of digital communication platforms like Line and WhatsApp, together with individuals' readiness to adopt sustainable practices, significantly contribute to long-term sustainability outcomes. These results affirm the centrality of digital transformation in empowering local communities, as online platforms not only facilitate more efficient communication and resource sharing but also create new opportunities for entrepreneurial collaboration and market expansion. Importantly, the evidence also highlights the role of readiness and willingness as critical enablers of resilience, underscoring that sustainable development in community enterprises is contingent not merely on resources but also on the collective mindset and adaptability of stakeholders.

Beyond individual and technological factors, the study draws attention to the indispensable role of governance and institutional support. Sustainable community enterprises require structured frameworks that balance economic growth with social inclusion and environmental stewardship. Governmental and administrative bodies must therefore provide consistent support through policy incentives, regulatory clarity, and targeted capacity-building programs, while corporate actors should advance beyond conventional notions of corporate social responsibility by embedding sustainable leadership and green governance into their organizational ethos. Such collaboration between public, private, and community stakeholders can create a synergistic ecosystem that enhances economic resilience and social equity.

The contribution of this research lies in its integration of demographic, financial, and technological dimensions into a comprehensive framework for understanding sustainability in the social economy. The insights not only enrich the academic literature but also provide actionable guidance for policymakers, practitioners, and community leaders seeking to strengthen grassroots enterprises in developing nations. Ultimately, advancing sustainability in community enterprises necessitates a holistic and multi-stakeholder approach—one that intertwines local participation, digital innovation, institutional support, and ethical corporate practices. By doing so, community enterprises can serve as engines of inclusive growth, social well-being, and long-term resilience, thereby playing a pivotal role in achieving the Sustainable Development Goals and shaping a more equitable future for society.

6.1 Research implications

The findings of this study carry several important implications for policymakers, practitioners, and corporate stakeholders. For policymakers, the results highlight the need to strengthen governance frameworks and provide targeted support for community enterprises through financial assistance, regulatory facilitation, and training programs that enhance digital literacy and readiness. For community leaders, the evidence highlights the importance of utilizing digital communication platforms to promote collaboration, disseminate knowledge, and expand market access. At the corporate level, businesses are encouraged to integrate sustainable leadership and corporate social responsibility into their strategies, aligning with community-based initiatives to ensure inclusive growth and long-term resilience. Taken together, these implications suggest that a multi-stakeholder approach—combining government, community, and corporate efforts—is essential to advancing the sustainability of community enterprises and the social economy in developing contexts.

6.2 Limitations

Despite its contributions, this study has limitations. The use of convenience sampling, while effective for reaching a large number of respondents, restricts the generalizability of the findings to the broader Thai population. Furthermore, the study's reliance on self-reported data may be subject to biases such as social desirability or selective recall, which could affect the accuracy of responses. The quantitative design also limited the ability to explore deeper contextual and qualitative dimensions of community enterprise sustainability, such as cultural influences, leadership dynamics, or community-specific challenges. These limitations underscore the need for cautious interpretation of the results, while acknowledging the value they bring to the existing body of knowledge.

6.3 Future Research

Future research should build upon these findings by employing mixed-method approaches that integrate both quantitative and qualitative data to capture a more comprehensive understanding of community enterprise sustainability. Longitudinal studies would be particularly valuable for examining how demographic, financial, and digital factors evolve over time and influence sustainability outcomes at different stages of enterprise development. Comparative studies across regions or countries could also offer insights into how cultural, institutional, and policy differences shape the trajectories of social economies in diverse contexts.

Moreover, future investigations should place greater emphasis on the role of green governance, exploring how environmental policies, regulatory frameworks, and eco-conscious leadership models can be systematically embedded into community enterprise operations. Equally, research should examine the development and implementation of technology policies that guide digital adoption, data governance, and cybersecurity practices in community enterprises. Such policies are vital for ensuring that digital tools are leveraged responsibly, inclusively, and sustainably. By addressing these dimensions alongside existing demographic and technological determinants, future research can provide a more holistic and policy-relevant framework. This will not only strengthen the theoretical discourse on sustainability but also generate actionable insights for policymakers, practitioners, and community leaders seeking to design resilient, digitally enabled, and environmentally responsible social economies.

ACKNOWLEDGEMENT

This study is the result of a collaborative partnership between Pathumthani University, Thailand, and Bangkokthonburi University, Thailand. The authors gratefully acknowledge the International College of Pathumthani University and the Research and Innovation Center of Pathumthani University for their unwavering support, guidance, and encouragement throughout the completion of this work.

DECLARATION

This study is not subject to competing interests, as stated by the authors.

REFERENCES

Ahmadin, M. (2022). Social research methods: Qualitative and quantitative approaches. *Jurnal Kajian Sosial Dan Budaya: Tebar Science, 6*(1), 104-113. http://ejournal.tebarscience.com/index.php/JKSB/article/view/103

Atiku, S. O., Jeremiah, A., Semente, E., & Boateng, F. (Eds.). (2024). *Sustainable development in industry and society 5.0: Governance, management, and financial implications*. IGI Global Scientific Publishing., DOI: 10.4018/979-8-3693-7322-4

Bansal, S., Garg, I., & Sharma, G. D. (2019). Social entrepreneurship as a path for social change and driver of sustainable development: A systematic review and research agenda. *Sustainability, 11*(4), 1091. DOI: 10.3390/su11041091

Boateng, E. Y., & Abaye, D. A. (2019). A review of the logistic regression model with emphasis on medical research. *Journal of Data Analysis and Information Processing*, *7*(4), 190–207. DOI: 10.4236/jdaip.2019.74012

Bouyahrouzi, E. M., Benmimoun, R., El Kihel, Y., & Bajjou, M. S. (2025). Integrating industry 4.0 technologies and maintenance 4.0 for sustainable manufacturing: A systematic literature review. *International Journal of Advanced Manufacturing Technology*. Advance online publication. DOI: 10.1007/s00170-025-16194-3

Chaves, R., & Monzón, J. L. (2012). Beyond the crisis: The social economy, prop of a new model of sustainable economic development. *Service Business*, *6*, 5–26. DOI: 10.1007/s11628-011-0125-7

Del-Aguila-Arcentales, S., Alvarez-Risco, A., Jaramillo-Arévalo, M., De-la-Cruz-Diaz, M., & de las Mercedes Anderson-Seminario, M. (2022). Influence of social, environmental and economic sustainable development goals (SDGs) over continuation of entrepreneurship and competitiveness. *Journal of Open Innovation, 8*(2), 73. DOI: 10.3390/joitmc8020073

Fatimah, Y. A., Govindan, K., Murniningsih, R., & Setiawan, A. (2020). Industry 4.0 based sustainable circular economy approach for smart waste management system to achieve sustainable development goals: A case study of Indonesia. *Journal of Cleaner Production, 269*, 122263. DOI: 10.1016/j.jclepro.2020.122263

Gomila, R. (2021). Logistic or Linear? Estimating causal effects of experimental treatments on binary outcomes using regression analysis. *Journal of Experimental Psychology. General, 150*(4), 700–709. DOI: 10.1037/xge0000920 PMID: 32969684

Guerrero, M., Liñán, F., & Cáceres-Carrasco, F. R. (2021). The influence of ecosystems on the entrepreneurship process: A comparison across developed and developing economies. *Small Business Economics, 57*(4), 1733–1759. DOI: 10.1007/s11187-020-00392-2

Hadi, N. U., Almessabi, B., & Khan, M. I. (2025). Leveraging industry 4.0 and circular open innovation for digital sustainability: The role of circular ambidexterity. *Journal of Open Innovation, 11*(2), 100545. DOI: 10.1016/j.joitmc.2025.100545

Jangjarat, K., Kraiwanit, T., Satityapong, N., Sonsuphap, R., & Phaksipaeng, I. (2023). The social economy in the digital era: A perspective on community enterprises in a developing economy. *Journal of Social Economics Research, 10*(1), 13–21. DOI: 10.18488/35.v10i1.3317

Jiatong, W., Li, C., Murad, M., Shahzad, F., & Ashraf, S. F. (2021). Impact of social entrepreneurial factors on sustainable enterprise development: Mediating role of social network and moderating effect of government regulations. *SAGE Open, 11*(3), 21582440211030636. DOI: 10.1177/21582440211030636

Khan, I. S., Ahmad, M. O., & Majava, J. (2021). Industry 4.0 and sustainable development: A systematic mapping of triple bottom line, circular economy and sustainable business models perspectives. *Journal of Cleaner Production, 297*, 126655. DOI: 10.1016/j.jclepro.2021.126655

Khan, M. I., Yasmeen, T., Khan, M., Hadi, N. U., Asif, M., Farooq, M., & Al-Ghamdi, S. G. (2025). Integrating industry 4.0 for enhanced sustainability: Pathways and prospects. *Sustainable Production and Consumption, 54*, 149–189. DOI: 10.1016/j.spc.2024.12.012

Khiawnoi, P., Nakudom, S. S., Noobutr, P., & Kaewzang, U. (2025). Empowering communities: The strategic role of community enterprise entrepreneurs in sustainable development. *Sustainability*, *17*(12), 5483. DOI: 10.3390/su17125483

Kim, D., & Lim, U. (2017). Social enterprise as a catalyst for sustainable local and regional development. *Sustainability*, *9*(8), 1427. DOI: 10.3390/su9081427

Li, W., Yi, P., Zhang, D., & Zhou, Y. (2020). Assessment of coordinated development between social economy and ecological environment: Case study of resource-based cities in Northeastern China. *Sustainable Cities and Society*, *59*, 102208. DOI: 10.1016/j.scs.2020.102208

Marín, L., Nicolás, C., & Rubio, A. (2019). How gender, age and education influence the entrepreneur's social orientation: The moderating effect of economic development. *Sustainability*, *11*(17), 4514. DOI: 10.3390/su11174514

Montiel, I. (2008). Corporate social responsibility and corporate sustainability: Separate pasts, common futures. *Organization & Environment*, *21*(3), 245–269. DOI: 10.1177/1086026608321329

Nguyen, H. T., Le, D. M. D., Ho, T. T. M., & Nguyen, P. M. (2021). Enhancing sustainability in the contemporary model of CSR: A case of fast fashion industry in developing countries. *Social Responsibility Journal*, *17*(4), 578–591. DOI: 10.1108/SRJ-03-2019-0108

Nogueira, E., Gomes, S., & Lopes, J. M. (2025). Unveiling triple bottom line's influence on business performance. *Discover Sustainability*, *6*, 43. DOI: 10.1007/s43621-025-00804-x

Núñez, R. B. C., Bandeira, P., & Santero-Sánchez, R. (2020). Social economy, gender equality at work and the 2030 agenda: Theory and evidence from Spain. *Sustainability*, *12*(12), 5192. DOI: 10.3390/su12125192

Obilor, E. I. (2023). Convenience and purposive sampling techniques: Are they the same? *International Journal of Innovative Social & Science Education Research*, *11*(1), 1–7. https://seahipaj.org/journals-ci/mar-2023/IJISSER/full/IJISSER-M-1-2023.pdf

Petcho, W., Szabo, S., Kusakabe, K., & Yukongdi, V. (2019). Farmers' perception and drivers of membership in rice production community enterprises: Evidence from the Central Region, Thailand. *Sustainability*, *11*(19), 5445. DOI: 10.3390/su11195445

Pfajfar, G., Shoham, A., Małecka, A., & Zalaznik, M. (2022). Value of corporate social responsibility for multiple stakeholders and social impact–Relationship marketing perspective. *Journal of Business Research, 143*, 46–61. DOI: 10.1016/j.jbusres.2022.01.051

Phuangsuwan, P., Limna, P., & Siripipatthanakul, S. (2024). Ethics in the social sciences research. *Advance Knowledge for Executives, 3*(4), 1-11. https://www.researchgate.net/publication/394223864

Pivoto, D. G., de Almeida, L. F., da Rosa Righi, R., Rodrigues, J. J., Lugli, A. B., & Alberti, A. M. (2021). Cyber-physical systems architectures for industrial internet of things applications in industry 4.0: A literature review. *Journal of Manufacturing Systems, 58*, 176–192. DOI: 10.1016/j.jmsy.2020.11.017

Pongracz, E. (2020). Social economy and social entrepreneurship–The manifestation of social responsibility in the era of globalization. In *SHS Web of Conferences*. EDP Sciences. https://doi.org/DOI: 10.1051/shsconf/20207404021

Ramadani, V., Agarwal, S., Caputo, A., Agrawal, V., & Dixit, J. K. (2022). Sustainable competencies of social entrepreneurship for sustainable development: Exploratory analysis from a developing economy. *Business Strategy and the Environment, 31*(7), 3437–3453. DOI: 10.1002/bse.3093

Romprasert, S., & Trivedi, A. (2021). Sustainable economy on community enterprise and digital marketing. *ABAC Journal, 41*(1), 62–80. http://www.assumptionjournal.au.edu/index.php/abacjournal/article/view/4195

Ross, B., Pilz, L., Cabrera, B., Brachten, F., Neubaum, G., & Stieglitz, S. (2019). Are social bots a real threat? An agent-based model of the spiral of silence to analyse the impact of manipulative actors in social networks. *European Journal of Information Systems, 28*(4), 394–412. DOI: 10.1080/0960085X.2018.1560920

Shamsudin, M. F., Hassim, A. A., & Abd Manaf, S. (2024). Mastering probability and non-probability methods for accurate research insights. *Journal of Postgraduate Current Business Research, 9*(1), 38–53. https://www.abrn.asia/ojs/index.php/jpcbr/article/view/157

Siripipatthanakul, S., Phuangsuwan, P., & Limna, P. Muthmainnha, Siripipattanakul, S., Vui, C. N., & Jaipong, P. (2025). Artificial intelligence (AI) influencing sustainable governance: Governance adopting AI. In Swaroop, A., Virdee, B., Correia, S. D., & Polkowski, Z. (Eds.) *Proceedings of Data Analytics and Management. ICDAM 2024. Lecture Notes in Networks and Systems*. Springer. https://doi.org/DOI: 10.1007/978-981-96-3372-2_38

Srivastava, M., Aftab, J., & Tyll, L. (2025). The influence of artificial intelligence and additive manufacturing on sustainable manufacturing practices and their effect on performance. *Sustainable Futures : An Applied Journal of Technology, Environment and Society*, *10*, 100820. DOI: 10.1016/j.sftr.2025.100820

Suriyankietkaew, S., Krittayaruangroj, K., & Iamsawan, N. (2022). Sustainable leadership practices and competencies of SMEs for sustainability and resilience: A community-based social enterprise study. *Sustainability*, *14*(10), 5762. DOI: 10.3390/su14105762

Tariq, M. U. (2025). Strategic place marketing in cross-border regions: Enhancing territorial cooperation and economic growth. In Sousa, B. (Ed.), *Cross-Border Regions Cooperation and Implications for Organizations* (pp. 107–136). IGI Global Scientific Publishing., DOI: 10.4018/979-8-3373-1912-4.ch004

Terdpaopong, K., & Kraiwanit, T. (2021). The technology readiness of Thai governmental agency. *The Journal of Asian Finance, Economics and Business*, *8*(12), 431–441. https://scholar.kyobobook.co.kr/article/detail/4010028724143

Tien, N. H., Anh, D. B. H., Ngoc, N. M., & Do Thi, Y. N. (2019). Sustainable social entrepreneurship in Vietnam. *International Journal of Entrepreneurship*, *23*(3), 1–12. https://www.abacademies.org/articles/sustainable-social-entrepreneurship-in-vietnam-8433.html

Uakarn, C., Chaokromthong, K., & Sintao, N. (2021). Sample size estimation using Yamane and Cochran and Krejcie and Morgan and green formulas and Cohen statistical power analysis by G* Power and comparisons. *APHEIT International Journal of Interdisciplinary Social Sciences and Technology, 10*(2), 76-88. https://so04.tci-thaijo.org/index.php/ATI/article/view/254253

Wang, S., Liu, C., & Zhou, Z. (2024). Government-enterprise green collaborative governance and urban carbon emission reduction: Empirical evidence from green PPP programs. *Environmental Research*, *257*, 119335. DOI: 10.1016/j.envres.2024.119335 PMID: 38849001

Wang, W., & Wang, X. (2023). Does provincial green governance promote enterprise green investment? Based on the perspective of government vertical management. *Journal of Cleaner Production*, *396*, 136519. DOI: 10.1016/j.jclepro.2023.136519

Wongadisai, W., Chanchalor, S., & Murphy, E. (2020). Rural community enterprises in Thailand: A case study of participation. *World Review of Entrepreneurship, Management and Sustainable Development*, *16*(3), 276–298. DOI: 10.1504/WREMSD.2020.107255

Younis, N. M. M. (2023). Sustainability reports and their impact on firm value: Evidence from Saudi Arabia. *International Journal of Management and Sustainability, 12*(2), 70–83. DOI: 10.18488/11.v12i2.3275

Yu, P., Wong, S., & Prabhakar, A. (Eds.). (2025). *AI strategies for social entrepreneurship and sustainable economic development.* IGI Global Scientific Publishing., DOI: 10.4018/979-8-3693-6392-8

Yusif, S., & Hafeez-Baig, A. (2024). Impact of stakeholder engagement strategies on managerial cognitive decision-making: The context of CSP and CSR. *Social Responsibility Journal, 20*(6), 1101–1121. DOI: 10.1108/SRJ-05-2023-0295

Zong, Z., & Guan, Y. (2025). AI-driven intelligent data analytics and predictive analysis in Industry 4.0: Transforming knowledge, innovation, and efficiency. *Journal of the Knowledge Economy, 16*, 864–903. DOI: 10.1007/s13132-024-02001-z

KEY TERMS AND DEFINITIONS

Binary regression, also known as binary logistic regression: It is a statistical model used to predict the probability of a binary dependent variable (e.g., yes/no, success/failure) based on one or more independent variables. In contrast to linear regression, it employs a logistic function to convert the outcome into probabilities, ensuring that fitted values are constrained between 0 and 1, with the probability rising as the linear predictor increases.

Community Enterprise: Community Enterprise is a well-established organization with a very great reputation and good feedback from clients. However, for all our engagement, we have had limited capacity over the years to research, define, and describe our impact. Community enterprise has arisen as a revolutionary force in local economic systems, creating new avenues for sustained growth and exemplifying the significant impact of information technology (IT) on the digital economy. By merging traditional community values with contemporary technical innovations, community enterprises enhance local markets and foster resilience against global economic volatility.

Convenience sampling: It is a non-probability sampling method in which researchers select individuals who are most readily available, rather than employing random selection. This strategy is frequently employed for rapid preliminary investigations or when temporal and budgetary constraints exist; nonetheless, it poses the risk of yielding a biased sample that fails to accurately represent the larger population, thereby constraining the capacity to generalize the results.

Corporate Social Responsibility (CSR): CSR is a business strategy in which firms incorporate social and environmental issues into their operations and engage with stakeholders, striving to be socially accountable for their effects on the community, environment, and society, beyond just profit maximization. Typical CSR programs encompass philanthropic contributions, environmental stewardship, ethical labor practices, and community enhancement, which bolster a company's brand, draw talent, and promote long-term sustainability for both the enterprise and society.

Questionnaire Survey: A questionnaire is a document containing written inquiries used for data collection. A survey, on the other hand, encompasses the comprehensive process of constructing and administering the questionnaire, gathering the data, and subsequently evaluating and interpreting the responses to derive insights about a larger population. Consequently, a questionnaire serves as a tool within the broader survey technique.

Social Economy for Sustainability: The social economy represents an innovative industry that integrates social innovation with community-focused problem-solving, resulting in hybrid company models that prioritize social aims over financial profit while ensuring economic sustainability. This paradigm shift highlights the capacity of community enterprises to serve as catalysts for inclusive growth, striking a balance between profit, social justice, sustainability, and equality. Simultaneously, the academic domain has witnessed an increase in research on sustainable development, with a particular focus on the three pillars of sustainability: social, economic, and environmental.

Sustainability: Sustainability entails fulfilling current requirements without jeopardizing the capacity of future generations to meet their own demands, often encapsulated as the "Triple Bottom Line," which encompasses environmental, economic, and social dimensions. It entails harmonizing human welfare with the meticulous preservation of natural resources to guarantee sustained ecological integrity, economic sustainability, and social equality.

Chapter 2
Technology-Driven Organization Policies:
E-Waste Governance Framework

Astri Dwi Andriani
https://orcid.org/0000-0003-3951-4829
Putra Indonesia University, Indonesia

Supaprawat Siripipatthanakul
https://orcid.org/0000-0001-6671-2682
Bangkokthonburi University, Thailand

ABSTRACT

This chapter presents a systematic review that shows a technology-oriented e-waste governance framework emphasizes the use of technology to oversee the entire lifecycle of electronics—from manufacturing to disposal—through policies such as Extended Producer Responsibility (EPR), fostering a circular economy, and employing technology for collection and recycling. Essential elements encompass implementing Extended Producer Responsibility (EPR) to ensure manufacturer accountability, promoting a circular economy that emphasizes repairability and recyclability, and utilizing modern technologies, such as artificial intelligence, for electronic waste processing. This strategy integrates legislative measures, technological advancements, and economic incentives to reduce environmental damage and reclaim valuable resources.

1. INTRODUCTION

In recent decades, digital transformation has accelerated at an unprecedented pace, not only in how organizations operate but also in the functioning of society

DOI: 10.4018/979-8-3373-6821-4.ch002

as a whole. Elements of modern business ecosystems include smart devices, artificial intelligence, cloud computing, and the Internet of Things (IoT). But behind this wave of innovation lies a less ostentatious truth: electronic waste, or e-waste, is piling up in ever-larger mountains as technology advances. According to the Global E-Waste Monitor 2024, the amount of e-waste generated worldwide was 62 million metric tons in 2022, and the projected amount by 2030 is 82 million metric tons, unless significant action is taken (United Nations Institute for Training and Research [UNITAR], 2024). Worryingly, 22.3% of e-waste is collected and recycled through environmentally friendly processes, while the remainder is sent to informal disposal systems that often use unsafe methods, such as open burning and acid leaching (UNITAR, 2024). As an academic who has witnessed not only educational and technological institutions but also the effects of digital acceleration, he has personally observed how uncontrolled use of this material can become the driver of innovation, rather than the source of ecological injustice.

While the importance of sustainability improvements is widely recognized, many technology organizations have yet to establish the governance systems needed to manage their electronic resources throughout their lifecycle—acquisition, use, decommissioning, and disposal. While corporate social responsibility (CSR) policies or informal recycling initiatives are viable, not many organizations incorporate e-waste management into their overall sustainability and digital transformation policies. There is empirical evidence on the situation of e-waste in developing countries, such as Southeast Asia. A significant percentage of obsolete electronic devices end up in informal markets, where they are dismantled without safety measures, releasing toxic substances such as lead, mercury, and cadmium into the environment (Balde et al., 2024; Forti et al., 2020). However, this is not a technical problem, but a very profound ethical issue. Organizations that accelerate the use of technology without planning for its disposal, or even the waste it generates, are likely undermining ecological integrity and humanity.

The lack of adequate e-waste management within organizations constitutes a multidimensional gap in policy and management. From a regulatory perspective, most national and regional e-waste policies are largely ineffective and uncoordinated. For example, a few Southeast Asian countries have adequate Extended Producer Responsibility (EPR) systems (Kojima & Yoshida, 2022). Regarding communication and stakeholder engagement, e-waste management is characterized by limited awareness within the company, low transparency, and limited employee, supplier, and community involvement. Many organizations view obsolete technology as a maintenance issue rather than a culturally recognized sustainability issue. Personally, I view this as a failure of vision and governance, a symptom that digital advancement is outpacing our capacity to manage the side effects as a moral and institutional community.

This chapter is important because it bridges the gap between technology policy, sustainability, and organizational ethics. When technology policy focuses solely on innovation, it risks ignoring the environmental and social implications of technological development. Conversely, these sustainability initiatives, which lack regulatory coherence and organizational accountability, tend to be superficial rather than transformative. Therefore, organizational ethics must be revised to incorporate environmental stewardship and intergenerational responsibility. As an academic and practitioner at the intersection of education, governance, and technology, I view this fusion not as an ideological goal but as a moral duty —an invitation for institutions to redefine progress from a sustainable perspective.

The primary objective of this chapter is to outline a Technology-Driven E-Waste Governance Framework that balances technological innovation, environmental accountability, and regulatory compliance. The framework will be constructed from a synthesis of recent literature, a theoretical overview, and comparative cases from corporate, educational, and governmental organizations. It seeks to offer a conceptual and practical model for incorporating sustainability into the core of digital transformation. The framework also identifies communication and stakeholder engagement as drivers of accountability and resilience in e-waste governance.

The rapid pace of digital transformation offers immense opportunities; without careful regulation, it can create what scholars refer to as the dark side of digitalization (Rana et al., 2023). Reconceptualizing technology policy through the lens of green governance and ethical responsibility enables organizations to develop digital ecosystems that are not only efficient and innovative but also resilient, circular, and sustainable for future generations.

The outline of this chapter helps the reader understand the context and its applications in analyzing the two sides of technological advancement and the unintended effects of e-waste on digital innovation policies. This chapter discusses methods of green governance in technology-based organizations, policy frameworks, and compliance cultures. It summarizes theoretical approaches to technology governance, communication, and sustainability research. It presents the Technology-Driven E-Waste Governance Framework and comparative case studies that showcase experiences from the business, education, and public sectors. This chapter discusses policy and practice implications and concludes with a call for a sustainable and ethical digital future. Finally, the chapter is written not only from an academic perspective but also from a human one, affirming that technological evolution should not be achieved at the expense of the environment.

2. LITERATURE REVIEW

2.1 Technology Policy and E-Waste: The Two-Sided Sword of Digital Advancement

The use of technology policies has traditionally been considered the engine of modernization and competitiveness, enabling societies to innovate, grow, and adapt to the digital economy. To accelerate industrial transformation, governments worldwide invest billions of dollars annually in digital infrastructure, cloud computing, and artificial intelligence initiatives (Organization for Economic Cooperation and Development [OECD], 2023). However, as these technologies expand, they also generate vast new quantities of e-waste. The Global E-Waste Monitor 2024 estimates that, in 2022, more than 62 million metric tons of e-waste were generated worldwide, while only 13.8 million tons were reported to have been recycled (UNITAR, 2024). The remainder of this waste is typically disposed of in informal sectors, where the dismantling process releases toxic heavy metals into the soil and waterways, making technological advancement one of the sources of environmental decentralization. In this way, technology policy, as a means of innovation, has become a silent cause of the planet's growing environmental crisis.

In theory, technology policy structures seek to generate efficiency in both the economy and society. However, in reality, they often prioritize digital use and economic productivity over environmental responsibility. For example, the Digital Economy Plan implemented by several ASEAN countries primarily focuses on fostering innovation ecosystems, digital trade, and increased productivity, with little consideration of environmental sustainability indicators (ASEAN, 2021). The consequence of this imbalance is what researchers call the policy-innovation paradox, in which states and companies rush to innovate without corresponding investments in waste management and environmental regulations (Rana et al., 2023). As a result, e-waste is not only a side effect of innovation but also a failure of the governance system, an unintended outcome of ambitious, ecologically blind technological policy. The double face of digital advancements is also defined by the rapid obsolescence of devices, which is, in many cases, exacerbated by intentional obsolescence schemes and consumer cultures. The lifespan of smartphones has been reduced to less than 3 years, and corporate IT infrastructure, such as servers and routers, is often renewed every 5 years to remain operational (Statista, 2023). The result is an endless cycle of hardware that sustains the e-waste stream. In developing economies such as Indonesia, Malaysia, and the Philippines, e-waste, both imported and domestic, is arriving at a faster rate than recycling and recovery facilities can handle (Kojima & Yoshida, 2022). Therefore, the lack of adequate policy alignment between digital transformation and environmental sustainability creates a vicious cycle in which the use of technology takes precedence over its responsible disposal.

Furthermore, the e-waste trade is global, making governance even more difficult. Despite the Basel Convention (1989), which aims to control transboundary flows of hazardous waste, illegal transport of used electronics persists, often under the guise of secondhand goods. According to estimates by the United Nations Environment Program (UNEP, 2023), the proportion of electronics exported to Africa and Asia can reach 30% of products that are non-functional or irrecoverable upon delivery. These flows alleviate another weakness of technological governance: policies aimed at innovation and economic development lack transnational accountability mechanisms to address the environmental effects of globalization. In this case, technological policy operates in silos, measuring progress in gigabytes and growth rates, and the silent accumulation of waste is not reflected in the balances.

From a human perspective, this paradox of innovation and degradation is particularly poignant. Technology is thought to provide empowerment, efficiency, and connectivity worldwide, but it also shifts some costs to vulnerable communities burdened by informal recycling and exposure to toxic substances. Informal recycling centers in Ghana (Agbogbloshie) and Indonesia (Tangerang) have been reported to expose workers (usually children) to hazardous chemicals as they dismantle discarded electronics for small profits (Forti et al., 2020; Nnorom & Osibanjo, 2021). These human narratives are moral lessons: if technology policy is not used to protect both digital rights and environmental justice, it cannot be considered successful. Ignoring the social aspect of e-waste creates another ecological imbalance, akin to the digital divide.

Therefore, technology policy reform to address e-waste governance must be integrative—that is, it must balance incentives toward innovation and environmental responsibility. Researchers have proposed integrating Green Digital Governance principles into national technology strategies, focusing on lifecycle management, circular economy design, and producer responsibility (Bocken et al., 2022; Li et al., 2023). The concepts of Extended Producer Responsibility (EPR), eco-design requirements, and right-to-repair legislation are increasingly recognized as effective policy instruments for bridging the gap between innovation and sustainability. However, institutional communication, organizational culture, and cross-sector cooperation are vital conditions for the success of these tools. Essentially, e-waste governance is not a technical or legal exercise, but a cultural one; that is, organizations need to internalize sustainability as a value of digital transformation.

In conclusion, the digital age represents a paradise of wealth and waste. Technology policy, previously considered an exclusive instrument of competitiveness, must be transformed into a multidimensional governance system that combines digital innovation with environmental conservation. The difficulty in rejecting technology is not the problem; rather, it is the self-definition of progress. With the transition to the Fourth Industrial Revolution, the question is no longer whether digitalization will persist or how to achieve sustainability. This chapter presents the thesis that the solution lies in establishing coherent e-waste governance frameworks that balance the three aspects of innovation, communication, and environmental management. The following chapter delves deeper into this topic.

2.2. Green Governance of Technology-Based Organizations: Policy Frames and Culture of Compliance

The debate about sustainability in technology organizations has evolved from a simple rhetorical defense to a strategic imperative. With the growing relevance of environmental, social, and governance (ESG) indicators in determining organizational legitimacy, the concept of green governance has emerged as a framework that integrates environmental responsibility into decision-making. The World Economic Forum (2023) reports that more than three out of five global investors prioritize ESG performance among the issues affecting capital allocation decisions, and that organizations that fail to demonstrate sustainable governance face reputational and financial losses. However, the implementation of green governance, particularly in e-waste management, remains uneven. Most organizations profess their commitment to sustainability and apply it despite fragmented policies and poor compliance cultures, which could be defined as a green gap between declaration and practice, according to scholars (Husted & de Sousa-Filho, 2017).

Green governance is essentially a form of corporate governance that integrates environmental considerations into accountability and transparency frameworks (Lock & Seele, 2021). It places sustainability at the heart of corporate social responsibility (CSR) initiatives and strategies, rather than relegating them to the background. This shift is particularly complex in technology-focused organizations, as it is highly dynamic due to rapid technology and product obsolescence and the globalization of supply chains. The framework for good green governance should be a clear policy framework that expresses responsibilities for e-waste reduction, life-cycle assessment, and compliance reporting. The Circular Economy Action Plan, managed by the European Commission (2020), is a good example, as it focuses on the principle of extended producer responsibility (EPR) and integrates product design, manufacturing, and post-consumer recovery. However, most organizations in developing economies lack the institutional capacity to translate these structures into local action, leading to policy diffusion and limited internalization.

Communication is key to bridging this gap between policies and practices. Organizational communication theories highlight that monitoring and reporting are not the only aspects of governance; rather, they involve constructing meaning and legitimacy among stakeholders (Christensen & Langer, 2020). Within the framework of e-waste governance, communication is considered a connective tissue between technical processes and ethical principles, between compliance mechanisms and cultural change. For example, when internal communication about sustainability is viewed as a dialogue, rather than an order, it creates a sense of ownership among staff and promotes behavioral change (Cornelissen, 2023). External communication, including sustainability reporting and digital disclosures, increases organizational transparency and stakeholder trust. In this way, green governance becomes performative governance: it is not only about what organizations do, but also how they report on their commitment and progress.

Another important dimension is the culture of compliance. Compliance, traditionally perceived as the legal obligation to follow the law, has transformed into a more dynamic phenomenon that also involves ethical commitment and active participation (Kaptein, 2022). In technology-based companies, developing such a culture involves integrating sustainability values into daily activities and strategic priorities. Studies by Eccles et al. (2020) reveal that companies with a high ethical standard and sustainability-oriented leadership are 40% more likely to achieve sustainability in their long-term performance. Therefore, the issue of compliance should be internalized as a culture. In the context of e-waste management, this is not limited to disposal procedures but rather to establishing a shared organizational awareness that integrates responsible waste management into the corporate identity.

In terms of policy, green governance requires multilevel cooperation between government, industry, and academia. Policymakers provide the regulatory framework, organizations implement compliance, innovate, and lead, and universities provide research, education, and advocacy. For example, the Green Industry Platform, promoted by the United Nations Industrial Development Organization (UNIDO), illustrates how cross-sector partnerships and the capacity to drive change toward greater sustainability can be established through the coordination of technological modernization and environmental responsibility (UNIDO, 2022). This triple connection strengthens systems thinking—the core of sustainable governance—by enabling all institutional actors to contribute to the circular circulation of resources and knowledge.

To humanize green governance, it is necessary not to treat it as an act of bureaucratic compliance, but rather as a cautionary tale about how the organization relates to the planet. Technology-based organizations are not ideals; they are individuals with values, attitudes, and decisions that influence the environment. With the growing popularity of digital innovation as the foundation for the future of work, education, and communication, it is necessary to understand that sustainability is no longer a limiting factor but a driver of creative change. Personally, I believe that the ecological wisdom of technology is the moral compass of the 21st-century organization, guiding technological intelligence toward ecological wisdom.

In conclusion, green governance offers the ethical and functional underpinnings of sustainable digital transformation. It requires logical policy frameworks, open communication, and a culture of regulatory compliance that extends beyond it. Organizations that were previously sources of disruption have been forced to become architects of technology-driven regeneration. To achieve this, they must ensure their digital strategies align with green management, where innovation does not trump integrity. The next section of this chapter discusses the theoretical perspectives on which this synthesis is based, integrating the conceptual frameworks of communication theory, governance studies, and technology-driven e-waste governance models into a unified framework.

2.3 Theoretical Perspectives Research: Integration of Communication, Governance, and Sustainability Theories

E-waste governance is interdisciplinary and therefore requires a theoretical framework that transcends the analysis of a single discipline. The issues of organizational behavior, communication, technology, and sustainability are complex and cannot be fully explained by a single theory. Therefore, the chapter utilizes four interrelated theoretical perspectives—including Technology Policy and Governance Theory, Organizational Communication Theory, Sustainability and Green Governance Models, and Resilience and Systems Thinking—to create a comprehensive conceptual perspective through which the analysis and proposal of the Technology-Driven E-Waste Governance Framework were conducted. It is not only analytically in-depth but also normative: the question this integration answers is not only how e-waste governance works, but how it should work in an ethically responsible digital age.

The first conceptual pillar is Technology Policy and Governance Theory. Technological governance encompasses the institutional relationships, norms, and policy tools that influence the management and distribution of technological innovation (Kuhlmann et al., 2019). It addresses not only technology in terms of control but also its ethical and social coordination. In this context, e-waste governance can be understood as a subset of the broader debate on responsible innovation, a movement that decentralizes technology, making it inclusive, sustainable, and reflexive (Owen et al., 2013). However, policy research has long observed a disjunction between the acceleration logic of digital innovation and the deliberative logic of environmental policy (Sovacool & Hess, 2017). This is what Beck (1992) has come to call the so-called risk society, where technological advances pose — and will continue to pose — new environmental and ethical hazards faster than institutions can regulate them.

To address this deficiency, more agile and proactive forms of governance are needed; that is, policies must adapt to technological changes rather than respond to them.

The second theoretical axis, Organizational Communication Theory, situates governance in the social dynamics of meaning, dialogue, and legitimacy. Traditional policy frameworks tend to view communication as a means of implementation, whereas modern literature treats it as a constitutive process of governance (Christensen & Langer, 2020). In e-waste management, communication is a medium, a message, a way of structuring the organization of sustainability and its perception and realization by organizations. The Communicative Constitution of the Organization (CCO) perspective (Putnam & Nicotera, 2010) affirms that organizations do not just talk about governance; they constitute it through their communication practices. This implies that policy adherence, ethical awareness, and trust among positive stakeholders are socially constructed through everyday discourse, reporting, and accountability. For technology-driven organizations, good e-waste governance therefore relies on open communication systems that align institutional language with sustainable behavior.

The third pillar, Sustainability Models and Green Governance, underpins the discussion of ecological and ethical responsibility. The Triple Bottom Line framework (Elkington, 1997) remains central, as it assumes that an organization's success must be achieved through a balance between people, planet, and profit. However, more recent models, such as the Circular Economy and cradle-to-grave design philosophies, build on this reasoning by focusing on resource regeneration and product lifecycle responsibility (Geissdoerfer et al., 2017). By applying these models to e-waste governance, technology is no longer understood as a consumable good but as an open system of material and meaning circulation. This compels organizations to use life-cycle assessment (LCA) tools, eco-design principles, and closed-loop supply chains. These strategies align with the ethical imperative of stewardship: the idea that innovation should not deplete, but rather replenish, ecological capital (Dyllick & Muff, 2016). In this sense, e-waste governance is a framework of moral responsibility because it transforms.

The theoretical synthesis is completed with Resilience and Systems Thinking, which offers a dynamic perspective on governance as an interdependent system. According to Donella Meadows (2008), systems thinking emphasizes feedback loops, interrelationships, and a system's self-healing through learning. In e-waste governance, this translates into perceiving organizations not as islands, but as part of broader ecological and sociotechnical networks. In this sense, resilience does not simply mean the capacity to absorb external or regulatory shocks, but also the ability to adapt to them (Folke et al., 2010). Adaptive governance systems promote iterative learning, that is, policies that change as new information, technologies, and risks emerge. This reflexivity is crucial for technology-based organizations, where change is so rapid that even the most static and compliant systems become obsolete before they can be formulated.

By combining these theoretical perspectives, a multidimensional understanding of e-waste governance is achieved. Technology Policy Theory describes the structural context and power relations; Organizational Communication Theory describes the sensemaking practices that maintain compliance and legitimacy; Sustainability Models define moral and environmental objectives; and Systems Thinking endows the framework with adaptability, transparency, and resilience. Their combination constitutes an interdisciplinary architecture, a theoretical infrastructure upon which the proposed Technology-Driven e-Waste Governance Framework will be built. Through this synthesis, the abstract concept of governance becomes a practice of life, expression, and moral guide.

At the individual level, we consider this theoretical integration both an academic requirement and a moral obligation. Humanized theory is a kind of guide, a method for directing knowledge toward the good. The challenge facing both academics and practitioners is how to ensure that technology governance is not simply a process of risk reduction, but rather a process of reframing the very idea of progress, environmental stewardship, stewardship of generations, and stewardship of the totality of innovation. These theoretical insights are consolidated into a viable model, as outlined in the next section: the process of translating political ideals and communication ethics into a structured model for sustainable e-waste management in technology-based organizations.

2.4. E-Waste Governance Framework: Theory and Principles of Policy Formulation

To create an effective e-waste governance system, an integrated understanding of the interrelationships among technology policy, communication, and sustainability within complex organizational systems is necessary. Most countries have established e-waste regulations, but few structures are designed to address the organizational aspect—that is, how internal governance processes, leadership ethics, and communication channels implement these regulations. To address this gap, this chapter proposes the Technology-Driven e-Waste Governance Framework (TD-EWGF) to balance innovation and environmental protection. This framework operates at three interacting levels: policy architecture, organizational communication and culture, and adaptive sustainability systems. Each level builds on the theoretical insights presented above and aims to shape responsible digital transformation for policymakers and organizational leaders.

The TD-EWGF, at the policy architecture level, begins by integrating sustainability into the organization's technology policy. Traditional approaches to technology focus on innovation, efficiency, and profitability, but a governance-based approach adds lifecycle accountability, resource recovery, and the integration of the circular economy to these priorities (Ghisellini, Cialani, & Ulgiati, 2016). A technological asset should not be understood as a short-term operational instrument, but as a tangible entity with a quantifiable ecological footprint throughout its lifecycle—from design, acquisition, and use to refurbishment and decommissioning. This methodology aligns with the principle of Extended Producer Responsibility (EPR), which holds manufacturers, importers, and users of electronic equipment accountable for their environmental impacts (Widmer et al., 2005). Integrating EPR into internal governance systems is critical to ensure that sustainability is not a one-time feature but a structural aspect that influences procurement practices, supplier relationships, and waste management measures.

The second level, communication and organizational culture, recognizes that governance is not merely a functional aspect of policy, but rather a collectively shared meaning and action. Organizational communication theories emphasize that sustainability compliance and awareness are socially constructed through dialogue, stories, and reporting (Christensen & Langer, 2020). In the TD-EWGF, communication functions in three ways, including informational transparency, participatory engagement, and ethical framing. Informational transparency: The open disclosure of e-waste data, including total volume produced, processing, and final disposal, thereby fostering trust and accountability among stakeholders. Participatory engagement involves various actors, such as employees, suppliers, regulators, and local communities, in decision-making and oversight. The ethical framing, in turn, redefines e-waste as an ethical challenge and an opportunity for innovation and leadership, rather than an administrative burden. Regarding the practice of green governance in organizations, my experience in universities and companies shows that organizations that humanize sustainability by involving everyone in communication processes are more likely to internalize green governance as part of their cultural identity, rather than as part of their compliance checklist.

Systems thinking and resilience theory also translate into the design of governance at the third level, called adaptive sustainability systems. The TD-EWGF focuses on feedback, lifelong learning, and data-driven adaptation, rather than on conventional compliance processes (Meadows, 2008; Folke et al., 2010). This includes adopting real-time digital observations, such as Internet of Things (IoT) sensors to track asset usage, blockchain-based certification to track e-waste recycling, and AI-enhanced analytics to predict patterns in e-waste generation (Baldé et al., 2024). Thanks to these technologies, organizations can identify inefficiencies early, predict policy changes, and redefine their purchasing or recycling strategies. This concept of adaptive governance is not limited

More importantly, the ethical triad of accountability, transparency, and inclusion underpins the TD-EWGF. Accountability emphasizes organizations' ethical and social responsibilities to ensure that technological advancement contributes to the health of the planet rather than destroying it (Dyllick & Muff, 2016). Transparency provides the communication infrastructure that legitimizes sustainability claims and reduces the risk of greenwashing. Inclusion will imply that governance structures are developed in collaboration with diverse stakeholders, especially informal actors in the recycling industry, who are often marginalized yet important to the processing of e-waste globally (Lepawsky, 2018). These three principles can be implemented to enable organizations to transform governance from a process of command and control to one of dialogic learning and shared responsibility. From a human perspective, this represents a paradigm shift: sustainability is not just a duty, but a caring, growth-oriented mindset within organizational ethics.

Five operational pillars are also proposed for implementing the framework's principles in practice. (1) Policy Integration: integrating e-waste governance into digital transformation strategies; (2) Stakeholder Collaboration: aligning industry, academia, and government; (3) Capacity Building: training workers and suppliers in sustainable technology practices; (4) Monitoring and Accountability: using data analytics to generate transparent reporting; and (5) Innovation Incentives: promoting eco-design and green entrepreneurship. These pillars are intended to be sector-specific and scalable. For example, an educational institution may focus on circular procurement and student awareness programs, while a technology company may focus on eco-innovation and supply chain certification. This flexibility makes the TD-EWGF theoretically and practically implementable.

The Technology-Driven E-Waste Governance Framework can define what it means to be a sustainable organization in the age of digital technology. It suggests that sustainability cannot be seen as a limitation on innovation, but as its ethical development, a new intelligence that combines ecological foresight with technological ingenuity. With the increasing digitalization of the global economy, e-waste governance will serve as an indicator of whether our technological advances are guided by long-term foresight or immediate expediency. Personally and as an academic, this framework is a call to action: to reconsider the morality of digital change, to speak truthfully, and to create institutions as strong as the technologies that created them. Then, and only then, can we trust that digital innovation will not cause ecological deterioration, but rather regenerate the environment.

2.5. Case Studies: Comparative Experience in Business, Educational, and Governmental Agencies

In the context of comparative institutional experience, the practical value of the Technology-Driven e-Waste Governance Framework (TD-EWGF) is best appreciated. E-waste governance cannot be approached as a homogeneous problem; it is filtered through diverse institutional logics, leadership cultures, and accountability. This section analyzes the framework's operation in three contrasting environments—including businesses, educational institutions, and government agencies—to shed light on the disparities and similarities in the application of sustainable technology governance. It aims to demonstrate that the principles of accountability, transparency, and inclusion will not change, but their application must depend on the cultural and structural DNA of each type of organization.

2.5.1. Corporate Sector: Technology Companies and the Circular Economy

In the corporate environment, the e-waste challenge is compounded by high product turnover, consumer preference for novelty, and complex international supply chains. Large technology companies have begun incorporating circular-economy principles into their governance frameworks, offering a partial example of the TD-EWGF model. For example, Apple Inc. reported recovering over 48,000 metric tons of e-waste in 2023, which was used to manufacture newer products such as aluminum, gold, and rare earths (Apple Environmental Progress Report, 2024). This practice aligns with the framework's pillars of policy integration and oversight. Similarly, Dell Technologies has implemented closed-loop recycling of post-consumer plastics and metals used in the manufacture of discarded electronics (Dell Sustainability Report, 2023). These examples demonstrate that companies implement governance based on data-driven observation, lifecycle accountability, and innovation incentives. However, despite these achievements, research indicates that less than 20% of global companies have effective e-waste reporting systems (Baldé et al., 2024). The difference between leaders and laggards demonstrates that technological competence alone is insufficient for governance; rather, ethical commitment and cross-sector cooperation are required.

2.5.2. Educational Institutions: Building a Sustainable Technological Culture

Universities and schools have a unique position in the digital ecosystem. They are major purchasers of electronic equipment and powerful promoters of morality.

However, e-waste management in the sector has not been fully implemented and is mainly limited to ad hoc recycling initiatives rather than tailored policy interventions. A good example is the National University of Singapore (NUS), which established a Zero Waste Task Force in 2022 with the goal of reducing e-waste by 30% by 2025 through digital inventory audits, forging partnerships with responsible disposal providers, and engaging students (NUS Sustainability Report, 2023). This strategy involves capacity building and engagement processes within the TD-EWGF. Pilot projects for e-waste literacy campaigns at Putra University, Indonesia, yielded positive results: over 70% of staff and students reported developing a better understanding of sustainable IT use through communication-based interventions (Andriani & Prasetyo, 2024). The results validate the idea that communication is not simply a process of transmission, but an agent of transformation, developing a sense of shared stewardship and moral imagination. Therefore, schools and universities can be crucial green governance laboratories for applying the theory at the civic level.

2.5.3. Government Agencies: Public Accountability and Regulatory Leadership

Government institutions are the primary aggregators and regulators of electronic equipment. However, the shortcomings of regular e-waste governance are often due to bureaucratic inertia, disjointed procurement systems, and limited budget cycles. A converse example is the Estonian Ministry of Environment, which has a publicly available asset management system to track the lifecycle of government electronic devices from purchase to disposal, with blockchain-verified transparency of disposal (European Commission, 2023). This is consistent with the current situation in the Estonian Ministry of Environment.

2.5.4. Intersectoral Learning and Comparative Perspectives

Case comparisons can vary from various structural perspectives. First, the most decisive factor is always leadership commitment: sustainability initiatives with clear executive-level mandates demonstrate much greater compliance and innovation outcomes (Eccles, Ioannou, & Serafeim, 2014). Second, transparency in communication (data disclosure, sustainability reporting, and stakeholder discussions) is the legitimacy mechanism that distinguishes between real and symbolic governance. Third, adaptive capacity across sectors is enhanced through the use of digital technology, IoT tracking, artificial intelligence analytics, and cloud-based reporting. Finally, the contextualization of cultures is inevitable; what is effective in business settings may not be directly applicable to education and government unless it is culturally contextualized and participatory. Therefore, the TD-EWGF is not a general-purpose model, but rather a meta-structure that enables local adaptation through feedback, learning, and shared governance.

2.5.5. Humanizing Governance: An Ethical Digital Civilization

E-waste governance is not just about data and systems; it is ultimately a story of human beings. Describe the negotiation between societies' ecological responsibility and technological ambition. The turning point in any institution—corporate, educational, or governmental—lies not in the sophistication of its policies but in the conviction of its people. According to a NUS sustainability officer, he found that every old laptop is, in fact, a narrative about our principles, as much as it is about our technology. This observation aligns with the TD-EWGF: norm as communication, sustainability as empathy, and innovation as care. From this perspective, e-waste governance is presented not as a control that requires management, but as the moral story of how human beings learn to live with what they create.

2.6 Implication of the Policy and Practice: Sustainable E-Waste Governance Implementation Plans

The Technology-Driven E-Waste Governance Framework (TD-EWGF) requires institutional commitment and systemic coordination for its implementation. It is not enough to offer convincing theoretical models; sustainability lies in translating principles into viable, repeatable policies and routines. This section identifies the policy and practical implications of implementing the TD-EWGF, focusing primarily on strategic alignment, regulatory synergy, leadership accountability, and stakeholder inclusion. The objective is to provide guidance on developing resilient digital ecosystems that balance technological advancements with environmental stewardship and social justice.

2.6.1. Integration of Strategic Policies

The initial and most important implication is that it must be combined with a strategic policy. In most organizations, technology policies, sustainability goals, and procurement systems are managed separately rather than working together (OECD, 2023). To achieve e-waste governance, environmental objectives must be incorporated into digital transformation policies. Therefore, digital blueprints at the national or institutional level should include quantifiable e-waste reduction targets, lifecycle standards, and indicators for resource recovery. Governments can implement green public procurement (GPP) systems by selecting suppliers that participate in circular-economy certification programs or take-back programs (European Commission, 2023). Likewise, institutions should institutionalize sustainability audits to assess digital efficiency and environmental impact. The TD-EWGF is transforming environmental responsibility into competitiveness and reputation by aligning e-waste management with performance metrics and innovation.

2.6.2 Regulatory Coherence and International Cooperation

The second implication relates to regulatory coherence and international cooperation. National regulations alone are insufficient given the extent of globalization in e-waste supply chains. A laptop discarded in Jakarta can be disassembled in Ghana or recycled in Shenzhen. This international movement requires converging standards and common liability regimes. The Basel Convention on the Control of Transboundary Movements of Hazardous Wastes provides a foundation, but its implementation is inconsistent (UNEP, 2023). Therefore, policymakers must adopt polycentric forms of governance (Ostrom, 2010), in which various centers of power, such as the national, regional, corporate, and community levels, coordinate through joint data and monitoring systems. International organizations such as UNEP, ITU, and UNU could lead an initiative to establish an e-Waste Data Alliance (EAA) to monitor flows, establish best practices, and generate collective intelligence. This is because such global collaboration would not only operationalize the TD-EWGF but also transform it into a planetary governance system.

2.6.3. Accountability and Leadership in the Organization

At the institutional level, the governance of EAAs is determined by leadership ethics and accountability structures. Executive accountability has been found to correlate with strong sustainability outcomes (Eccles, Ioannou, & Serafeim, 2014). To achieve this, organizations should institutionalize the role of Chief Sustainability Officer (CSO) or a similar position within their IT departments, and establish e-waste policies with defined lines of accountability and budgetary oversight. Sustainability performance measures should include recycling rates, carbon savings through device reuse, and compliance with e-waste disposal requirements (Dyllick & Muff, 2016). Furthermore, public reporting should adopt descriptive, evidence-based formats, including ESG (environmental, social, and governance) indicator dashboards that link e-waste data to overall climate commitments. The TD-EWGF instills an organizational culture of moral coherence, in which the distinction between doing good and doing good no longer holds, by introducing accountability into leadership systems.

2.6.4 Stakeholder Engagement and Communication Infrastructure

Another vital implication is the stakeholder engagement and communication infrastructure that supports it. A form of governance based solely on regulation will not hold; instead, persuasion, participation, and shared meaning are necessary. The TD-EWGF proposes multi-level stakeholder platforms in which producers, recyclers, regulators, and consumers communicate through transparent digital channels. For example, blockchain-based traceability systems can enable end-users to confirm that their returned devices were recycled responsibly (Baldé et al., 2024). Educational campaigns should promote digital responsibility literacy among businesses, universities, and public agencies to empower people to reduce e-waste. The vision of waste as a resource must be regularly disseminated through these channels and change people's perceptions of waste: that it is not about disposal, but about responsible management. In this sense, communication is not an auxiliary instrument, but the connective tissue of government: it infuses policies with human meaning.

2.6.5. Technological Innovation for a Sustainable Digital Future

The fifth policy-implication innovation focuses on technology as a driver of sustainability. While digitalization is a source of e-waste, it offers solutions for managing it. AI can also be used to optimize device lifecycles, anticipate hardware wear and tear, and schedule maintenance or reuse (Forti et al., 2024). Internet of Things (IoT) sensors can monitor equipment usage, and machine learning models can predict the possibility of material recovery. Governments and research institutions are encouraged to invest in green innovation ecosystems: incubators that support startups dedicated to e-waste recycling technology, eco-design, and biodegradable electronics. These types of efforts will be aligned with the United Nations Sustainable Development Goals (SDG 12: Responsible Consumption and Production), which promote technological advancements that do not degrade natural resources but rather foster their sustainability. Therefore, the TD-EWGF promotes an innovation paradigm in which sustainability has become the very logic of invention.

2.6.6. Developing Organizational Strength and Resilience.

Finally, the TD-EWGF recommends incorporating resilience and future-readiness as requirements within organizational design. In a rapidly changing world of obsolescence, leadership must transform into a dynamic rather than a conformist approach. To address emerging e-waste challenges, institutions must develop scenario planning and systems-mapping capabilities to predict potential future issues, such as those related to artificial intelligence chips, electric vehicles, and 6G networks.

Interdisciplinary research collaborations among academia, government, and industry can serve as learning systems that continually revise policies in response to data feedback and technological insights (Folke et al., 2010). Humanistically, resilience is not a technical strength, but the moral strength to care for the future. A strong organization, therefore, has imaginative governance, the ability to generate creativity in crises, and responsible waste management.

2.6.7. Toward a Global Ethic of Responsible Innovation

The TD-EWGF, in essence, seeks to develop what might be called a global ethic of responsible innovation: a moral theory that views technological advancement as responsible stewardship rather than domination. Managing e-waste is a civilizational challenge rather than simply a managerial one, and involves empathy, dialogue, and foresight. By reimagining technology as a co-evolutionary ally of nature, organizations have the opportunity to enact a new form of intelligence—ecological intelligence—that prioritizes regeneration as much as invention. This is what sustainable digital transformation means for both policymakers and practitioners: creating technologically and ethically sustainable futures. In the age of innovation and responsibility, e-waste governance is not only an environmental imperative but also a story of hope.

3. METHODOLOGY

According to Siripipatthanakul et al. (2025, a-h); Limna & Siripipattanakul (2025); Phuangsuwan et al. (2025); Jaipong & Vui (2025); Kok & Sriboonruang (2025); the PRISMA method was modified in this chapter for meta-analysis integrated with content and thematic analysis, following four steps as follows;

Step 1: **Identifying the Problem** - Initial entries can be identified through manual searches, literature databases, or other approaches. The inquiry commenced with a random selection of the topic. The article contains 154 papers.

Step 2: **Screening** - Remaining records are identified after duplicates are removed. This phase assesses the pertinence of record titles and abstracts to the study's subject or objectives. The second phase entails the deliberate selection of publications from Google Scholar, Scopus, and Web of Science. This phase selected 78 papers.

Step 3: **Eligibility** - This section outlines the extent of the study that is feasible after screening. In this phase, essential information articles are assessed for eligibility. In Phase Three, 55 papers were selected for systematic examination through a selective sampling process. The chosen papers employed AI, Grammarly, Turnitin, and Word Cloud terminology mainly from 2020 to 2025.

Step 4: **Inclusions -** This phase provides a comprehensive analysis of systematic review studies. These studies meet the criteria for further inquiry. The flowchart depicts the degree of research discontinuation at each phase and the underlying factors. Prevalent grounds for exclusion encompass irrelevance to the research subject, inadequate study design, insufficient data, or unsuitable inclusion criteria. The fourth phase was executed by three academic specialists employing PRISMA. The content analysis validated the systematic review's precision.

4. RESULTS

4.1 Grammarly Results

The appropriateness of grammar is more than 95%, and the content similarity is about 1%.

4.2 Word Cloud Result

Figure 1. Word Cloud Result

https://www.wordclouds.com/

4.3 Theoretical Framework

Figure 2. Theoretical Framework

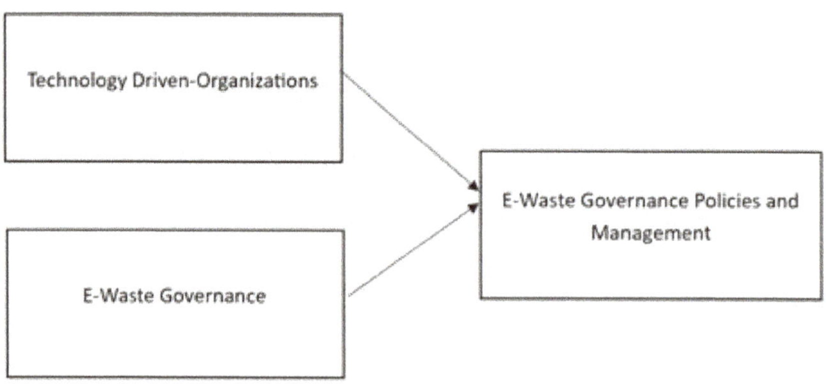

5. DISCUSSIONS

This chapter supports Grandhi et al. (2024), who found that, despite 57 million tons of e-waste generated in 2021, only 9 million were recycled, a major issue. Problems, including unlawful recycling, landfill disposal, and exports of e-waste, affect the world. E-waste reduction requires ongoing, well-structured efforts, supported by appropriate policy interventions, to minimize its generation and conserve the environment. E-waste management enterprises can profit from these policies by ensuring proper disposal, public awareness, and the repair and refurbishment of gadgets, thereby achieving higher recycling rates. Extended Producer Responsibility (EPR), WEEE directives, regional laws, and state-level laws in several nations have shown effectiveness with incentive recycling programs and stronger hazardous material rules. The material focuses on global E-waste trends and emphasizes the need for robust laws to mitigate environmental impacts and reduce electronic costs through efficient recycling.

The results confirm Zainuddin et al. (2024) that technology is essential to e-waste management because it offers creative ways to collect, recycle, and dispose of old electronics. Advanced technology can address environmental and health concerns associated with electronic waste and promote a cyclical, sustainable paradigm for device consumption. As societies realize the importance of sustainable e-waste management, research on technology and e-waste management is rising worldwide. Researchers are developing new technologies to improve recycling, reduce environmental impact, and promote a circular economy, making our planet more sustainable. It explains technology and e-waste management trends for study. The integration of technology and e-waste management shows the importance of scientific contributions. Many terms are used in the subject, but "electronic waste," "e-waste," and "metal recovery" are the most common.

The findings support Oyebode (2024) that E-waste has increased due to the fifth industrial revolution (5IR). Electronic waste must be managed to reduce its environmental and public health impacts. 5IR is essential for managing computer and electronic waste. The risks of e-waste, the economic benefits of recycling, and global best practices for managing and forecasting it are all factors. An integrated strategy is key to the success of e-waste management. The plan should include policy and technology. Developing e-waste recycling technology and enhancing recycling infrastructure are technological measures. Policy efforts include enacting laws to ensure safe e-waste disposal and promoting public awareness campaigns regarding recycling. A sustainable development. Formal collection and documentation systems can track the flow of e-waste and evaluate management operations. Collaboration between industry, government, and civic society was stressed to control electronic waste.

The result confirms Golpeo et al. (2025) that e-waste management is essential for resource recovery and environmental protection. This study analyzes how a circular economy framework and intelligent technology might improve urban electronic waste management. E-waste reduction, reuse, and recycling techniques vary across homes, LGUs, HEIs, and transporters. Maintaining and repairing gadgets is the norm across many industries, whereas renting is overlooked. Households prioritize equipment maintenance, carriers encourage trade-ins, and higher education institutions buy used and reuse obsolete technologies. Transporters emphasize proper disposal, higher education institutions emphasize modification, and households/local governments emphasize sustainable procurement. Intelligent technologies, such as automated sorting and AI-enhanced material recovery, improve efficiency and support circular economy principles. Governments must improve education, infrastructure, and regulations to support sustainable e-waste management. Stakeholder collaboration, technological innovation in recycling, and extended producer responsibility (EPR) policies can boost resource recovery and reduce environmental damage.

6. CONCLUSION

6.1 The Road to a Sustainable Digital Ecosystem

Technology is the creator of contemporary civilization and the source of new ecological concerns in the developing digital age. The acceleration of innovation has not only increased economic productivity and social connectivity but also generated ever-increasing amounts of e-waste, a physical manifestation of the contradiction between improvement and depletion. This chapter examines the paradox within the Technology-Driven e-Waste Governance Framework (TD-EWGF), arguing that the e-waste problem is not primarily environmental but organizational, communicative, and ethical. The boardroom, policy design, and raising awareness among digital citizens are the path to a sustainable digital ecosystem, but not the landfill.

The central focus of this chapter is the declaration of ethical regulation and the environmental roots of technological innovation. An unsustainable digital transformation such as this represents a lack of collective vision, a mortgaged future surrendered to the speed of action. This TD-EWGF proposal reinterprets governance as a process of alignment between policy coherence, communicative transparency, and adaptive learning. Consider e-waste management not post-production, but design-in, as a principle embedded in the DNA of digital organizations. This reengineering of end-of-pipe management toward lifecycle integration represents a paradigm shift in governance, moving from compliance-oriented to purpose-oriented governance (Dyllick & Muff, 2016).

However, this balance is the goal that requires multi-stakeholder cooperation and systemic accountability. Governments must be proactive in legislation, companies must operate conscientiously, and individuals must be mindful. Empirical data indicate that countries that have adopted e-waste policies, such as Japan, Germany, and South Korea, have achieved material recovery rates above 80%, indicating that sustainability is not an ideal, but a feasible reality (Baldé et al., 2024). However, achieving the same level of success globally will require implementing polycentric governance frameworks, open data systems, and effective cross-sector partnerships. The TD-EWGF presents a roadmap for these interdependencies and builds shared responsibility in governance through distributed governance intelligence.

Notably, this framework values the central role of communication as a means and mechanism for sustainability. It is not policies that govern, but people through dialogue, mutual understanding, and engagement. By clearly sharing their sustainability goals, organizations create stories that build culture, foster compliance, and innovation. In this sense, governance is both technical and rhetorical: it is persuasive and therefore presented in advance, before being regulated. Ethical narrative is a vital part of a sustainable digital ecosystem, as is regulatory sophistication (Heath & Palenchar, 2021). Communication becomes the ethical architecture of governance, enabling empathy amid complexity.

The technological frontier must shift from extraction to regeneration. With ecological ethics, it is possible to transform resource efficiency, material tracking, and circularity through AI, blockchain, and the Internet of Things (Forti et al., 2024). However, when innovation is not linked to moral reasoning, it can lead to greater inequality and waste. The key to the future of digital ecosystems lies in integrating environmental intelligence into technological systems: sustainability is not a byproduct but an algorithm. This integration will redefine competitiveness: organizations that design for durability and recyclability will dominate markets determined by conscious consumers and green capital flows.

6.2 Technology Organization Policies

Philosophically, this chapter suggests a new social contract between humanity and technology. It is a promise of care: the recognition that technological advances, to be meaningful, must also preserve life. The TD-EWGF embodies this spirit, combining the logic of innovation and the ethics of responsibility. The framework creates a vision of a future in which devices are designed to be disassembled, organizations are held accountable for their lifecycles, and policies respond to the new technological reality. A sustainable digital ecosystem is not a goal in itself, but rather an ethical development that involves a technology-driven e-waste governance framework that formulates comprehensive policies and regulations, typically focused on Extended Producer Responsibility (EPR), and utilizes digital tools to oversee the complete lifecycle of Electrical and Electronic Equipment (EEE). This method employs technologies such as IoT for asset tracking, data analytics to assess generation and recycling rates, and digital platforms to improve transparency and stakeholder collaboration, thereby fostering a circular economy by encouraging eco-design, efficient collection, and sustainable material recovery.

The issue of E-waste in developing countries stems from the importation of electronic and used goods from wealthy nations, coupled with a lack of effective E-waste management strategies. Insufficient protections, regulations, and policies. The legislation, technology, and enforcement regarding the proper disposal of electronic trash have resulted in significant human health and environmental issues in various nations. Concerns stem not only from the substantial amount of E-waste brought into developing nations but also from the extensive array of hazardous compounds associated with it. In light of the gravity of human health and environmental issues,

An initiative has been launched to address the existing and prospective E-waste-related issues in developing countries. Participants in the session have discussed and recognized prevalent scenarios and issues. An accord has been established to formulate a guideline for E-waste management within the developing country context, serving as a foundation for collaboration among nations, aimed at the future creation of an appropriate framework, supporting tools, and initiatives for E-waste management in the area.

REFERENCES

Andriani, A. D., & Prasetyo, B. (2024). Building digital sustainability awareness through organizational communication in higher education. *Journal of Environmental Communication Studies*, *12*(3), 201–219.

Apple Inc. (2024). *Environmental progress report 2024.* https://www.apple.com/environment/pdf/Apple_Environmental_Progress_Report_2024.pdf

ASEAN. (2021). *ASEAN digital masterplan 2025.* Association of Southeast Asian Nations. https://asean.org

Baldé, C. P., Forti, V., Gray, V., Kuehr, R., & Stegmann, P. (2024). *The Global E-Waste Monitor 2024.* United Nations Institute for Training and Research (UNITAR), International Telecommunication Union (ITU), and International Solid Waste Association (ISWA). https://ewastemonitor.info

Beck, U. (1992). *Risk society: Towards a new modernity.* SAGE Publications.

Bocken, N. M. P., Short, S. W., Rana, P., & Evans, S. (2022). Sustainable business models and circular economy transitions. *Journal of Cleaner Production*, *372*, 133764.

Christensen, L. T., & Langer, R. (2020). Corporate communication and sustainability: An organizational legitimacy perspective. *Journal of Business Ethics*, *162*(2), 415–428.

Cornelissen, J. (2023). *Corporate communication: A guide to theory and practice* (7th ed.). SAGE Publications.

Dell Technologies. (2023). *Sustainability and impact report 2023.* https://www.dell.com/learn/us/en/uscorp1/sustainability

Dyllick, T., & Muff, K. (2016). Clarifying the meaning of sustainable business: Introducing a typology from business-as-usual to true business sustainability. *Organization & Environment*, *29*(2), 156–174.

Eccles, R. G., Ioannou, I., & Serafeim, G. (2014). The impact of corporate sustainability on organizational processes and performance. *Management Science*, *60*(11), 2835–2857.

Eccles, R. G., Ioannou, I., & Serafeim, G. (2020). The impact of corporate sustainability on organizational processes and performance. *Management Science*, *66*(9), 3975–3998.

Elkington, J. (1997). *Cannibals with forks: The triple bottom line of 21st-century business*. Capstone Publishing.

European Commission. (2020). *A new circular economy action plan for a cleaner and more competitive Europe*. Publications Office of the European Union. https://eur-lex.europa.eu

European Commission. (2023). *E-governance and sustainability in the European digital agenda*. European Union Publications.

Folke, C., Carpenter, S. R., Walker, B., Scheffer, M., Chapin, T., & Rockström, J. (2010). Resilience thinking: Integrating resilience, adaptability and transformability. *Ecology and Society*, *15*(4), 20.

Forti, V., Baldé, C. P., & Kuehr, R. (2020). *E-waste statistics: Guidelines on classification, reporting and indicators*. United Nations University.

Forti, V., Baldé, C. P., & Kuehr, R. (2024). Artificial intelligence and e-waste management: The next frontier of digital sustainability. *Journal of Cleaner Production*, *441*, 141294.

Geissdoerfer, M., Savaget, P., Bocken, N. M., & Hultink, E. J. (2017). The circular economy: A new sustainability paradigm? *Journal of Cleaner Production*, *143*, 757–768.

Ghisellini, P., Cialani, C., & Ulgiati, S. (2016). A review on circular economy: The expected transition to a balanced interplay of environmental and economic systems. *Journal of Cleaner Production*, *114*, 11–32.

Golpeo, J. M., Andres, G. L., Faronal, I. N., & Casuat, C. D. (2025). Sustainable E-Waste Management Through a Circular Economy Framework: Integrating Smart Technologies for Resource Recovery in Urban Environments. In *E3S Web of Conferences* (Vol. 643, p. 01005). EDP Sciences., DOI: 10.1051/e3sconf/202564301005

Grandhi, S. P., Dagwar, P. P., & Dutta, D. (2024). Policy pathways to sustainable E-waste management: A global review. *Journal of Hazardous Materials Advances*, *16*, 100473. DOI: 10.1016/j.hazadv.2024.100473

Heath, R. L., & Palenchar, M. J. (2021). *Strategic issues management: Organizations and public policy challenges* (3rd ed.). SAGE Publications.

Husted, B. W., & de Sousa-Filho, J. M. (2017). The impact of sustainability governance, country stakeholder orientation, and country risk on environmental, social, and governance performance. *Journal of Cleaner Production*, *155*, 93–102.

Jaipong, P., & Vui, C. N. (2025). Sustainable Private Hospitals Based on Prices, Services, and Facilities. In *Sustainable Development for Hospitals and Clinics* (pp. 97–126). IGI Global Scientific Publishing., DOI: 10.4018/979-8-3693-9800-5.ch004

Kaptein, M. (2022). Understanding unethical behavior by unraveling ethical culture. *Human Relations*, *75*(2), 191–216.

Kojima, M., & Yoshida, A. (2022). E-waste management in East and Southeast Asia: Current policy and future perspectives. *Journal of Environmental Management*, *310*, 114754.

Kok, S. L. S., & Sriboonruang, P. (2025). Healthcare Digitization and Sustainability of Penang Healthcare Providers: A Case of Medical Tourism. In *Sustainable Development for Hospitals and Clinics* (pp. 127-162). IGI Global Scientific Publishing. DOI: DOI: 10.4018/979-8-3693-9800-5.ch005

Kuhlmann, S., Stegmaier, P., & Konrad, K. (2019). The tentative governance of emerging science and technology—A conceptual introduction. *Research Policy*, *48*(5), 1091–1097.

Lepawsky, J. (2018). *Reassembling rubbish: Worlding electronic waste*. MIT Press.

Li, Y., Song, B., & Zhu, L. (2023). Integrating green governance into digital policy: Lessons from the EU circular economy. *Environmental Policy and Governance*, *33*(2), 153–169.

Limna, P., & Siripipattanakul, S. (2025). Sustainable Clinic Management Based on Practices and Human Resource Management (HRM). In *Sustainable Development for Hospitals and Clinics* (pp. 65–96). IGI Global Scientific Publishing. DOI: DOI: 10.4018/979-8-3693-9800-5.ch003

Lock, I., & Seele, P. (2021). Green governance: Integrating sustainability into corporate communication and accountability. *Business Strategy and the Environment*, *30*(5), 2324–2338.

Meadows, D. H. (2008). *Thinking in systems: A primer*. Chelsea Green Publishing.

National University of Singapore (NUS). (2023). *NUS sustainability report 2023*. https://sustainability.nus.edu.sg

Nnorom, I. C., & Osibanjo, O. (2021). Overview of e-waste management practices and regulations in developing countries. *Waste Management & Research*, *39*(5), 619–630.

OECD. (2023). *E-waste management and policy coherence: Bridging technology and environment. OECD Environment Working Papers, No. 194.*

Organisation for Economic Co-operation and Development (OECD). (2023). *Digital economy outlook 2023*. OECD Publishing.

Ostrom, E. (2010). Polycentric systems for coping with collective action and global environmental change. *Global Environmental Change, 20*(4), 550–557.

Owen, R., Stilgoe, J., Macnaghten, P., Gorman, M., Fisher, E., & Guston, D. H. (2013). *Responsible innovation: Managing the responsible emergence of science and innovation in society*. John Wiley & Sons.

Oyebode, O. J. (2024, April). Management of Computer and Electronic Waste through the Fifth Industrial Revolution for the Cleaner Environment. In *2024 International Conference on Science, Engineering and Business for Driving Sustainable Development Goals (SEB4SDG)* (pp. 1-14). IEEE. DOI: DOI: 10.1109/SEB4SDG60871.2024.10629821

Phuangsuwan, P., Rungsiyanont, S., & Sriboonruang, P. (2025). Sustainable Development in Hospitals for Elderly Dysphagia. In *Geriatric Dentistry. In Sustainable Development for Hospitals and Clinics* (pp. 225–252). IGI Global Scientific Publishing., DOI: 10.4018/979-8-3693-9800-5.ch008

Putnam, L. L., & Nicotera, A. M. (2010). Communicative constitution of organization is a question: Critical issues for addressing it. *Management Communication Quarterly, 24*(1), 158–165.

Rana, N. P., Luthra, S., Mangla, S. K., Islam, R., & Roderick, S. (2023). Exploring the dark side of digital transformation: Environmental challenges and sustainability paradoxes. *Technological Forecasting and Social Change, 188*, 122329.

Siripipatthanakul, S., Chaiprakarn, S., Rungsiyanont, S., & Phuangsuwan, P. (2025, a). A Case Study of Sustainable Development in Hospitals in Thailand Using ChatGPT. In *Sustainable Development for Hospitals and Clinics* (pp. 1-32). IGI Global Scientific Publishing. DOI: DOI: 10.4018/979-8-3693-9800-5.ch001

Siripipatthanakul, S., Phuangsuwan, P., & Jaipong, P. (2025, e). Mobile Application Security: A Case Study of Bank of Ayudhya PCL.(KMA), Thailand. In *International Conference on Data-Processing and Networking* (pp. 479-493). Springer, Singapore. https://doi.org/DOI: 10.1007/978-981-96-3102-5_36

Siripipatthanakul, S., Phuangsuwan, P., & Jaipong, P., & Muthmainnha. (2025, f, September). Wi-Fi Networks and Security: A Systematic Review. In *International Conference on Artificial Intelligence and Networking* (pp. 81-92). Singapore: Springer Nature Singapore. https://doi.org/DOI: 10.1007/978-981-96-2015-9_6

Siripipatthanakul, S., Phuangsuwan, P., Siripipattanakul, S., & Jaipong, P. (2025, c). Sustainable Tourism Based on Challenges, Tourist Experiences, and Host Communities: A Case Study of Ayutthaya, Thailand. In *Best Practices for Human Resources Management in Tourism and Hospitality* (pp. 361-394). IGI Global Scientific Publishing. DOI: DOI: 10.4018/979-8-3693-8779-5.ch014

Siripipatthanakul, S., Phuangsuwan, P., Siripipattanakul, S., Limna, P., Jaipong, P., & Chaiprakarn, S. (2025, g, September). Sustainable Development Adopting Digital Intelligence Among University Students. In *International Conference on Artificial Intelligence and Networking* (pp. 211-220). Singapore: Springer Nature Singapore. https://doi.org/DOI: 10.1007/978-981-96-2015-9_14

Siripipatthanakul, S., Phuangsuwan, P., Siripipattanakul, S., Limna, P., Jaipong, P., & Chaiprakarn, S. (2025, h, September). Privacy and Information Security for Sustainable Community Among TikTok's Users. In *International Conference on Artificial Intelligence and Networking* (pp. 221-233). Singapore: Springer Nature Singapore. https://doi.org/DOI: 10.1007/978-981-96-2015-9_15

Siripipatthanakul, S., Rakpathum, W., Siripipattanakul, S., Kaewnaknaew, C., Phuangsuwan, P., & Jaipong, P. (2025, d). The Future Artificial Intelligence (AI) for the Hotel and Hospitality Industry: The Case of Hotel's Human Resource Management (HRM). In *Best Practices for Human Resources Management in Tourism and Hospitality* (pp. 395-422). IGI Global Scientific Publishing. DOI: DOI: 10.4018/979-8-3693-8779-5.ch015

Siripipatthanakul, S., Rungsiyanont, S., & Siripipattanakul, S. (2025, b). Sustainable Dentistry Using Artificial Intelligence (AI) for Digital Education and Business Among University Dental Hospitals. In *Sustainable Development for Hospitals and Clinics* (pp. 33-64). IGI Global Scientific Publishing. DOI: DOI: 10.4018/979-8-3693-9800-5.ch002

Sovacool, B. K., & Hess, D. J. (2017). Ordering theories: Typologies and conceptual frameworks for sociotechnical change. *Social Studies of Science*, *47*(5), 703–750. PMID: 28641502

United Nations Environment Programme (UNEP). (2023). *Illegal e-waste trade and environmental governance.* UNEP Report Series.

United Nations Environment Programme (UNEP). (2023). Global review of the Basel Convention and e-waste enforcement. UNEP.

United Nations Industrial Development Organization (UNIDO). (2022). *Green Industry Platform: Advancing sustainable industrialization.* UNIDO Reports.

United Nations Institute for Training and Research (UNITAR). (2024). *The Global E-Waste Monitor 2024: Facts and Figures.* https://ewastemonitor.info

Widmer, R., Oswald-Krapf, H., Sinha-Khetriwal, D., Schnellmann, M., & Böni, H. (2005). Global perspectives on e-waste. *Environmental Impact Assessment Review, 25*(5), 436–458.

World Economic Forum. (2023). *Global Risks Report 2023.* WEF Publications. https://www.weforum.org

Zainuddin, S. A., Abdullah, T., Ibrahim, M. A. H., Nawi, N. C., Nasir, N. A. M., Latef, A. S. A., & Musa, Z. C. (2024). Analyzing Global Research Trends in Technology and E-Waste Management: A Bibliometric Perspective. *Technology-Driven Business Innovation: Unleashing the Digital Advantage, 2*, 443–453. DOI: 10.1007/978-3-031-62656-2_39

KEY TERMS AND DEFINITIONS

E-waste Governance Framework: An e-waste governance framework comprises rules, laws, and regulations that direct the treatment of electronic waste, with the objective of safeguarding human health and the environment. Essential elements encompass delineating obligations for stakeholders, such as manufacturers and recyclers; instituting regulations for collection, transportation, and disposal; and advocating for the recovery of valuable resources through programs like the circular economy. Efficient frameworks also tackle technology as a supporting role, the cross-border transfer of electronic trash, and promote public awareness and engagement.

Extended Producer Responsibility (EPR): Extended Producer Responsibility (EPR) is a policy framework that holds producers accountable for the environmental consequences of their products over their entire lifecycle, encompassing post-consumer phases such as disposal and recycling. Extended Producer Responsibility (EPR) transfers the physical and/or financial obligations of waste management from governmental entities to producers, incentivizing them to create more environmentally sustainable products and finance efficient collection and recycling infrastructure.

Governance Policies: Governance policies are the regulations, practices, and procedures that guide and regulate an organization, guaranteeing effective, ethical management in the best interest of its stakeholders. They are crucial for establishing clear expectations, ensuring transparency, reducing risk, and informing decision-making, often encompassing board responsibilities, risk management, financial accountability, and business ethics.

Technology-Driven Organizations: Technology-driven corporations are enterprises that integrate technology as a fundamental component of their strategy and operations to enhance efficiency, foster innovation, and stimulate growth. These firms have a culture that prioritizes digital transformation, data-driven decision-making, and agility, enabling them to swiftly adjust to market fluctuations and secure a competitive edge. They contrast with firms that merely use technology as a supporting role, incorporating it comprehensively across all divisions to inform their vision and strategy.

Waste Management: Waste management encompasses the processes and actions necessary to oversee waste from its origin to its ultimate disposal. This encompasses the collection, transportation, treatment, and disposal of trash, along with the oversight and control of the waste management process, as well as waste-related legislation, technologies, and economic frameworks.

Chapter 3
Integrating Green HRM Policies With AMO and SCT Theory to Promote Sustainable Performance and Employee Well-Being

Syamsul Hadi
https://orcid.org/0000-0001-7226-4978
Universitas Sarjanawiyata Tamansiswa, Indonesia

Kusuma Chandra Kirana
Universitas Sarjanawiyata Tamansiswa, Indonesia

Ambar Lukitaningsih
https://orcid.org/0009-0002-9847-9242
Universitas Sarjanawiyata Tamansiswa, Indonesia

Utik Bidayati
Universitas Ahmad Dahlan, Indonesia

Dasmadi Dasmadi
STIE APRIN, Indonesia

ABSTRACT

This chapter covers the evolution, mediation, moderation, and research directions of green HRM. PRISMA-based systematic literature reviews excluded 549 papers, leaving 140. Analysis of themes involved comparisons. GHRM fosters financial,

DOI: 10.4018/979-8-3373-6821-4.ch003

operational, and innovative performance and well-being (mental health, contentment, engagement, retention) through employee involvement, green behavior, green innovation, and organizational culture. AI, context, and digital technology improve these encounters. This unique analysis presents GHRM research from 2006 to 2025, integrates ecology-inclusivity (DEI) into Ability, Motivation, Opportunity, and Social Exchange Theory, and prioritizes research. The theoretical implications support GHRM as a social-exchange-based mechanism that uses the Ability-Motivation-Opportunity pathway to turn HR skills into sustainable advantages and requires a model that balances economic, social, and environmental aims.

1. INTRODUCTION

Green Sustainable Human Resource Management (GHRM) has become a concern for organizations worldwide, with a direct impact on organizational performance in both the private and non-profit sectors. Furthermore, it can contribute to good corporate governance by harmonizing employee welfare. Clearly, this indicates a massive shift in awareness towards sustainability, particularly highlighting the long-term welfare of employees amid rapid technological development (Dyageti & Kiran, 2025; Yin, 2023). "Humans will inevitably lose when competing in a race against horses, but in terms of acceleration, how can we ride horses so that we can go faster?" That is the quote that accurately describes current technological developments. It is no surprise that HRM management has evolved so rapidly from traditional practices towards green sustainability, which no longer ignores ecological responsibility and social inclusivity. This is currently challenging, as various generations with unique perspectives often cause friction within organizations (Bratton et al., 2025). The vision always disseminated by the United Nations in the objectives of *sustainable development goals* (SDGs) makes us believe that every organization must look for balance, social issues, and environmental concerns, which can be quoted as internal elements of company governance and external things related to politics and norms within the economic process (Madero-Gómez et al., 2023). These challenges are like art, where there are ways to overcome them. However, in practice, HRM has evolved in line with the company's sociocultural context, leading to green HRM that integrates environmental factors into human resource management to improve organizational performance and foster environmentally friendly behavior (Faisal, 2023; Noor Faezah et al., 2024). Several studies indicate that companies capable of sustainably managing the challenges of implementing Green HRM can maintain, and even increase, employee productivity. In fact, they can increase employee awareness to always be involved according to their abilities by a significant percentage of 25%

A review conducted by Madero-Gómez et al. (2023), for example, shows gaps across studies, such as how HRM practices across research subjects have different impacts on organizational sustainability, which ultimately affects organizational performance and employee welfare, and also highlights several contexts that have a negative impact. This shows that the art of implementing GHRM varies significantly across organizations with unique sociocultural characteristics. In addition, indirect aspects also vary. Researchers are aware that these differences are reasonable and therefore valid, allowing for differences in different contexts, especially those related to mediating aspects such as the role of involvement, motivation, green behavior, and employee resilience (Gyensare et al., 2024; Lu et al., 2023; Nakra et al., 2024; Winarno & Silvianita, 2024). In addition to mediating aspects, moderation also contributes to the diversity of these relationships, such as artificial intelligence and commitment as moderating factors (Housheya & Atikbay, 2025). Not only that, but the research object also becomes a debate over how *green HRM* practices affect financial performance, or even through employees' attitudes and behaviors (Permadi et al., 2024). As highlighted by Purnasari et al. (2025), behavior plays a crucial role in improving financial literacy.

Furthermore, the most fundamental issue remains the gap between the ecological and inclusive dimensions of sustainable HRM practices, which hinders and even complicates comprehensive implementation to support sustainability itself. This may have been highlighted by many organizations, so that organizations are not yet confident and fully utilizing the potential of implementing sustainable HRM as a planned and strategic driving force to continue to be long-term oriented in improving performance and increasing employee resilience in various dynamic situations today (Caputo et al., 2025; Rosyafah et al., 2025). Various development sciences, particularly the relationship between GHRM and performance improvement and well-being, clearly show that they have not been comprehensively integrated into a single theory, such as social exchange and ability motivation opportunity.

This study views sustainable HRM as a strategic professional internal management system that integrates various aspects such as the environment, society, and economy into policy formulation and how HR implementation encourages all available resources to focus on the main factors of organizational sustainability, which are indeed related to sustainable performance and maintaining the welfare of employees as the leading actors in organizational operations, as well as being a source of capital that needs to be increased (Agrawal et al., 2024; Bratton et al., 2025). The mechanism of Green HRM, employee welfare and Organizational performance based on *Ability-Motivation-Opportunity (AMO)* theory, where improved intent in an organization will result from supporting its' ability, keep motoring it and provide opportunities use of exchange social using *social exchange theory* (SCT), whereby when employees receive fair support, they will respond with positive behavior and not just economic transactions (Wang et al., 2025). This theory is essential for dissecting how sustainable HRM practices can enhance organizational performance and employee well-being (Abu-Mahfouz et al., 2023; Din et al., 2023). In this context, Green HRM asserts that success in an organization depends on the extent to which it creates a reciprocal relationship with its employees, ultimately strengthening the company's long-term sustainability orientation.

Given the gaps in previous research and the urgency of this study, this study aims to synthesize empirical research and theoretical approaches related to the concept and practice of green HRM across the existing literature and to examine their impact on organizational performance and employee welfare. Thus contributing to an understanding of how HRM mechanisms, particularly in terms of performance and sustainable welfare, and the positive impact of implementing green HRM. In addition, this study highlights the integration of mediation and moderation to improve organizational performance, providing insights into various integrations (Nakra et al., 2024). Moreover, this research would have managerial implications, specifically for Managers particularly in Human Resource to further enhance their understanding and insight on how they can structure support mechanism when implementing focused/oriented polices towards sustainability awareness such that the will naturally find it easier form a sustainable organization since it is now evident that leaving out being sustainable will significantly hamper the long term success of an organization (Bratton et al., 2025; Kosasih, 2024), of course, by combining ecological and social perspectives in every policy established.

2. LITERATURE REVIEW: RESEARCH THEORY

2.1 Ability Motivation Opportunity (AMO) Theory

The Ability Motivation Opportunity (AMO) theory was first developed by which explains how individual performance in an organization can be improved through three main approaches, namely ability, motivation, and opportunity. These dimensions have been used in various studies, particularly in human resource management (HRM) and organizational behavior, where the primary focus of every organization is to improve performance sustainably in a dynamic, changing environment. Various contexts, such as the academic field, show that the three AMO dimensions can contribute to performance improvement, particularly by increasing knowledge and research in academic circles, where ability and motivation have a positive impact on performance (Benítez-Núñez et al., 2024). This theory emphasizes that when managers seek to implement effective HR management, they must apply a balanced combination of abilities, intrinsic and extrinsic motivation, and fair opportunities for employees, so that there is synergy in sustainably supporting effective organizational management. Applying only one of these elements will not result in successful transformation, let alone one oriented towards long-term sustainability (Alkhalaf & Al-Tabbaa, 2024; Liu et al., 2024). However, each company has its own unique values/culture, and almost every organization, whether national or international, even at the SME level, is different, although there are many similarities in general. This environment also affects the application of the AMO approach. (Gooderham et al., 2022) . This theory reinforces the application of GHRM, which can directly improve organizational performance and enhance employee well-being by continuously developing capabilities, motivating, and providing fair opportunities.

2.2 Social Exchange Theory **(SCT)**

The role of the organization is always associated with the reciprocity it provides to employees, and vice versa. Relating to the present study, social exchange theory (SET) are very relevant in integrating AMO theory on performance improvement where this theory assumes that if organisation supports employees' growth both through range of training, coaching as well as motivating in terms of fair compensation, beneficial outcome and fair behaviour within environment organizational support instruction or normative; then employee will feel obligated to reciprocate back as much to company such as improve their performance, increase role awareness by the high extra-role behavior during work time not only all over supporting the organization but also involve into psychological contracts with an organizations a well (Ahmad et al., 2023; Blau, 2017; Emerson, 1976). Whilst implementing HRM practices across organisations, the concern should be for employees' wellbeing, including the right to their position and job, opportunities for development, and remuneration and rewards following due process. This can be considered a form of staff support, which, in turn, enhances loyalty. Thus, it is clear that this improves employee performance, which, in turn, enhances organizational sustainability.

Furthermore, because both parties feel satisfied, well-being will arise (Parent-Lamarche et al., 2023; Rajâa & Mekkaoui, 2025). However, it should be noted that the impact of social exchange theory depends heavily on consistent procedures and fairness in its implementation. It is easy to imagine that when an organization makes promises but does not prioritize fairness in their implementation, employees will feel disappointed and lose trust in the organization, which will reduce reciprocity and even have the opposite effect, thereby reducing performance and impacting well-being (Kilroy et al., 2023). The exchange of resources between organizations and employees must be managed wisely to create equitable justice that can be maintained and developed in the increasingly dynamic and uncertain future.

3. METHODOLOGY

This study uses a simple and systematic method that includes stages in a comprehensive search for relevant literature using various reliable databases through inclusion and exclusion criteria to ensure the relevance and quality of articles, as well as thematic analysis to map the direction of research and in-depth understanding in the context of sustainable HRM (Hadi et al., 2020, 2025).

Explore: Deepening the research question: How does sustainable HRM affect performance and well-being?

The PRISMA approach was applied to streamline the process, based on the studies by Asrifan et al. (2025), Siripipatthanakul (2025), and Hadi et al. (2025). The content analysis is based on the studies of Siripipatthanakul et al. (2025; a-h); and this is reflected in Table 1 below:

Table 1. Stage selection of articles

Stages Selection	**Number of Articles**	**Description**
Identification	488	The article was obtained from search results based on research questions regarding the impact of sustainable HRM on performance and well-being.
Citation Chaining	61	Addition of articles through *backward* and *forward citation chaining* processes
Total articles obtained	54	Total articles before the duplication removal process
Deletion of duplicate articles	1	There is one duplicate article that needs to be deleted
Total articles after deletion	548	Articles that passed the initial selection stage after article deletion.
Highly Relevant Articles	140	Total articles that passed the selection and were deemed most appropriate and used in this deep review

The next stage involves synthesizing each article to identify and analyze them to answer the research questions, explore developments in the literature, and provide recommendations for further research, particularly in the field of sustainable HRM.

4. RESULTS & DISCUSSION

4.1 Descriptive analysis

The descriptive results of this literature study show that most of the research used quantitative methods, followed by systematic literature reviews, and that the results reflect a multidisciplinary approach. Some of the results can be summarized as follows:

1. *Organizational performance:* More than 100 articles show that sustainable HRM practices have a positive effect on financial, operational, and innovation performance across sectors, including manufacturing, hospitality, and public administration. In addition, these studies emphasize mediating variables related to green innovation and green employee behavior in improving sustainable organizations and increasing competitive advantage (Awwad Al-Shammari et al., 2022; Gazi et al., 2025; Kharel et al., 2025). Several articles also highlight various challenges, particularly in measuring performance impact, and recommend the use of longitudinal research to determine not only cross-sectional but also long-term effects, thereby ensuring more accurate data (Kosasih, 2024; Yuxiaosi, 2024).
2. *Employee well-being:* 80 articles show that sustainable HRM practices can improve employee well-being, including mental health, job satisfaction, engagement, and retention (Jaskeviciute et al., 2021; Sharma et al., 2023). Employee well-being is also associated with organizational trust, inclusion, and work-life balance initiatives, which positively impact employee loyalty and productivity (Al-Sulaiti, 2024; Usha & Ramkumar, 2025). Some studies point to the multidimensionality of wellbeing, with weak direct relationships in some cases, and show job resources and statutory culture as the two leading enablers (Tortia et al., 2022).
3. *Mediating and Moderating:* variables such as employee engagement, green motivation, organizational culture, and leadership style can act as mediating or moderating factors that influence the effectiveness of sustainable Green HRM practices on positive organizational outcomes (Gilal et al., 2024). The integration of digital technology and artificial intelligence (AI) can also serve as a moderator, strengthening employee engagement and the effectiveness of sustainable HRM (Herlina & Iskandar, 2025). Additionally, psychological contracts and adaptability serve as moderating factors, strengthening the relationship between sustainable HRM and employee well-being (Nakra et al., 2024; Rafique et al., 2024).

4. *Context and Industry:* This study spans various research sectors that enrich the implementation of sustainable green HRM. Some areas, such as manufacturing, hospitality, education, and the public sector, are among the most studied (Järlström et al., 2023; Vovk & Vovk, 2024). This research is broadly scoped to include Asia, Europe, Africa, and the Middle East, examining how cultural and regulatory factors impact sustainable HRM practices (Ofosuhene, 2024). Size and type of organization have also been found to influence the adoption of HRM and the success level of sustainable HRM, where SMEs have different challenges than large organizations (Kutaula et al., 2025; Zihan & Makhbul, 2024)

4.2 The evolution of GHRM

The findings of this review of various articles can be classified as an evolutionary trajectory in research, from a basic understanding of GHRM to a more comprehensive conceptualization. After that, GHRM began to expand with various theoretical foundations, leading to empirical evidence emerging across sectors about GHRM itself, and finally, it began to integrate technology, inclusiveness, and GHRM implementation strategies. This evolution can be seen in the following Table 2:

Table 2. The evolution of Green Human Resource Management

Year Range	Research Direction	Description
2006–2011	Fundamentals of HRM and Sustainability	Early research examined the relationships among HRM, company performance, and employee welfare and identified the complexities and challenges of linking HR practices to sustainable organizational outcomes. During this period, the concept of *Green HRM* was introduced, emphasizing the important role of HR in environmental sustainability and worker health.
2017–2019	GHRM Framework and CSR Integration	The work in this period examined CSR and sustainable HRM, emphasizing sustainable business models and the strategic integration of human resource strategies to motivate and retain human capital.
2020–2022	GHRM and Theoretical Advances	The focus of such research is to study the development of Green HRM practices (recruitment, training, and performance appraisal) and their influence on environmental performance and organizational sustainability. Social Exchange Theory and Ability-Motivation-Opportunity (AMO) are employed as some of the theories that describe mediating mechanisms between employee behaviors and organizational outcomes
2023–2024	Empirical Evidence from Various Sectors on GHRM	Academic research supports the GHRM's positive impact on employee engagement, wellbeing, organizational commitment, and sustainable performance across sectors (including manufacturing, hospitality, and education). The study also examines the mediating role of green innovation, employees' motivation, and organizational culture, as well as barriers to implementation
2025	Technology Integration, Inclusivity, and Strategic GHRM	Recent studies emphasize the relationships among artificial intelligence (AI), digitalization, and inclusive leadership within the GHRM system. The focus is increasingly shifting toward strategically aligning GHRM with global sustainability goals, improving employee well-being, organizational agility, and meeting social and environmental demands.

4.3 Thematic Review

The literature on current HRM practices reveals several interesting themes, particularly its role in improving performance and well-being to support environmentally friendly organizational sustainability. In addition, mediating and moderating factors are also highlighted in this theme. The themes resulting from the integration of 140 articles can be seen in the following Table 3:

Table 3. The themes resulting from the articles' integration

Thematic	Theme Description
Green Human Resource Management (GHRM) and Organizational Performance	64 articles indicate that GHRM is a driver in improving innovation, social, and economic performance across various industries. Green recruitment, training, performance management, and employee engagement have been proven to improve sustainability and competitive advantage (Akbar et al., 2024; Ghosh & Haque, 2024; Kharel et al., 2025; Yin, 2023)
Employee wellbeing and GHRM	52 articles indicate that GHRM enhances employee wellbeing by supporting mental health, promoting work-life balance, and implementing inclusive policies. Wellbeing also acts as a mediator of employee performance and retention (Ahmed et al., 2025; Dyageti & Kiran, 2025; Madero-Gómez et al., 2023; Mahdia, 2024; Sharma, S. Prakash, et al., 2023)
The Mediating Role of Employee Engagement and Green Behavior	44 articles indicate that employee engagement and pro-environmental behavior are important mediators of the influence between GHRM and organizational outcomes. Where employees who are actively involved in green initiatives demonstrate higher commitment to sustainable performance (Ibanez et al., 2024; Muhammad Arslan Amjad et al., 2025; Wang et al., 2025)
Integration of Ecological Sustainability and Social Inclusivity	42 articles show the integration of GHRM with *Diversity, Equity, and Inclusion (DEI)* principles in enhancing employee engagement, resilience, and long-term sustainable performance (Agrawal et al., 2024; Soekotjo et al., 2024)
The Role of Technology and AI in GHRM	21 articles explore the role of AI, big data analytics, and digital tools in improving GHRM implementation through decision-making, employee wellbeing, and organizational efficiency (Hashmi & Ghai, 2025; Herlina & Iskandar, 2025; Nurimansjah, 2023; Sugiharto et al., 2025)
GHRM and Corporate Social Responsibility (CSR)	19 articles show that GHRM plays a role in strengthening corporate social responsibility through ethical governance, community participation, and social-environmental balance, which enhance the reputation and long-term success of organizations (Chawla et al., 2025; Winarno & Silvianita, 2024)
Contextual and Industry-Specific Application	18 articles show that GHRM can be applied in various sectors such as hospitality, manufacturing, education, and MSMEs, with adjustments to the cultural context and local regulations (Al-Sulaiti, 2024; Masood, 2024; Ofosuhene, 2024; Vovk & Vovk, 2024)
Green Innovation and Sustainable Performance	17 shows that green innovation acts as a mediator in the relationship between GHRM and organizational performance. Leadership and employee motivation are key factors in fostering environmentally friendly innovation (Gazi et al., 2025; Naseer Muhammad et al., 2025; Zihan & Makhbul, 2024)

continued on following page

Table 3. Continued

Thematic	Theme Description
Theoretical Frameworks and Models of GHRM	15 articles indicate that theories such as *the Resource-Based View (RBV), Ability-Motivation-Opportunity (AMO),* and *Social Exchange Theory* are used to explain the mechanisms of GHRM and its impact on organizational outcomes (Anshima et al., 2025; Kutaula et al., 2025; Wu, 2024)
Retention, Commitment, and Employee Satisfaction	14 articles show that GHRM improves retention and job satisfaction by building a supportive work environment and organizational values that align with individual values (Din et al., 2023; Dira et al., 2024; S. Siva et al., 2024)
Organizational Culture and Leadership in GHRM	14 articles indicate that *transformational* and *inclusive* leadership styles are effective in strengthening GHRM to enhance employee engagement and pro-environmental behavior (Chennai et al., 2025; Gargee Banerjee, 2025; Permadi et al., 2024)
Challenges and Barriers to GHRM Implementation	12 articles identify barriers, including a lack of awareness, resistance to change, resource constraints, and minimal managerial support. Therefore, effective leadership, training, and clear metrics are needed to overcome these barriers (Dev, 2025; Kosasih, 2024; Roucham et al., 2023)
Measurement and Analytics in GHRM	10 articles indicate that HR analytics and metrics are used to measure the efficiency of sustainability programs, survival/hand operation/ inactivity , and environmental performance while enhancing evidence-based decision-making (Catherine Ezeafulukwe et al., 2022; Kaaria, 2024)

Based on a thematic analysis of 140 articles, the most dominant theme related to GHRM as the primary driver of organizational sustainability was followed by employee well-being, engagement, and green behavior, which served as important mediating variables. Moreover, a new trend has emerged regarding the importance of technology, artificial intelligence, and data analytics as tools to reinforce GHRM practices. We emphasize here that the success of GHRM depends largely on organizational culture and the appropriateness of industrial and regional conditions. Although there is already substantial evidence, challenges remain, including limited resources and resistance to change.

4.4 The Impact of GHRM on Performance and Well-being

The reviewed literature agrees that GHRM practices have a positive impact on organizational performance and employee well-being, which is often mediated by factors such as employee engagement, organizational culture, and green motivation. However, there are some differences in views regarding the strength of certain mediating variables, the role of technology, and contextual factors such as industry and research location, possibly due to differences in methodological, sectoral, cultural, or organizational contexts across studies; see Table 4.

Table 4. The agreement and divergence of GHRM on performance and well-being

Comparison Criterion	Studies in Agreement	Studies in Divergence
Organizational Performance Outcomes	Most research has unveiled a positive effect of Sustainable HRM practices on financial, operational, and innovation performance, leading to enhanced productivity, brand reputation, and competitive advantage (Akbar et al., 2024; Dyageti & Kiran, 2025; Ghosh & Haque, 2025; Montalvo-Falcón et al., 2023; Yadav et al., 2024)	Authors are reporting that the balance between work-life policies does not influence economic performance (Caputo et al., 2025), and green training gives different results depending on the context (Yin, 2023)
Employee wellbeing Indicators	GHRM can increase employees' prueti/wellbeing by providing mental health support, job satisfaction, and engagement in the organization that help/and a higher retention rate as well as a low level of turnover (Ahmed et al., 2025; Dyageti & Kiran, 2025; Garcia, 2025; Madero-Gómez et al., 2023; Mahdia, 2024; Sharma, S. Prakash, et al., 2023)	Some research reveals a small or mediated connection between HRM and wellbeing, with mixed relationships to work stress, where additional demands may negatively impact wellbeing, albeit enhanced performance (Tortia et al., 2022; Vanhala & Tuomi, 2006)
Mediating and Moderating Variables	Organizational outcomes are influenced by employee engagement, green motivation, organizational culture, and leadership that serve as mediating/moderating variables between GHRM and organizational outcomes (Anshima et al., 2025; Dira et al., 2024; Gargee Banerjee, 2025; Ibanez et al., 2024; Muhammad Arslan Amjad et al., 2025; Ogbonnaya & Aryee, 2022)	Organizational outcomes are influenced by employee engagement, green motivation, organizational culture, and leadership that serve as mediating/moderating variables between GHRM and organizational outcomes (Alam et al., 2023), and the moderating effect of AI integration is contingent (Hashmi & Ghai, 2025; Herlina & Iskandar, 2025)
Sustainability Integration Approaches	Several works take an integrative approach that unites ecology, social, and economic factors in HRM, promoting a more comprehensive sustainability perspective that integrates diversity, equity, and inclusion (DEI) and corporate social responsibility (CSR). (Agrawal et al., 2024; Bratton et al., 2025; Soekotjo et al., 2024) (Agrawal et al., 2024; Gunawan & Mikhail, 2025; Soekotjo et al., 2024)	Other works separate between green and social HRM initiatives (or examine ecological or economical aspects separately) for a more partial nature of integration, according to the literature , gerrymandering so far (Massoudie et al., 2024; Winarno & Silvianita, 2024)
Contextual and Industry Variations	The potential impact of Sustainable HRM in industries including: engineering, hospitality, education, SMEs, and the public sector has been documented with positive results globally (Ghosh & Haque, 2025; Järlström et al., 2023; Massoudie et al., 2024; Ofosuhene, 2024)	Several reviews emphasize context-related blockages, such as resource shortages, cultural variations, and commitment of officials, shaping the success of implementation (Barakat et al., 2024; Dev, 2025; Kosasih, 2024; Mathew et al., 2025)

4.5 Theoretical Implications

1. The synthesis of research findings reinforces the theoretical basis that GHRM practices positively influence organizational performance and employee well-being by integrating ecological and social dimensions. This supports frameworks such as Ability-Motivation-Opportunity (AMO), which explains how HRM capabilities can drive sustainable competitive advantage and employee engagement (Anshima et al., 2025; Kharel et al., 2025; Rahman & Saba, 2025)
2. The mediating role of employee engagement, green behavior, and well-being in the relationship between Sustainable HRM and organizational outcomes is consistent. This demonstrates the relevance of Social Exchange theory as a lens for understanding these dynamics (Gilal et al., 2025; Muhammad Arslan Amjad et al., 2025)

3. The emerging conceptual model integrates ecological sustainability with social inclusivity, for example, through the integration of GHRM with the principles of Diversity, Equity, and Inclusion (DEI), emphasizing a holistic approach to GHRM that strengthens organizational resilience and employee well-being (Agrawal et al., 2024; Soekotjo et al., 2024)
4. The role of green innovation as a mediator between HRM practices and organizational sustainability performance is increasingly recognized, showing that HRM not only supports environmental goals but also drives innovation processes that are important for long-term sustainability (Muhammad et al., 2025; Zihan & Makhbul, 2024)
5. The integration of technology, including Artificial Intelligence and HR analytics, within the GHRM framework is theoretically believed to enhance decision-making, employee well-being, and organizational adaptability, marking a new direction for the development of HRM theory that is more inclusive of digital transformation (Hashmi & Ghai, 2025; Nurimansjah, 2023; Shettigar & Chakraborty, 2025).
6. This literature also challenges traditional HRM paradigms by emphasizing a dual focus on employee well-being and organizational sustainability, and calls for a more comprehensive model that balances economic, social, and environmental goals (Bratton et al., 2025; Ogbonnaya & Aryee, 2022; Tortia et al., 2022)

4.6 Practical Implications

1. It is recommended that organizations adopt GHRM practices with green recruitment, training, performance appraisal, and employee engagement to improve environmental performance and enhance the well-being of employees, which would add to the efficiency of the organization (Alegbesogie, 2023; Kamboj & A, 2024; Yin, 2023)
2. Enhanced employee engagement and wellbeing from GHRM practices affect retention, motivation, and productivity....[more] It contributes to long-term success in the relationship between business and SDGs (Ahmed et al., 2025; Madero-Gómez et al., 2023; Sharma, S. Prakash, et al., 2023)
3. Leadership commitment, management support, and creating a green organizational culture have been identified as significant impediments in the practice of GHRM to overcome constraints with respect to resources as well as resistance (Chatterjee et al., 2025; Dev, 2025; Roucham et al., 2023)
4. Industry-level approaches accomplished, for example, by the adoption of customized GHRM practices in the manufacturing and hospitality as well as public administration sectors, illustrate the relevance of contextualization in terms of

organizational size, culture, and environmental orientation (Gilal et al., 2025; Vovk & Vovk, 2024)

5. The digital and AI aspects of GHRM practices will yield tangible value by addressing issues of employee well-being and workload relief, in line with corporate sustainability. Hence, organizations are recommended to adopt human resource technology solutions (Hashmi & Ghai, 2025; Herlina & Iskandar, 2025; Shamshuddin et al., 2025)
6. Policymakers and HR practitioners should promote lifelong learning to develop sustainability competencies, green innovation, and inclusive practices, in order to build a sustainable workforce that can drive organisational transformation (Manao & Hadi Senen, 2024; Masood, 2024; S. Siva et al., 2024).

4.7 Gap and Future Agenda

In this section, the gaps collected and separated by research area — both in terms of their impact, integration with various contexts, mechanisms of relationship with independent and dependent variables, mediation, and even moderation — will be explained in detail in the following Table 5.

Table 5. Gap and future research direction

Gap Areas	Future Research Directions	Research Priorities
Longitudinal studies on the impact of GHRM A cross-sectional research design predominates, making it challenging to understand the causal implications and long-term impacts of sustainable HRM on organisational performance and employee wellbeing.	Conduct longitudinal and experimental research that follows evolution, and also establish cause-and-effect relationships between GHRM practices and their outcomes. Also, deal with the current methodological constraints (Kosasih, 2024; Mahdia, 2024; Masood, 2024)	High
Integration of contextual and cultural factors Scant empirical evidence towards the moderating effect of organizational culture and leadership style, as well as regional context, on the relationship between sustainability HRM and its outcomes	Creating cross-cultural comparative and industry-level research to investigate the modulated effect of culture and leadership in GHRM implementation. Transferability and practical relevance (Chatterjee et al., 2025; Din, Yang, Khan, Mohi Ud Din, et al., 2023; Gilal et al., 2025; Permadi et al., 2024)	High
Mechanisms linking GHRM to well-being Insufficient examination of mediators like organizational trust, psychological safety, and overall dimensions of well-being in GHRM effects	Exploring psychosocial mediators—trust, safety, and holistic wellbeing—through mixed-methods. Filling the gap in the literature on complex employee outcomes (Jaskeviciute et al., 2021; Mahdia, 2024; Tortia et al., 2022)	High
The role of technology and AI in GHRM The limited role of AI and digital tools in improving sustainable GHRM practices and employee engagement.	Empirical investigation of AI-led GHRM interventions, ethical implications, employee wellbeing, and sustainability. To assess the goods and bads (Hashmi & Ghai, 2025; Herlina & Iskandar, 2025; Nurimansjah, 2023; Shamshuddin et al., 2025)	Medium

continued on following page

Table 5. Continued

Gap Areas	Future Research Directions	Research Priorities
Balancing sustainability objectives in HRM We have little empirical knowledge as to how organizations deal with trade-offs among ecological, social, and economic sustainability in HRM.	Investigating decision-making and frameworks in HRM that facilitate effective TBL trade-offs/balancing. (Dev, 2025; Tak et al., 2024)	Medium
Employee green behavior as a mediator/ moderator The mediating or moderating role of employees' green behavior and motivation in mediating/moderating the relationship between GHRM and performance has been explored in only a few prior studies.	Multilevel, longitudinal investigation to clarify anatomic mediating processes of green behavior and motivation in the relationship between GHRM practices, their contingent factors (i.e., green values), and organization-level outcome requires further empirical validation (AlKetbi & Rice, 2024; Gilal et al., 2025; Muhammad Arslan Amjad et al., 2025; Zhang & Peng, 2023)	High
GHRM in SMEs and developing economies Insufficient studies on the challenges, practices, and outcomes of sustainable HRM in SMEs, as well as in developing countries.	Investigating the obstacles, enablers, and implications for performance of GHRM adoption in SMEs operating within diverse developing economies: A mixed-methods approach. Fostering inclusive sustainability and applied-based guidance (Kutaula et al., 2025; Roucham et al., 2023; Zihan & Makhbul, 2024)	High
Measurement and standardization of sustainable GHRM constructs Differences and discrepancies in how GHRM is operationalized and how performance measures are used prevent comparisons across studies.	Construction and validation of measurement tools and reporting constructs for GHRM practices & outcomes (Anshima et al., 2025; Viitasaari et al., 2022; Yuxiaosi, 2024)	Medium
The mediating role of innovation in sustainable HRM outcomes Insufficient empirical attention has been paid to how green innovation serves as a mediator in the relationship between SHRM and organizational performance.	Examining the mechanisms of GHRM on green innovation and performance through a longitudinal design. Its mediating role needs further empirical evidence (Kharel et al., 2025; Naseer Muhammad et al., 2025)	Medium
Employee involvement in GHRM Finally, we need to better understand how employee engagement can mediate the effects of SHRM across sectors and cultures.	Carry out sectoral and cross-cultural research to investigate organizational mechanisms for engagement and to create GHRM interventions that enhance it. Activism is a significant mediator; however, its contextual nuances must be further explored (Dira et al., 2024; Ibanez et al., 2024; Muhammad Arslan Amjad et al., 2025)	High

5. CONCLUSION

The literature on green human resource management (GHRM) shows a significant positive impact on organizational performance and employee well-being. In general, research shows that applying sustainability principles to HRM practices can improve financial, operational, and innovation outcomes while promoting holistic employee health, satisfaction, and engagement. Mediating factors such as employee engagement, green motivation, and green employee behavior play an important role in linking HR practices to organizational sustainability outcomes. Green innovation and pro-environmental behavior are also significant game changers that enhance competitive edge and long-term sustainability. Moreover, inclusive HR practices that align with diversity, equity, and inclusion (DEI) principles complement positive environmental efforts in supporting employee well-being and organizational social responsibility. The literature also underscores the significance of organizational culture and leadership style, such as inclusive leadership and management support, as moderators of how GHRM practices work. The use of digital technologies, such as AI and HR analytics, is being more acknowledged as a catalyst leading to higher sustainability implementation, greater employee commitment, and promoting evidence-based decision-making." However, research on the link between technology and sustainable HRM remains relatively scarce, leaving room for further inquiry.

Green HRM (Human Resource Management) policies are strategies, methods, and systems that aim to foster environmental sustainability within a business. Essential policies include environmentally focused recruitment to attract eco-aware applicants, educational initiatives to inform employees about sustainability, and performance management that integrates environmental goals and incentives for sustainable practices. Additional methods include minimizing paper consumption, encouraging telecommuting, using energy-efficient devices, and cultivating a widespread ethos of environmental stewardship.

Contextual factors, such as industry, firm size, region, and business culture, have been found to impact the adoption and implementation of GHRM. While manufacturing, hospitality, education, and public organization samples have been primarily studied, cross-cultural and longitudinal studies are yet needed to understand constraints and causal relationships in depth. From a methodological perspective, the dominance of cross-sectional designs limits the strength of causal conclusions, so future research is advised to use longitudinal and mixed-method approaches that incorporate employee perspectives alongside managerial views. Conceptually, most research draws on theories such as the Ability-Motivation-Opportunity (AMO) and Social Exchange Theories. However, there is still room to develop a more integrative theoretical framework that explicitly combines ecological, social, and technological dimensions. There are practical challenges as well: notably , addressing sustainability in HR operations (operationalizing sustainability), balancing economic objectives with social and environmental goals, and overcoming barriers to implementing TMSHR, such as resource limitations and low organizational awareness. This feedback loop reiterates HRM as more than just an administrative task, but rather a strategic imperative for developing adaptive and innovative organizations. Despite these advances, future research should focus on longitudinal validation, new technologies and their consequences for SHRM, multilevel analysis, and the development of contextual frameworks grounded in the theoretical foundations of sustainable HRM.

6. RESEARCH IMPLICATIONS

This chapter enhances the theoretical understanding of sustainable performance by innovatively integrating Social Cognitive Theory (SCT) with the Ability-Motivation-Opportunity (AMO) framework. Initially, it strengthens the link between Green Human Resource Management (GHRM) and sustainable performance by demonstrating how green competency behavior enhances employee skills and how green motivation strengthens the Ability-Motivation-Opportunity (AMO) framework. This aligns focus on the interdependent interaction between environmental and personal factors. This chapter expands the use of the AMO concept by situating it within the context of environmental sustainability, suggesting that green skills and motivation are crucial for leveraging the opportunities presented by GHRM practices by emphasizing the impact of intrinsic motivators, such as GP, on pro-environmental behavior, thereby reinforces the idea that organizational frameworks and individual enthusiasm collaborate to yield sustainable outcomes.

REFERENCES

Abu-Mahfouz, S., Halim, M. S. A., Bahkia, A. S., & Noryati Alias, P. D. in M., & Tambi, A. M. (2023). Sustainable human resource management practices in organizational performance: The mediating impacts of knowledge management and work engagement. *Journal of Entrepreneurship. Management and Innovation*, *19*(2), 57–97. DOI: 10.7341/20231922

Agrawal, N., Beriwal, K., & Daga, N. (2024). Sustainability Through Human Resource Management: A Conceptual Framework. In R. Sharma, I. Bhardwaj, S. Grima, T. Sachdeva, K. Sood, & E. Özen (Eds), *Sustainable Development Goals: The Impact of Sustainability Measures on Wellbeing* (Vol. 113B, p. 0). Emerald Publishing Limited. https://doi.org/DOI: 10.1108/S1569-37592024000113B003

Ahmad, R., Nawaz, M. R., Ishaq, M. I., Khan, M. M., & Ashraf, H. A. (2023). Social exchange theory: Systematic review and future directions. *Frontiers in Psychology*, *13*, 1015921. DOI: 10.3389/fpsyg.2022.1015921

Ahmed, M. U., Farhad, F., & Hussain, M. G. (2025). *The Effect of Sustainable Practices on Organizational Success and Worker Well-Being*. Business, Economics and Management. https://doi.org/DOI: 10.20944/preprints202507.1755.v1

Akbar, A., Ahmad, S., Khalid, M., Aslam, M. F., & Bhatti, M. A. A. (2024). Analyzing the Effect of Green HRM on Organizational Performance. [BBE]. *Bulletin of Business and Economics*, *13*(2), 864–869. DOI: 10.61506/01.00402

Al-Sulaiti, K. (2024). Sustainability and Sustainable HRM Case Study: Sustainable HRM in Light of Ministry of Culture, Qatar. *WSB Journal of Business and Finance*, *58*(1), 19–28. DOI: 10.2478/wsbjbf-2024-0003

Alam, M. N., Campbell, N., Das, S., Hashim, F., Ur Rehman, I. H., & Iqbal, J. (2023). Green Training and Development Revolutionizing Organizational Performance: The Moderating Role of Green Employee Involvement in the Bangladeshi Pharmaceutical Industry. *International Business Research*, *16*(9), 36. DOI: 10.5539/ibr.v16n9p36

Alegbesogie, A. I. (2023). The Impact of Green Human Resource Management Practices on Organisational Performance. *Vilnius University Proceedings*, *37*, 6–11. DOI: 10.15388/VGISC.2023.1

AlKetbi, A., & Rice, J. (2024). The Impact of Green Human Resource Management Practices on Employees, Clients, and Organizational Performance: A Literature Review. *Administrative Sciences*, *14*(4), 78. DOI: 10.3390/admsci14040078

Alkhalaf, T., & Al-Tabbaa, O. (2024). The effect of ability, motivation and opportunity (AMO) on SMEs' innovation performance. *Creativity and Innovation Management, 33*(1), 21–38. DOI: 10.1111/caim.12578

Amjad, M. A., Khoso, R. A., Soomro, M. A., & Khan, A. A. (2025). Engaging Minds and Greening Workspaces: Investigating the Mediating Role of Employee Engagement in Strengthening the Relationship Between Green HRM Practices and Organizational Sustainability Outcomes. *Journal of Management & Social Science, 2*(3), 186–208. DOI: 10.63075/9j2zwv02

Anshima, A., Bhardwaj, B., Sharma, D., & Chand, M. (2025). Ensuring sustainability through green HRM practices: A review, synthesis and research avenues. *Journal of Organizational Effectiveness: People and Performance, 12*(1), 57–90. DOI: 10.1108/JOEPP-10-2023-0493

Asrifan, A., Siripipatthanakul, S., Alaqad, M.H., Saodi, S., Sadaruddin, S. (2025). Privacy and Security Management Practices for Organizations. Open source preview, 2025, pp. 151–182

Awwad Al-Shammari, A. S., Alshammrei, S., Nawaz, N., & Tayyab, M. (2022). Green Human Resource Management and Sustainable Performance With the Mediating Role of Green Innovation: A Perspective of New Technological Era. *Frontiers in Environmental Science, 10*. Advance online publication. DOI: 10.3389/fenvs.2022.901235

Banerjee, G. (2025). The impact of Employee Engagement and Organizational Environmental Performance: A Literature Review. *Journal of Information Systems Engineering & Management, 10*(35s), 991–1001. DOI: 10.52783/jisem.v10i35s.6179

Barakat, S., Alneyadi, R., Alderei, M., Bani Malek, S., Alahbabi, F., & Jasim, M. (2024). The Impact Of Sustainable Hrm Practices On Job Performance In Uae Organizations. *International Journal of Management Trends: Key Concepts and Research, 3*(1), 29–38. DOI: 10.58898/ijmt.v3i1.29-38

Benítez-Núñez, C., Díaz-Díaz, N. L., Ballesteros-Rodríguez, J. L., & de Saá-Pérez, P. (2024). Explaining academic researchers' performance from the Ability–Motivation–Opportunity (AMO) perspective. *Studies in Higher Education, 49*(12), 2423–2440. DOI: 10.1080/03075079.2024.2307965

Blau, P. M. (2017). *Exchange and Power in Social Life* (1st ed.). Routledge., DOI: 10.4324/9780203792643

Bratton, A., Renwick, D. W. S., Paillé, P., Matthews, B., Leidner, S., Karamali, E., & Holland, P. (2025). Guest editorial: Sustainable human resource management and organizational performance: new definitions, navigating tensions, and global insights. *Journal of Organizational Effectiveness: People and Performance*, *12*(1), 1–13. DOI: 10.1108/JOEPP-03-2025-569

Caputo, A., Tron, A., Sassetti, S., Giakoumelou, A., & Colantoni, F. (2025). Managing for resilience through social sustainable human resource approach: The effect on organisational performances. *Management Decision*. Advance online publication. DOI: 10.1108/MD-05-2024-1144

Chatterjee, S., Chaudhuri, R., & Vrontis, D. (2025). Creating organizational value and sustainability through green HR practices: An innovative approach with the moderating role of top management support. *Business Ethics, the Environment & Responsibility*, *34*(1), 17–31. DOI: 10.1111/beer.12569

Chawla, M., Sharma, P., Gautam, S., & Popescu, C. R. Gh. (2025). Sustainability in Human Capital Management: Shaping the Future With Technology. In T. Kapoor, A. K. Singh, K. Sahay, H. Kaur, & R. Verma (Eds), *Sustainable Management Practices for Employee Retention and Recruitment* (pp. 265–290). IGI Global. https://doi.org/ DOI: 10.4018/979-8-3693-8679-8.ch013

Chennai, Manimalar, Dr. R. Pavithra, S., & B, Preethi. (2025). Green HRM and Inclusive Leadership: A Dual Strategy for Sustainable Growth. *International Scientific Journal of Engineering and Management*, *04*(09), 1–9. DOI: 10.55041/ ISJEM.ESEH025

Dev, A. (2025). ECO-HR: Integrating Sustainability into workforce management" "Reducing HR's Environmental Footprint. *International Journal For Multidisciplinary Research*, *7*(2), 41484. DOI: 10.36948/ijfmr.2025.v07i02.41484

Din, A. U., Yang, Y., Khan, K., Din, Q. M. U., Golo, M. M. A., & Khan, I. U. (2023). Role of employee commitment and individual value in green HRM practices and sustainable firm performance. *Second International Conference on Sustainable Technology and Management (ICSTM 2023)*, *12804*, 404–413. https://doi.org/DOI: 10.1117/12.3004029

Din, A. U., Yang, Y., Khan, K., Mohi Ud Din, Q., Mansoor Ahmed Golo, M., & Khan, I. U. (2023). Role of employee commitment and individual value in green HRM practices and sustainable firm performance. In Qu, X. (Ed.), *Second International Conference on Sustainable Technology and Management (ICSTM 2023)* (p. 67). SPIE. https://doi.org/DOI: 10.1117/12.3004029

Dira, A. F., Noor, G. P., Bangun, M. F. A., Winardi, N. M. A., Kamal, F., & Utomo, K. P. (2024). The Role of Employee Engagement in Green HRM to Create Sustainable Humanist Performance. *EKOMBIS REVIEW: Jurnal Ilmiah Ekonomi Dan Bisnis, 12*(1). Advance online publication. DOI: 10.37676/ekombis.v12i1.4856

Dyageti, S. R., & Kiran, K. C. U. (2025). THE ROLE OF HR IN DRIVING ORGANIZATIONAL SUSTAINABILITY THROUGH EMPLOYEE WELL BEING PROGRAMS. *International Journal of Advances in Business and Management Research, 03*(Supplementary), 78–85. DOI: 10.62674/ijabmr.2025.v3isupp1.010

Emerson, R. M. (1976). Social Exchange Theory. *Annual Review of Sociology*, •••, 335–362.

Ezeafulukwe, C., Okatta, C. G., & Ayanponle, L. (2022). Frameworks for sustainable human resource management: Integrating ethics, CSR, and Data-Driven Insights. *World Journal of Advanced Research and Reviews, 13*(3), 583–592. DOI: 10.30574/wjarr.2022.13.3.0151

Faisal, S. (2023). Green Human Resource Management—A Synthesis. *Sustainability, 15*(3), 2259. DOI: 10.3390/su15032259

Garcia, A. G. (2025). The impact of sustainable practices on employee well-being and organizational success. *Brazilian Journal of Development, 11*(3), e78599. DOI: 10.34117/bjdv11n3-054

Gazi, Md. A. I., Masud, A. A., Emon, Md., Ibrahim, Md., & Senathirajah, A. R. bin S. (2025). The triadic relationship between green HRM, innovation, and pro-environmental behaviour: A study of their interactions and impacts on employee productivity and organizational sustainability. *Environmental Research Communications, 7*(1), 015016. DOI: 10.1088/2515-7620/ada676

Ghosh, A., & Haque, S. (2024). The impact of GHRM and EGB on organization's financial and operational performance – an empirical analysis on Indian ICT sector. *Journal of Organizational Effectiveness: People and Performance, 12*(1), 105–126. DOI: 10.1108/JOEPP-08-2023-0359

Ghosh, A., & Haque, S. (2025). The impact of GHRM and EGB on organization's financial and operational performance – an empirical analysis on Indian ICT sector. *Journal of Organizational Effectiveness: People and Performance, 12*(1), 105–126. DOI: 10.1108/JOEPP-08-2023-0359

Gilal, F. G., Nawaz, A., Riaz, A., Channa, K. A., & Jahanzeb, A. (2024). Achieving sustainability through green human resource management practices and environmental orientation fit: A multilevel modeling. *Evidence-Based HRM: A Global Forum for Empirical Scholarship, 13*(2), 299–318. https://doi.org/DOI: 10.1108/EBHRM-09-2022-0223

Gilal, F. G., Nawaz, A., Riaz, A., Channa, K. A., & Jahanzeb, A. (2025). Achieving sustainability through green human resource management practices and environmental orientation fit: A multilevel modeling. *Evidence-Based HRM: A Global Forum for Empirical Scholarship, 13*(2), 299–318. https://doi.org/DOI: 10.1108/EBHRM-09-2022-0223

Gooderham, P. N., Pedersen, T., Sandvik, A. M., Dasí, À., Elter, F., & Hildrum, J. (2022). Contextualizing AMO Explanations of Knowledge Sharing in MNEs: The Role of Organizational and National Culture. *Management International Review, 62*(6), 859–884. DOI: 10.1007/s11575-022-00483-0

Gunawan, W. B., & Mikhail, L. (2025). Strategic and sustainable human resource management: Twin weapon for achieving competitive advantage in organization. *Priviet Social Sciences Journal*, *5*(6), 45–58. DOI: 10.55942/pssj.v5i6.401

Gyensare, M. A., Adomako, S., & Amankwah-Amoah, J. (2024). Green HRM practices, employee well-being, and sustainable work behavior: Examining the moderating role of resource commitment. *Business Strategy and the Environment, 33*(4), 3129–3141. DOI: 10.1002/bse.3642

Hadi, S., Lukitaningsih, A., Kirana, K.C., ... Hidayat, R., Siripipatthanakul, S. (2025).Privacy and Security Management Practices for Organizations. Open source preview, 2025, pp. 89–116

Hadi, S., Purnasari, E., Hidayat, R., Lukitaningsih, A., Kirana, K. C., & Siripipatthanakul, S. (2025). Challenges and Management Practices in Protecting Employee Personal Data in the Digital Era. In *Privacy and Security Management Practices for Organizations* (pp. 89–116). IGI Global Scientific Publishing., DOI: 10.4018/979-8-3693-9720-6.ch004

Hadi, S., Tjahjono, H. K., & Palupi, M. (2020). *Systematic Review: Meta Sintesis Untuk Riset Perilaku Organisasional.* Viva Victory Abadi.

Hashmi, S., & Ghai, R. K. (2025). *Applying Artificial Intelligence in Sustainable HR Practices: A Pathway to Organizational Endurance*. https://doi.org/DOI: 10.20944/preprints202507.1204.v1

Herlina, M. G., & Iskandar, K. (2025). Integrating Sustainable HRM, AI, and Employee Well-Being to Enhance Engagement in Greater Jakarta: An SDG 3 Perspective. *E3S Web of Conferences, 601*, 00020. https://doi.org/DOI: 10.1051/e3sconf/202560100020

Housheya, N., & Atikbay, T. (2025). The Role of Green HRM in Promoting Green Innovation: Mediating Effects of Corporate Environmental Strategy and Green Work Climate, and the Moderating Role of Artificial Intelligence. *Sustainability*, *17*(16), 7238. DOI: 10.3390/su17167238

Ibanez, A. A. E., Khalefa, M., & Aldulaimi, M. (2024). Green Human Resource Management Practices and Employee Engagement: A Case Study of Leading Sustainable Organizations. *International Journal of Green Management and Business Studies*, *4*(2). Advance online publication. DOI: 10.56830/IJGMBS12202401

Järlström, M., Saru, E., & Pekkarinen, A. (2023). Practices of Sustainable Human Resource Management in Three Finnish Companies: Comparative Case Study. *South Asian Journal of Business and Management Cases*, *12*(1), 31–51. DOI: 10.1177/22779779231154656

Jaskeviciute, V., Stankeviciene, A., Diskiene, D., & Savicke, J. (2021). The relationship between employee well-being and organizational trust in the context of sustainable human resource management. *Problems and Perspectives in Management*, *19*(2), 118–131. DOI: 10.21511/ppm.19(2).2021.10

Kaaria, A. G. (2024). Essential Human Resource Metrics and Analytics for Sustainable Work Environments: Literature Mapping and Conceptual Synthesis. *East African Journal of Business and Economics*, *7*(1), 241–262. DOI: 10.37284/eajbe.7.1.1976

Kamboj, J., & A, E. (2024). Mapping the green human resource management practices: A systematic scoping review and its implications for employees' well-being. *Human Systems Management*, *43*(6), 932–970. DOI: 10.3233/HSM-230183

Kharel, S., Chhetri, A. D., & Sharma, S. (2025). Advancing Organizational Sustainability Through Green Hrm: The Mediating Roles of Green Innovation and Employee Environmental Performance. *Journal of Balkumari College*, *14*(1), 1–9. DOI: 10.3126/jbkc.v14i1.80752

Kilroy, J., Dundon, T., & Townsend, K. (2023). Embedding reciprocity in human resource management: A social exchange theory of the role of frontline managers. *Human Resource Management Journal*, *33*(2), 511–531. DOI: 10.1111/1748-8583.12468

Kosasih, K. (2024). The Role of Human Resource Management in Sustainable Development: A Systematic Literature Review. [DEMAND]. *Economics Studies and Banking Journal*, *1*(2), 61–74. DOI: 10.62207/q4jya069

Kutaula, S., Chowdhury, S., Gillani, A., Budhwar, P. S., & Dey, P. K. (2025). Linking HRM with Sustainability Performance Through Sustainability Practices: Unlocking the Black Box. *British Journal of Management*, *36*(2), 615–632. DOI: 10.1111/1467-8551.12861

Liu, J., Wang, Z., Li, C., & Xu, R. (2024). Exploring pathways to digital transformation: fsQCA analysis based on the AMO framework. *PLoS One*, *19*(12), e0315249. DOI: 10.1371/journal.pone.0315249

Lu, Y., Zhang, M. M., Yang, M. M., & Wang, Y. (2023). Sustainable human resource management practices, employee resilience, and employee outcomes: Toward common good values. *Human Resource Management*, *62*(3), 331–353. DOI: 10.1002/hrm.22153

Madero-Gómez, S. M., Rubio Leal, Y. L., Olivas-Luján, M., & Yusliza, M. Y. (2023). Companies Could Benefit When They Focus on Employee Wellbeing and the Environment: A Systematic Review of Sustainable Human Resource Management. *Sustainability*, *15*(6), 5435. DOI: 10.3390/su15065435

Mahdia, A. (2024). The Role of Human Resource Management in Employee Well-Being and Mental Health: A Systematic Literature Review. [PRODUCTIVITY]. *Management Studies and Business Journal*, *1*(3), 286–303. DOI: 10.62207/phea4z38

Manao, A., & Hadi Senen, S. (2024). Sustainable Human Resource Management: Building an Adaptive and Inclusive Organizational Culture. *Dinasti International Journal of Education Management And Social Science*, *6*(2), 863–871. DOI: 10.38035/dijemss.v6i2.3673

Masood, R. Z. (2024). Sustainable Hrm Practices And Innovation In Healthcare Smes: Mediating Role Of Green Innovation Using Pls-Sem Analysis. *ShodhKosh: Journal of Visual and Performing Arts*, *5*(1). Advance online publication. DOI: 10.29121/shodhkosh.v5.i1.2024.3728

Massoudie, J., Afghan, H., Zamani, M., Samadi, M., & Gul, S. (2024). The Role of GHRM Practices Towards Organizational Commitment: A Mediation Analysis of Green Health and Safety Management. *Kardan Journal of Economics and Manangement Sciences*, *3*. Advance online publication. DOI: 10.31841/KJEMS.2024.168

Mathew, Dr. J. M., Fasina, M., Biju, P. J., & Hamsa, E. (2025). Role of HR in Promoting Sustainability and Corporate Social Responsibility. *International Journal of Accounting and Economics Studies, 12*(SI-1), 371–376. DOI: 10.14419/sxztgj60

Montalvo-Falcón, J. V., Sánchez-García, E., Marco-Lajara, B., & Martínez-Falcó, J. (2023). Green human resource management and economic, social and environmental performance: Evidence from the Spanish wine industry. *Heliyon, 9*(10), e20826. DOI: 10.1016/j.heliyon.2023.e20826

Muhammad, N., Laghari, A., Amin, D. H., Bangash, S. A., Ishfaq, U., & Nosh, N. (2025). The Green Innovation Engine: How GHRM Practices Influence Sustainable Performance in Manufacturing Industries. *Social Science Review Archives, 3*(1), 417–435. DOI: 10.70670/sra.v3i1.321

Muhammad, N., Laghari, A., Amin, D. H., Bangash, S. A., Ishfaq, U., & Nosh, N. (2025). The Green Innovation Engine: How GHRM Practices Influence Sustainable Performance in Manufacturing Industries. *Social Science Review Archives, 3*(1), 417–435. DOI: 10.70670/sra.v3i1.321

Nakra, N., Kashyap, V., & Arora, R. (2024). Interactive effects of sustainable HRM, career adaptability and career sustainability on employee well-being. *Employee Relations, 46*(8), 1706–1722. DOI: 10.1108/ER-01-2024-0021

Noor Faezah, J., Yusliza, M. Y., Chaudhary, R., Ramayah, T., & Fawehinmi, O. (2024). Green human resource management and employee ecological behaviour: The mediating role of green commitment. *Social Responsibility Journal, 20*(6), 1157–1176. DOI: 10.1108/SRJ-07-2023-0399

Nurimansjah, R. A. (2023). Dynamics of Human Resource Management: Integrating Technology, Sustainability, and Adaptability in the Modern Organizational Landscape. *Golden Ratio of Mapping Idea and Literature Format, 3*(2), 104–123. DOI: 10.52970/grmilf.v3i2.324

Ofosuhene, D. (2024). Adoption Of Green Human Resource Management Practices And Organisational Wellbeing In The Ghanaian Service Sector. *European Journal of Human Resource Management Studies, 7*(2). Advance online publication. DOI: 10.46827/ejhrms.v7i2.1672

Ogbonnaya, C., & Aryee, S. (2022). HRM Practices, Employee Well-Being, and Organizational Performance. In Brough, P., Gardiner, E., & Daniels, K. (Eds.), *Handbook on Management and Employment Practices* (Vol. 3, pp. 369–391). Springer International Publishing., DOI: 10.1007/978-3-030-29010-8_17

Parent-Lamarche, A., Dextras-Gauthier, J., & Julien, A.-S. (2023). Toward a new model of human resource management practices: Construction and validation of the High Wellbeing and Performance Work System Scale. *Frontiers in Psychology, 14*. Advance online publication. DOI: 10.3389/fpsyg.2023.1151781

Permadi, I. K. O., Fadah, I., & Rizki, V. L. (2024). Green Hrm and Employee Performance: How Organizational Culture Plays A Mediating Role. *Journal of Economics. Finance And Management Studies, 07*(10). Advance online publication. DOI: 10.47191/jefms/v7-i10-44

Purnasari, E., Hermuningsih, S., Hidayat, R., Hadi, S., & Kumar, A. (2025). *Antecedents and Consequences of Financial Literacy: A Comprehensive Systematic Literature Review of the Past Decade*. 3–14. https://doi.org/DOI: 10.2991/978-94-6463-698-7_2

Rafique, S., Tanveer, Y., Sadiq, S., & Tariq, A. (2024). Mediation of Green Perceived Organizational Support between Green Human Resource Management Practices and Employee Performance. *The Asian Bulletin of Big Data Management, 4*(3), 173–191. DOI: 10.62019/abbdm.v4i3.228

Rahman, M. A., & Saba, N. A. (2025). Driving Sustainable Business Performance through Green HRM: Examining the Mediating Roles of Green Technology Innovation and Green Organizational Culture. *Golden Ratio of Human Resource Management, 5*(2), 501–516. DOI: 10.52970/grhrm.v5i2.1326

Rajâa, O., & Mekkaoui, A. (2025). Revealing the impact of social exchange theory on financial performance: A systematic review of the mediating role of human resource performance. *Cogent Business & Management, 12*(1), 2475983. DOI: 10.1080/23311975.2025.2475983

Rosyafah, S., Hidayat, W. G. P. A., & Nitawati, E. Y. (2025). Sustainable HRM in the face of climate change: Building organizational resilience through green human capital practices. *Annals of Human Resource Management Research, 5*(2), 113–123. DOI: 10.35912/ahrmr.v5i2.2966

Roucham, B., Lefilef, A., & Mahmoudi, H. (2023). The Forward-Looking Studies on Green Human Resources Management: A Content Analysis. *MANAGEMENT AND ECONOMICS REVIEW, 8*(3), 259–275. DOI: 10.24818/mer/2023.10-01

Security: A Case Study of Bank of Ayudhya PCL.(KMA), Thailand. In *International Conference on Data-Processing and Networking* (pp. 479-493). Springer, Singapore. https://doi.org/DOI: 10.1007/978-981-96-3102-5_36

Shamshuddin, S., Baburao, C., Shaik, H., & Prasad, Y. V. V. V. S. S. S. V. (2025). Sustainable Employee Engagement and Retention: A Review of Green HRM and Digital Integration for Modern Workplaces. In Pujari, P., Khan, S. A., Kumar, A., & Naim, A. (Eds.), (pp. 357–380). Advances in Logistics, Operations, and Management Science. IGI Global., DOI: 10.4018/979-8-3693-8492-3.ch017

Sharma, S., & Prakash, S., & Devchand. (2023). Employee Well-Being And Sustainable Hr Practices: A Pathway To Long-Term Organizational Success. *Shodh-Kosh: Journal of Visual and Performing Arts*, *4*(2), 1602–1615. DOI: 10.29121/shodhkosh.v4.i2.2023.2706

Shettigar, Dr. R., & Chakraborty, Prof. A. (2025). Sustainable Business Strategies: An HR-Centred Approach To Resilience And Sustainable Growth. *International Journal of Environmental Sciences*, •••, 551–561. DOI: 10.64252/zkhtn743

Siripipatthanakul, S. (2025). Privacy and Security Management Practices for Organizations. Open source preview, 2025, pp. 1–306

Siripipatthanakul, S., Chaiprakarn, S., Rungsiyanont, S., & Phuangsuwan, P. (2025, a). A Case Study of Sustainable Development in Hospitals in Thailand Using ChatGPT. In *Sustainable Development for Hospitals and Clinics* (pp. 1-32). IGI Global Scientific Publishing. DOI: DOI: 10.4018/979-8-3693-9800-5.ch001

Siripipatthanakul, S., Phuangsuwan, P., & Jaipong, P. (2025, e). Mobile Application

Siripipatthanakul, S., Phuangsuwan, P., & Jaipong, P., & Muthmainnha. (2025, f, September). Wi-Fi Networks and Security: A Systematic Review. In *International Conference on Artificial Intelligence and Networking* (pp. 81-92). Singapore: Springer Nature Singapore. https://doi.org/DOI: 10.1007/978-981-96-2015-9_6

Siripipatthanakul, S., Phuangsuwan, P., Siripipattanakul, S., & Jaipong, P. (2025, c). Sustainable Tourism Based on Challenges, Tourist Experiences, and Host Communities: A Case Study of Ayutthaya, Thailand. In *Best Practices for Human Resources Management in Tourism and Hospitality* (pp. 361-394). IGI Global Scientific Publishing. DOI: DOI: 10.4018/979-8-3693-8779-5.ch014

Siripipatthanakul, S., Phuangsuwan, P., Siripipattanakul, S., Limna, P., Jaipong, P., & Chaiprakarn, S. (2025, g, September). Sustainable Development Adopting Digital Intelligence Among University Students. In *International Conference on Artificial Intelligence and Networking* (pp. 211-220). Singapore: Springer Nature Singapore. https://doi.org/DOI: 10.1007/978-981-96-2015-9_14

Siripipatthanakul, S., Phuangsuwan, P., Siripipattanakul, S., Limna, P., Jaipong, P., & Chaiprakarn, S. (2025, h, September). Privacy and Information Security for Sustainable Community Among TikTok's Users. In *International Conference on Artificial Intelligence and Networking* (pp. 221-233). Singapore: Springer Nature Singapore. https://doi.org/DOI: 10.1007/978-981-96-2015-9_15

Siripipatthanakul, S., Rakpathum, W., Siripipattanakul, S., Kaewnaknaew, C., Phuangsuwan, P., & Jaipong, P. (2025, d). The Future Artificial Intelligence (AI) for the Hotel and Hospitality Industry: The Case of Hotel's Human Resource Management (HRM). In *Best Practices for Human Resources Management in Tourism and Hospitality* (pp. 395-422). IGI Global Scientific Publishing. DOI: DOI: 10.4018/979-8-3693-8779-5.ch015

Siripipatthanakul, S., Rungsiyanont, S., & Siripipattanakul, S. (2025, b). Sustainable Dentistry Using Artificial Intelligence (AI) for Digital Education and Business Among University Dental Hospitals. In *Sustainable Development for Hospitals and Clinics* (pp. 33-64). IGI Global Scientific Publishing. DOI: DOI: 10.4018/979-8-3693-9800-5.ch002

Siva, S., & Saravanakumar, A. Piyali Biswas, Shreya Verma, & Uma Shankar. (2024). Developing Sustainable HRM Practices for Retention and Satisfaction in Organizations. *Nanotechnology Perceptions*. Advance online publication. DOI: 10.62441/nano-ntp.vi.4017

Soekotjo, S., Sosidah, S., Kuswanto, H., Setyadi, A., & Pawirosumarto, S. (2024). *A Conceptual Framework for Sustainable Human Resource Management: Integrating Ecological and Inclusive Perspectives*. https://doi.org/DOI: 10.20944/preprints202412.2299.v1

Sugiharto, B., Hiya, N., Simamora, H., Gaol, N. L. D. H. L., & Andriani, M. (2025). Strategic Human Resource Mapping and Sustainable HRM: Balancing Productivity and Employee Well-being. *Golden Ratio of Mapping Idea and Literature Format*, *5*(2), 21–52. DOI: 10.52970/grmilf.v5i2.1362

Tak, T., Agarwal, V., Mishra, S., Nilkant, D., & Ahluwalia, G. (2024). An examination of sustainable human resource management for achieving financial, social, and environmental goals. *Multidisciplinary Reviews*, *6*, 2023ss090. https://doi.org/DOI: 10.31893/multirev.2023ss090

Tortia, E. C., Sacchetti, S., & López-Arceiz, F. J. (2022). A Human Growth Perspective on Sustainable HRM Practices, Worker Well-Being and Organizational Performance. *Sustainability*, *14*(17), 11064. DOI: 10.3390/su141711064

Usha, M., & Ramkumar, N. (2025). Happy Workers, Healthy Business: The Impact of Sustainable Human Resource Management and Workplace Happiness on Employee Engagement of Women Employees in Tea Plantations. *Sustainability*, *17*(3), 1047. DOI: 10.3390/su17031047

Vanhala, S., & Tuomi, K. (2006). HRM, Company Performance and Employee Well-being. *Management Review*, *17*(3), 241–255. DOI: 10.5771/0935-9915-2006-3-241

Viitasaari, M., Järlström, M., & Saru, E. (2022). HRM Practices in Sustainability Reports of the World's Most Sustainable Corporations. *Proceedings - Academy of Management*, *2022*(1), 15957. DOI: 10.5465/AMBPP.2022.15957abstract

Vovk, I., & Vovk, Y. (2024). Sustainable personnel management in the hospitality industry: Enhancing organizational performance through employee engagement and commitment. *Economics. Management and Sustainability*, *9*(2), 44–58. DOI: 10.14254/jems.2024.9-2.4

Wang, Y., Ummar, R., Qureshi, T. M., Ul Haq, J., & Bonn, M. A. (2025). Employee Sustainability: How Green Practices Drive Employee Well-Being and Citizenship Behavior. *Sustainability*, *17*(3), 936. DOI: 10.3390/su17030936

Winarno, A., & Silvianita, A. (2024). The role of socially responsible human resource management in improving performance: An analysis of subjective well-being as a mediating variable. *Qubahan Academic Journal*, *4*(3), 454–468. DOI: 10.48161/qaj.v4n3a878

Wu, Y. (2024). The Sustainable Human Resource Management (HRM): A Systematic Literature Review. In K. Szegedi (Ed.), *Sustainability Can't Wait!: VII. BBU International Sustainability Student Conference The Book of Full Paper* (pp. 468–499). Budapesti Gazdasági Egyetem. https://doi.org/DOI: 10.29180/978-615-6886-04-0_20

Yadav, A., Jindal, L., Verma, R., & Vishwakarma, K. (2024). THE ROLE OF GREEN HUMAN RESOURCE MANAGEMENT – AN ASSESSMENT. *ShodhKosh: Journal of Visual and Performing Arts*, *5*(4). Advance online publication. DOI: 10.29121/shodhkosh.v5.i4.2024.1868

Yin, Q. (2023). The Impact of Green Human Resource Management on Organizational Performance. *Frontiers in Business, Economics and Management*, *11*(3), 112–115. DOI: 10.54097/fbem.v11i3.13198

Yuxiaosi, W. (2024). *The Sustainable Human Resource Management (HRM): A Systematic Literature Review*. Budapesti Gazdasági Egyetem., https://publikaciotar.uni-bge.hu/id/eprint/2408/

Zhang, X., & Peng, L. (2023). Exploring the relationship between green human resource management (GHRM) and organizational sustainable development performance. In Qu, X. (Ed.), *Second International Conference on Sustainable Technology and Management (ICSTM 2023)* (p. 80). SPIE. https://doi.org/DOI: 10.1117/12.3004951

Zihan, W., & Makhbul, Z. K. M. (2024). Green Human Resource Management as a Catalyst for Sustainable Performance: Unveiling the Role of Green Innovations. *Sustainability*, *16*(4), 1453. DOI: 10.3390/su16041453

KEY TERMS & DEFINITIONS

AMO Theory: AMO Theory, an acronym for Ability, Motivation, and Opportunity, is a paradigm that elucidates employee and organizational performance. It posits that performance is contingent on an employee's innate capability, their motivation to work, and the circumstances and resources available to them (opportunity). When all three elements are robust, performance is optimized; however, performance deteriorates if any one part is deficient.

Employee Performance: Employee performance assesses the extent to which an individual executes their work responsibilities, accomplishes tasks, and advances organizational objectives, taking into account aspects such as quality, quantity, and efficiency. It is affected by competencies, motivation, work environment, and leadership, and may be measured using metrics such as productivity, accuracy, and task completion rates, typically examined through performance evaluations, peer feedback, and goal monitoring.

Green HRM (Human Resource Management) policies: are strategies, practices, and systems designed to promote environmental sustainability within an organization. Key policies include *green recruitment* to attract eco-conscious candidates, *training programs* to educate employees on sustainability, and *performance management* that includes environmental objectives and rewards for eco-friendly behavior. Other practices involve reducing paper use, promoting remote work, adopting energy-efficient technologies, and fostering a general culture of environmental responsibility.

Green Human Resource Management: Green Human Resource Management (GHRM) is a strategy framework that incorporates environmental sustainability into all human resource operations to promote employee awareness of environmental issues and resource efficiency. Primary objectives include reducing the company's ecological footprint, fostering a sustainable culture, reducing costs, and enhancing employee motivation and engagement. Green Human Resource Management (GHRM) encompasses measures throughout the whole employee lifecycle, including sustainable recruitment, training, and performance management, to ensure the firm achieves its environmental objectives.

Organizational Performance: Organizational performance quantifies the efficacy with which a company attains its objectives, utilizing measurements such as financial and operational outcomes. It entails juxtaposing actual results with anticipated objectives and may be affected by a confluence of individual, group, and systemic elements. Monitoring performance enables firms to enhance plan execution, establish competitive advantages, and secure long-term sustainability.

SCT Theory: Social Exchange Theory (SET) posits that social behavior is the result of a cost-benefit analysis, where individuals seek to maximize rewards and minimize costs in their relationships. Individuals are thought to evaluate relationships based on perceived rewards versus costs and will continue, deepen, or end them accordingly. This theory applies to various relationships, from friendships and romantic partnerships to professional interactions.

Sustainable performance: defines an organization's ability to achieve its financial objectives while making a positive impact on the environment, employees' well-being, and society as a whole. To achieve sustainable performance, organizations need to integrate sustainability objectives into their overall strategy.

Chapter 4

Digital Infrastructure Transforming Healthcare, Cities, and Supply Chain: Green Governance Policies for Sustainable Resource Management

A. Sharmila
https://orcid.org/0009-0001-8110-2548
Saveetha Institute of Medical and Technical Sciences, Chennai, India

K. C. Raja Shree
https://orcid.org/0009-0009-1029-910X
Saveetha Institute of Medical and Technical Sciences, Chennai, India

G. S. Gigi
Saveetha Institute of Medical and Technical Sciences, Chennai, UK

ABSTRACT

Technology policy is essential for promoting sustainability and green governance in various sectors and countries. Green practices, renewable energy solutions, and resource-efficient systems are adopted through technology policy, which creates legislative frameworks, incentives, and innovation pathways. This study investigates how strategic technology policies support circular economy models, impact environmental management, and facilitate digital breakthroughs such as blockchain, AI, and IoT to improve sustainability outcomes. Additionally, it examines how well policy frameworks align with international environmental objectives, such as the Paris Agreement and the Sustainable Development Goals (SDGs) of the UN. The

DOI: 10.4018/979-8-3373-6821-4.ch004

results show that well-designed technology policies increase institutional capacity for green governance and hasten the shift to low-carbon economies, guaranteeing long-term ecological balance and socioeconomic resilience.

1. INTRODUCTION

The core networks, platforms, and technologies that enable the collection, storage, processing, and sharing of data across multiple systems and businesses are called digital infrastructure platforms. Technologies that enable the gathering, storage, processing, and sharing of data across multiple systems and businesses are referred to as digital infrastructure (Du et al., 2024). Resource-efficient and ecologically mindful technologies that maximize productivity, limit waste, and support long-term ecological and social objectives (Melville, 2010).

In the field of healthcare, digital infrastructure plays a crucial role in supporting smart hospitals and telemedicine platforms, helping to reduce unnecessary resource consumption, optimize energy usage, and minimize waste associated with medical supplies and pharmaceuticals. For instance, digital infrastructure aids smart hospitals and telemedicine platforms in reducing superfluous resource consumption, enhancing energy efficiency, and lessening waste from medical supplies and pharmaceuticals (Williams et al., 2019). IoT-enabled medical devices and automated inventory management systems empower hospitals to monitor usage patterns, forecast demand, and mitigate issues related to overstocking or expired medications, thereby improving operational efficiency and reducing their environmental impact. This approach enhances operational efficiency while lowering the environmental footprint associated with overstocking or expired medications (Ghobakhloo et al., 2021). The development of smart cities in urban planning relies heavily on integrated digital infrastructure to effectively oversee energy, water, waste, and transportation systems. This integrated digital infrastructure is essential for managing waste and transportation systems in the context of smart city development (Batty et al., 2012). Sensor networks, traffic management platforms, and digital twin models enable urban planners to simulate scenarios, anticipate resource demands, and implement proactive strategies to minimize energy consumption and emissions. These platforms and digital twin models enable urban planners to create simulations, predict resource needs, and take proactive measures to reduce energy use and emissions. Smart systems enhance the allocation of public resources, alleviate traffic congestion, and encourage sustainable mobility options, thereby fostering the development of resilient, adaptive, and environmentally friendly cities. Intelligent systems improve the distribution of public resources, decrease congestion, and advocate for sustainable mobility solutions, resulting in cities that are resilient, adaptive, and environmentally responsible (Albino et al., 2015).

Supply chains, often marked by their complexity and global interdependencies, gain substantial advantages from digital infrastructure, which enhances visibility, predictive analytics, and optimizes logistics and inventory management (Ivanov et al., 2019). As noted by Brandenburg et al. (2014), technologies such as blockchain, IoT, and AI enable comprehensive tracking, minimize material waste, and enhance accountability and transparency in sourcing and transportation. Similarly, blockchain, IoT, and AI support end-to-end tracking, reduce material waste, and bolster accountability and transparency in sourcing and transportation (Ghobakhloo et al., 2021). Furthermore, blockchain-enabled tracking guarantees transparency in sourcing, promotes responsible sourcing practices, and enhances sustainability compliance within multi-tiered supply networks. As systems become increasingly interconnected, the risk of security breaches affecting private data increases (Lodha et al., 2023). The challenges of standardization and interoperability complicate the integration of diverse systems, thereby heightening operational risks and inefficiencies. Issues related to standardization and interoperability render the integration of disparate systems increasingly challenging, thereby amplifying operational risks and inefficiencies (Ejiaku, 2014). To surmount these obstacles, it is essential to undertake coordinated efforts that encompass regulatory frameworks to safeguard data, financial incentives to encourage adoption, and standardization initiatives to facilitate seamless integration across platforms and sectors. Tackling these challenges necessitates collaborative efforts, including regulatory frameworks that protect data, financial incentives for adoption, and standardization initiatives to ensure seamless integration across platforms and sectors. The utilization of AI-driven optimization and real-time decision-making is crucial for enhancing operational efficiency across different industries. By analyzing extensive datasets derived from sensors, IoT devices, and enterprise systems, AI facilitates proactive resource management, predictive maintenance, and more informed decision-making (Varnosfaderani & Forouzanfar, 2024).

Expanding digital infrastructure on a global scale allows both developed and developing regions to reap the benefits of sustainable technology adoption, thereby fostering inclusive growth and advancing the realization of the Sustainable Development Goals worldwide. The policy recommendations for stakeholders underscore the essential involvement of policymakers, industry leaders, and urban planners in promoting the adoption of digital infrastructure. Strategic suggestions for stakeholders reiterate the vital function of policymakers, industry leaders, and urban planners in facilitating the uptake of digital infrastructure. To alleviate financial constraints faced by businesses, policymakers ought to devise and implement strategies that stimulate the development and expansion of digital infrastructure, concentrating on minimizing the tax burden, enhancing the quality of tax revenues, and fortifying regional financial ecosystems. For businesses, it is imperative that policymakers formulate and implement policies that promote the development and growth of digital infrastructure, with a focus on reducing fiscal pressures, improving the quality of fiscal revenue, and bolstering regional financial ecosystems (Laurenza et al., 2018).

Government and private R&D investments can foster technological innovation and facilitate the adoption of digital solutions in essential sectors, such as supply chains, healthcare, and urban planning. Both government and private sector investments in R&D can stimulate technological innovation and support the implementation of digital solutions in critical areas, including supply chains, healthcare, and urban planning (Zhuo et al., 2025). Urban planners are instrumental in incorporating digital infrastructure into urban development to establish smart and resilient cities. By integrating digital infrastructure into urban development, they aim to create smart, resilient cities. Through data analytics and digital twins, planners can enhance decision-making, improve energy and resource efficiency, and design systems that reduce environmental impact while improving livability (Pereira et al., 2024).

Real-time monitoring, predictive analytics, and data-driven decision-making enable improvements in operational efficiency while minimizing environmental impact. The use of analytics and data-driven decision-making enhances operational effectiveness and reduces environmental consequences (Ghobakhloo & Hong,2015). The integration of these technologies paves the way for a more resource-efficient, environmentally friendly, and sustainable future, while also fostering societal well-being and long-term economic development (Melville, 2010; Batty et al., 2012). By incorporating these technologies, an ecologically responsible and sustainable future is achievable, which further supports enduring societal well-being and economic advancement.

2. LITERATURE REVIEW

2.1 Digital Infrastructure

Digital infrastructure comprising Internet of Things (IoT) sensors, cloud and edge computing, artificial intelligence and machine learning (AI/ML), digital twins, and distributed ledgers (e.g., blockchain) is broadly recognised as the technological foundation for sustainable resource management across various sectors. This body of literature contends that these elements collectively facilitate real-time visibility, predictive optimisation, and reliable traceability, which are crucial capabilities for minimising waste, enhancing system efficiency, and bolstering resilience in socio-technical systems (Seng et al., 2022).

2.2 Healthcare and Digital Infrastructure

In healthcare, systematic reviews and empirical life-cycle assessment (LCA) studies reveal a complex interplay of advantages and disadvantages: digital health services, including telemedicine, remote patient monitoring, and electronic health records (EHRs), have the potential to reduce transport-related emissions and increase facility utilisation. However, these services also transfer specific environmental impacts to data centers, network infrastructure, and device lifecycles, underscoring the necessity of life-cycle assessment to substantiate sustainability claims. Consequently, researchers advocate for the development of standardised, sector-specific LCA frameworks and for investigations that measure both clinical outcomes and the comprehensive environmental impacts of digital health interventions (Lokmic-Tomkins et al., 2022).

2.3 Internet of Things (IoT) and Digital Infrastructure

The body of work concerning smart cities highlights the importance of IoT networks, cohesive urban data platforms, and digital twins as valuable tools for managing urban resources, including energy, water, waste, and mobility, through comprehensive monitoring, scenario simulation, and automated control. Digital twins are widely acknowledged for their contributions to resilience planning and to performing 'what-if' simulations; nevertheless, they also pose challenges regarding computation, data quality, and institutional integration for implementation at the city level (Huzzat et al., 2025).

Research in supply chain and logistics underscores the synergistic capabilities of IoT telemetry, AI forecasting, and blockchain provenance to reduce waste, improve routing, and support circular-economy initiatives. Extensive reviews reveal that the combination of blockchain and IoT improves traceability and trust in sensitive supply chains, particularly in the food and pharmaceutical sectors.

Nevertheless, researchers raise concerns about scalability, integration with current ERP systems, and the energy consumption associated with specific ledger architectures. It is frequently recommended that hybrid models, which incorporate permissioned ledgers and off-chain storage and emphasize energy-efficient protocols, be prioritized (Winkelmann et al., 2024). Cross-sector syntheses highlight several governance, equity, and environmental considerations that mitigate the technological potential. Firstly, deficiencies in interoperability and standards result in fragmented data ecosystems that hinder cross-domain optimisation (Kitchin & Moore-Cherry, 2021). Secondly, digital equity remains a persistent issue: factors such as connectivity, access to devices, and digital literacy influence who reaps the benefits of efficiency improvements and who risks being left out (Bansal, N., & Choudhary, H., 2024).

Thirdly, privacy, security, and governance in the public interest (including data trusts and procurement regulations) are consistently recognised as vital for reliable implementation (Gleeson & Walden, 2016). Lastly, the literature emphasises that the overall sustainability impact of digitalisation relies on lifecycle assessments of data centers, networks, and hardware, rather than solely on operational cost savings (Liew et al., 2025).

2.4 Supply Chain, Logistics, and Digital Infrastructure

Research in supply chain and logistics underscores the synergistic capabilities of IoT telemetry, AI forecasting, and blockchain provenance to reduce waste, improve routing, and support circular-economy initiatives. Extensive reviews reveal that the combination of blockchain and IoT improves traceability and trust in sensitive supply chains, particularly in the food and pharmaceutical sectors. Nevertheless, researchers raise concerns about scalability, integration with current ERP systems, and the energy consumption associated with specific ledger architectures. It is frequently recommended that hybrid models, which incorporate permissioned ledgers and off-chain storage and emphasise energy-efficient protocols, be prioritised (Winkelmann et al., 2024).

2.5 Digital Infrastructure's Contribution to Sustainability

According to Akhter and Sofi (2022), precision agriculture employs Internet of Things (IoT) sensors to track soil moisture, nutrient levels, and microclimatic factors. This facilitates more precise irrigation methods and minimizes water waste. By utilizing IoT-enabled monitoring devices, agricultural operations can significantly reduce environmental harm, improve crop yield efficiency, and optimize fertilizer use, all of which contribute to both economic and ecological sustainability (Zhang et al., 2023). Urban planners and policymakers can implement strategies that reduce resource waste, improve energy efficiency, and enhance citizens' quality of life, thanks to the actionable data provided by these real-time monitoring systems. Policymakers can implement interventions that reduce resource waste, improve energy efficiency, and enhance citizens' quality of life, aided by the actionable insights these real-time monitoring systems offer (Alshaflut, 2024). Big data analytics enhances real-time monitoring capabilities by providing the computational power needed to handle large volumes of data and generate predictive insights. Pereira et al. (2024) note that predictive analytics in smart cities enhances operational efficiency by recognizing patterns in energy consumption, peak demand times, and resource usage, which facilitates proactive planning that reduces environmental impact. By curtailing overproduction, reducing storage costs, and minimizing unnecessary emissions from transportation, the application of big data analytics in supply chain management supports demand forecasting, inventory optimization, and distribution efficiency, all of which contribute to environmental sustainability (Jain et al., 2017). By integrating real-time IoT data with Big Data analytics, decision-makers can leverage predictive analytics, facilitating a transition from reactive to proactive management strategies. By linking real-time IoT data with Big Data analytics, decision-makers gain predictive insights, enabling a shift from reactive to proactive management strategies (Lee & Lee, 2015).

Cloud environments empower organizations to optimize IT resource allocation through virtualization, workload balancing, and server consolidation, resulting in a significant reduction in energy consumption compared to traditional data centers. Green cloud computing strategies incorporate renewable energy sources, energy-aware scheduling algorithms, and intelligent load management to lower the carbon intensity of digital operations. The adoption of cloud infrastructure in large-scale organizations reduces the environmental footprint by minimizing physical hardware needs, decreasing power consumption, and optimizing cooling systems (Satpathy et al., 2025). IoT sensors and predictive analytics enable the creation of smart grids, thereby increasing the integration of renewable energy sources into the national grid, reducing transmission losses, and improving the efficiency of energy distribution (Al-kateeb & Abdullah, 2024). Additionally, cloud-based platforms enable companies to manage sustainability performance metrics, monitor carbon emissions in real time, and adhere to international frameworks such as the United Nations Sustainable Development Goals (Lee & Lee, 2015). Data privacy and security concerns are paramount, as IoT networks generate large amounts of sensitive data that, if compromised, could pose risks to individuals and organisations (Supriya & Padaki, 2016). Constrain adoption, particularly in resource-limited regions. Another crucial challenge concerns the environmental impact of electronic waste generated by the production, use, and disposal of digital materials, which can negate environmental benefits if not appropriately managed (Osibanjo & Nnorom, 2007). By integrating cloud computing, big data analytics, IoT, and smart technologies into resource management and operational procedures, digital infrastructure promotes sustainability in a transformative way. IoT and smart technologies enable accurate, real-time monitoring of material, energy, and water consumption, thereby minimizing waste and its negative environmental impacts (Lee & Lee, 2015; Zhang et al.,2023). Big data analytics facilitates predictive management, enabling proactive optimization of resources in urban planning, agriculture, and supply chains. Cloud computing provides scalable, energy-efficient platforms that reduce carbon footprint while meeting the digital demands of modern organizations (Alshaflut, 2024; Melville, 2010).

Figure 1. The Role of Digital Infrastructure in Sustainability

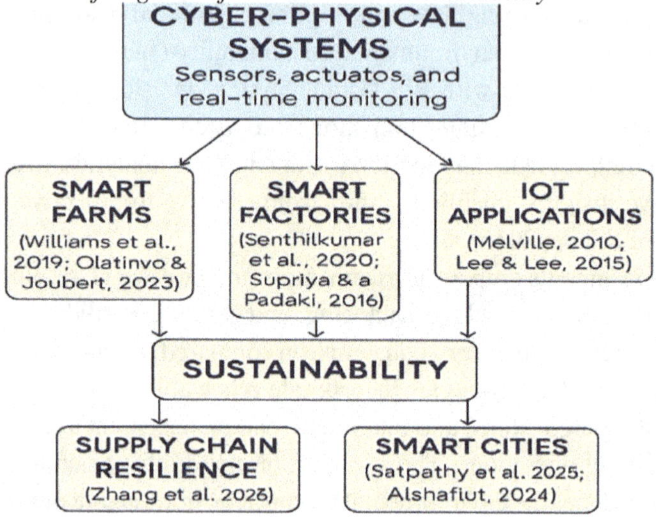

(Williams et al.,2019; Olatinwo & Joubert, 2023; Senthilkumar et al., 2020; Supriya & Padaki, 2016; Lee & Lee, 2015; Zhang et al., 2023; Satpathy et al., 2025; Alshaflut, 2024; Melville, 2010)

Digital infrastructure is crucial to promoting sustainability across sectors such as agriculture, energy, water management, and urban development. In industries such as agriculture, energy, water management, and urban development, digital infrastructure is crucial. Businesses and governments can reduce waste and carbon footprint, and improve resilience to environmental concerns by using real-time data, predictive models, and scalable infrastructure (Williams et al., 2019). In agriculture, for example, soil moisture sensors, weather monitoring devices, and nutrient analysers collect critical data that guides farmers in optimising irrigation schedules and applying fertiliser only when necessary. This reduces water consumption and prevents nutrient runoff into surrounding ecosystems, supporting sustainable agricultural practices (Wolfert et al.,2017)—IoT-enabled resource monitoring. Smart water meters and sensors can detect leaks and irregular consumption patterns, allowing municipalities to intervene early and avoid water waste. It demonstrates that such systems have significantly reduced unrevenue water losses in urban areas (Olatinwo & Joubert, 2023). In addition to increasing operational efficiency, these technologies enable stakeholders to adopt more environmentally responsible and sustainable practices (Senthilkumar et al.,2020).

In energy management, for example, utilities use predictive analytics to anticipate periods of peak demand and adjust energy supply accordingly. This proactive approach ensures that power plants operate at optimal efficiency and reduces unnecessary energy production, thereby minimizing greenhouse gas emissions (Ullah et al.,2018). Similarly, agricultural applications of big data analytics allow farmers to anticipate adverse weather events or pest outbreaks, thereby enabling preventive interventions that protect crops and conserve resources (Liakos et al., 2018). In logistics and supply chain management, big data solutions enable companies to streamline operations by analysing traffic patterns, delivery schedules, and fuel consumption data (Prange et al., 2018). Predictive analytics in waste management also offers substantial benefits: by analysing historical waste generation trends, municipalities can adjust collection schedules and allocate resources more efficiently, leading to improved efficiency and a lower environmental impact (Mishra et al., 2022).

Cloud computing, which provides scalable, adaptable, and energy-efficient IT infrastructure, is another revolutionary technology that promotes adaptable, energy-efficient computer infrastructure and is a game-changer for sustainability (Beloglazov et al., 2012). Cloud-based monitoring solutions in manufacturing save energy and reduce waste, and avoid costly downtime by tracking equipment performance and predicting repair needs. Waste and prevent expensive downtime by tracking equipment performance and forecasting repair plans. (Caggiano, 2018). In addition to improving energy efficiency, the move to cloud computing facilitates collaboration between stakeholders. Stakeholder collaboration: Businesses, governments, and academic institutions can access data. Computing resources through shared platforms, without using Cloud platforms, provide the processing capacity to manage and curate this data. At the same time, artificial intelligence (AI) algorithms improve decision-making by identifying trends and optimizing interventions, providing the processing capacity required to handle and retain this data. In contrast, artificial intelligence (AI) algorithms improve decision-making by spotting trends and maximising interventions (Albino et al.,2015). The synergy between these technologies allows cities to reduce their operational costs and emissions. Enables cities to lower operational costs and reduce emissions. Through the large-scale storage and analysis of this data by cloud platforms, farmers can obtain real-time insights and modify their methods as necessary (Liakos et al.,2018).

To maximize energy supply and demand management, energy utilities are combining cloud computing, predictive analytics, and smart meters. in the energy sector, combine cloud computing, predictive analytics, and smart meters (Mishra et al., 2022).

2.6 Smart Cities and Sustainable Urban Development: Combining Technology for a Greener Future

Integrating digital twin models, intelligent transportation systems, Internet of Things (IoT) systems, and platforms for citizen interaction is crucial for this transformation. By enhancing infrastructure, fostering active citizen participation, and optimizing resource management, these solutions collaborate to promote sustainable urban development (Zhuo et al.,2025).IoT-enabled sensor networks are transforming urban resource management by delivering real-time data that improves the efficiency of energy, water, and waste systems. In wastewater management, IoT-based systems enable real-time monitoring of water quality and flow, facilitating timely interventions and effective treatment processes (Alzahrani et al., 2023). For instance, AI-driven adaptive traffic lights can adjust green light durations based on traffic conditions. This innovation has led to reduced wait times and improved traffic flow at critical intersections (Ganin et al.,2019).

In a similar vein, Nagpur has commenced deploying AI-powered speed radar systems equipped with automatic number plate recognition (ANPR) cameras. These systems can detect and document speeding vehicles in real time, even at night, thereby enhancing road safety and compliance with traffic laws. Smart cities can significantly reduce greenhouse gas emissions by enhancing public transportation and optimizing traffic flow, ultimately making urban areas more sustainable and livable (Bestuzheva et al.,2023). By providing a dynamic and interactive platform, digital twins enable cities to anticipate challenges and develop practical solutions (Huzzat et al., 2025). Urban digital twins integrate data from various sources to simulate urban plans digitally prior to execution, delivering real-time and future forecasts that empower cities to make decisions grounded in evidence (Valadas, 2024).

This methodology improves the resilience and adaptability of urban systems, ensuring that cities can effectively address changing challenges and opportunities. Citizen engagement is essential for the success of smart city initiatives (Huzzat et al., 2025). Platforms that enhance communication between residents and city officials foster transparency, inclusivity, and active involvement in urban governance. By allowing citizens to express concerns, provide suggestions, and engage in decision-making processes, these platforms cultivate a sense of belonging and shared responsibility (Pereira et al., 2024).

Figure 2. Sustainable Urban Development and Smart Cities

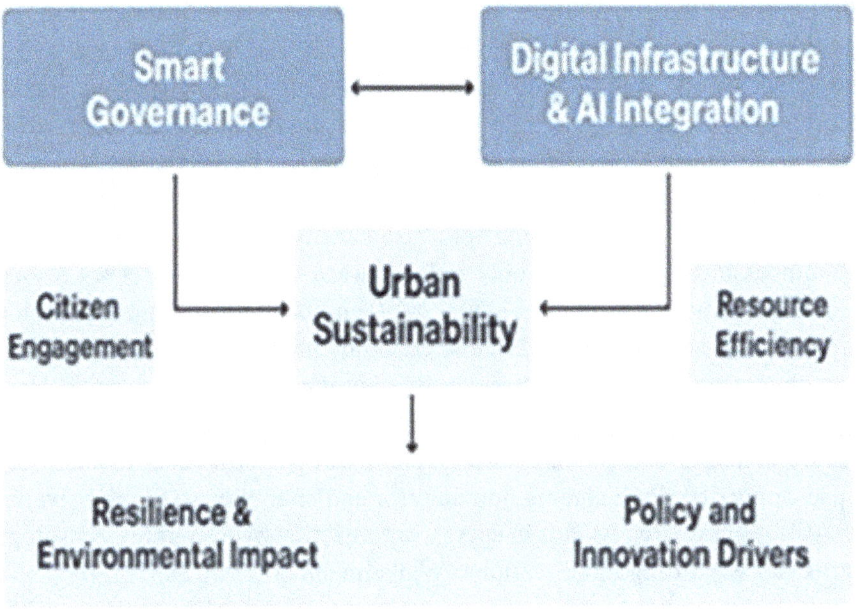

(Pereira et al., 2024; Moutinho et al.,2025; Huzzat et al., 2025; Bestuzheva et al., 2023; Ganin et al., 2019; Alzahrani et al., 2023; Zhuo et al., 2025)

2.7 Digital Supply Chains for Sustainable Resource Management

Digital supply chains, supported by technologies such as the Internet of Things (IoT), blockchain, artificial intelligence (AI), and predictive analytics, are transforming how businesses manage resources, minimize waste, and foster sustainability. These innovations are crucial for moving towards more sustainable and efficient supply chain practices (Zhang et al., 2022). Effective resource management in a sustainable context necessitates complete transparency within the supply chain. The ability to collect real-time data enables immediate identification of inefficiencies or disruptions, facilitating prompt corrective measures (Bharati & Podder, 2022). One of the numerous ways digital technology significantly aids circular economy initiatives is by enhancing supply chain transparency and traceability (Saberi et al., 2019). By accurately forecasting these elements, businesses can reduce overproduction, decrease waste, and ensure the efficient use of resources. For example, AI-powered tools can predict supply chain bottlenecks or changes in consumer demand, enabling companies to proactively adjust their operations (Zhang et al., 2022).

Furthermore, these analytics can identify areas with significant environmental impacts, such as energy-intensive operations or suppliers with high emissions, enabling targeted actions to mitigate adverse effects. IoT devices track the condition and location of products, while blockchain technology guarantees the authenticity and origin of materials, thereby promoting recycling and reuse (Huzzat et al., 2025). AI algorithms further enhance circularity by optimizing reverse logistics, forecasting product end-of-life, and proposing refurbishment or recycling alternatives. This all-encompassing approach not only reduces waste but also conserves resources and lowers carbon footprints. Jain et al. (2017) emphasize that digital technologies are crucial to various aspects of circular economy initiatives, including enhancing supply chain transparency and traceability. Automation, powered by AI and robotics, improves logistics efficiency by refining warehousing, transportation, and inventory management. In warehouses, robots can perform tasks such as sorting and packaging, thereby reducing human error and energy consumption (Ivanov et al., 2019). Additionally, AI aids in energy optimization by examining consumption patterns and suggesting modifications (Winkelmann et al., 2024).

Figure 3. Digital Supply Chains for Sustainable Resource Management

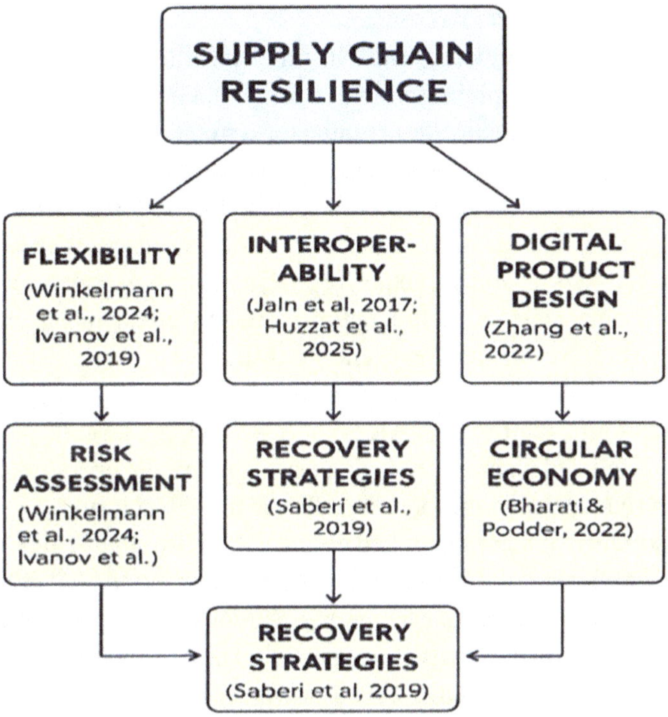

(Winkelmann et al., 2024; Ivanov et al.,2019; Jain et al., 2017; Huzzat et al., 2025; Zhang et al., 2022; Bharati & Podder, 2022; Saberi et al., 2019)

2.8 Obstacles and Hurdles in Digital Supply Chains for Sustainable Resource Management

Digital platforms enable cross-sector data interoperability, facilitating smooth communication and collaboration among systems and stakeholders, thereby addressing one of the most significant challenges in sustainable resource management: information fragmentation. By creating collaborative digital infrastructures, organisations can respond more effectively and cohesively to sustainability challenges, while improving resource utilisation and reducing environmental impact (Couldry et al., 2015).Data interoperability, which permits the sharing and collective analysis of information from healthcare systems, urban infrastructure, and supply chains, is essential for achieving comprehensive sustainability objectives. For example, merging environmental data with healthcare information allows for the assessment of how pollution and different environmental factors influence public health. This integration can inform targeted interventions, disease prevention strategies, and the efficient allocation of medical resources.

Similarly, linking supply chain data with urban infrastructure systems enables the optimisation of logistics, minimisation of operational redundancies, and reduction of carbon footprints, illustrating the potential for systemic environmental advantages across various sectors (Gamidullaeva et al., 2025).Shared digital platforms serve as centralised hubs that connect these sectors and facilitate coordinated efforts toward sustainability. In healthcare, such platforms integrate patient data with environmental and logistical datasets, enabling predictive analytics for disease outbreaks, resource planning, and emergency response optimisation. In urban management, digital platforms synchronise traffic flows, energy distribution, water usage, and waste management systems, improving the efficiency and resilience of city operations. In supply chains, digital platforms provide real-time tracking, inventory management, and predictive demand planning, reducing waste, increasing operational efficiency, and promoting circular economy practices. Research by (Couldry et al.,2015) emphasises the importance of these integrated digital systems in smart cities, highlighting their capacity to foster cross-sector collaboration, innovation, and governance, ultimately contributing to urban sustainability. Similarly, (Gamidullaeva et al.,2025) highlight the development of cross-sectoral digital platforms as tools for innovation ecosystem development, demonstrating how networked interactions among sectors can lead to innovative business models and end-to-end digital processes that are environmentally and economically sustainable. By enabling the aggregation and analysis of large-scale data across domains, these platforms allow decision-makers to assess sustainability metrics in real time, track progress toward environmental and social goals, and identify areas of inefficiency or risk. For instance, predictive analytics based on integrated healthcare and environmental data can help local governments identify regions most vulnerable to air pollution, enabling targeted public health interventions. In supply chains, overseeing the movement of materials and energy consumption among suppliers and distributors enables organizations to pinpoint areas of waste, enhance transportation routes, and lower greenhouse gas emissions. The capacity to visualize and analyse data across various sectors fosters an evidence-based strategy for sustainability, promoting more informed decisions that are in accordance with global sustainability frameworks like the United Nations Sustainable Development Goals (SDGs) (Couldry et al.,2015; Gamidullaeva et al.,2021). Infrastructure investment, including high-speed networks, cloud computing capabilities, and resilient digital architectures, is necessary to support large-scale cross-sector platforms, while stakeholder engagement, including collaboration among government agencies, private sector actors, and civil society, is essential for adoption and sustainability (Couldry et al., 2015).

The considerable influence of digital platforms on sustainability is evidently illustrated by their ability to promote innovation and collaborative problem-solving. By connecting data and processes across healthcare, urban management, and supply chains, these platforms enable the development of new services, business models, and operational practices that would not be possible in isolated systems. (Gamidullaeva et al., 2021) emphasize that cross-sector digital platforms enable networked interactions that facilitate the co-creation of solutions, such as predictive healthcare services informed by urban environmental monitoring or dynamic supply chain routing guided by real-time city traffic patterns. This integration not only enhances efficiency and resource utilization but also instigates systemic changes in how organizations approach sustainability, fostering a culture of continuous learning, collaboration, and innovation. In urban management, integrated platforms equip cities to dynamically modify energy distribution, traffic patterns, and waste management in response to real-time conditions, thereby improving both sustainability and quality of life. In supply chains, digital platforms enable swift adjustments to logistics and inventory strategies based on real-time demand and environmental factors, thereby minimizing resource wastage and lessening environmental impact (Couldry et al., 2015; Gamidullaeva et al., 2021). The incorporation of digital platforms within healthcare, urban management, and supply chains signifies an essential strategy for promoting sustainable development. By enabling cross-sector data interoperability, these platforms facilitate evidence-based decision-making, resource optimization, operational efficiency, and systemic innovation. They serve as centralized hubs for collaboration, allowing stakeholders to respond collectively to environmental, social, and economic challenges. Despite the ongoing challenges concerning data privacy, standardization, infrastructure, and stakeholder engagement, research conducted by (Couldry et al.,2015) and Gamidullaeva et al.,2021) highlights the capacity of digital platforms to facilitate transformative changes across various sectors. The ongoing advancement and implementation of these integrated digital infrastructures will be crucial in developing sustainable, resilient, and innovative systems for the future.

2.9 Prospective Pathways in Digital Supply Chains

The evolution of digital supply chains is increasingly shaped by artificial intelligence (AI), sustainable ICT practices, and scalable digital infrastructure, which collectively form the basis of modern supply chain management (Garay-Rondero et al., 2020). Within these technologies, artificial intelligence (AI) plays a vital role by enabling real-time decision-making and enhancing operational efficiency.

By analyzing large datasets generated by sensors, Internet of Things (IoT) devices, and enterprise systems, AI aids organizations in accurately predicting demand, enhancing inventory management, and boosting logistics efficiency (Satpathy et al.,

2025). (Culot et al., 2024) emphasize that AI-powered predictive analytics in supply chains helps companies anticipate potential disruptions, while simultaneously mitigating operational risks and enhancing overall resilience. Samuels & Singh (2025) argue that the strategic integration of AI is a crucial component for adaptable and responsive supply chain operations, enabling organizations to proactively respond to evolving market demands, shifting customer expectations, and external disruptions such as pandemics or geopolitical tensions. In the realm of supply chain operations, AI-driven forecasting systems empower businesses to foresee future demand fluctuations before they occur. These predictive systems leverage historical purchasing behaviors, market trends, and supply chain interruptions to generate precise forecasts, which consequently minimize excess inventory and operational waste. (Culot et al., 2024) points out that these AI-driven models also enable businesses to dynamically modify their procurement strategies, ensuring more effective resource allocation. Furthermore, algorithms powered by artificial intelligence enhance transportation logistics by pinpointing the most efficient delivery routes and minimizing fuel usage, thereby aiding in the reduction of greenhouse gas emissions. (Samuels & Singh, 2025) emphasizes that these innovations in logistics not only facilitate cost savings but also encourage eco-friendly practices, allowing organizations to comply with more stringent regulatory requirements and customer demands for sustainability.

In conjunction with AI, the adoption of sustainable ICT practices is essential for ensuring that supply chain operations do not contribute to environmental harm. Contemporary supply chains are heavily dependent on computing infrastructure, which requires substantial energy and generates considerable carbon emissions. Initiatives in green computing, such as the utilization of energy-efficient hardware, the optimization of data center cooling systems, and the implementation of renewable energy solutions, can greatly diminish energy consumption and the carbon footprints linked to digital supply chain operations. (Bharati & Podder, 2022).For example, some organizations have adopted liquid cooling technologies in their data centers, which reduces the energy required for air conditioning and cooling systems. These initiatives not only lower operational costs but also align with corporate sustainability goals, helping organizations meet their climate action commitments. Furthermore, sustainable ICT practices encompass more than just hardware enhancements. The design of efficient software and the optimization of algorithms are crucial in minimizing computational overhead and energy consumption. By refining software processes, organizations can ensure that their digital operations utilize fewer resources without sacrificing performance. Incorporating sustainability throughout the entire lifecycle of ICT systems from manufacturing and deployment to maintenance and disposal helps to reduce electronic waste and environmental impacts. Scalable digital infrastructure represents a crucial component of sustainable supply chain management, as it empowers organizations to swiftly adapt to evolving business requirements without incurring substantial additional costs or complexities. Cloud computing platforms, modular data centers, and standardized interoperable systems enable the seamless expansion and integration of digital operations. This scalable infrastructure permits organizations to modify capacity as necessary, offering the flexibility to respond to demand spikes or market disruptions without the need for significant investments in physical infrastructure. This is particularly advantageous for supply chains operating in volatile markets or regions with fluctuating consumer demand, as scalability ensures that companies can meet customer needs without overextending their resources (Zhang et al., 2022). Furthermore, standardization and interoperability are vital for facilitating smooth integration across various regions, industries, and technologies. By employing a unified set of protocols and data formats, businesses can more effectively share information and collaborate internationally, thereby streamlining global supply chain operations. Bharati & Podder (2022) highlight that a robust and interoperable digital infrastructure enhances efficient resource allocation by guaranteeing that data flows seamlessly. Such infrastructure enables organizations to base their decisions on data, improving resource utilization, reducing waste, and promoting better coordination among suppliers, manufacturers, and distributors.

The combination of artificial intelligence, sustainable information and communication technology practices, and scalable infrastructure fosters a comprehensive strategy for developing resilient and environmentally responsible supply chains. Optimization driven by AI guarantees that supply chain operations are increasingly predictive, efficient, and cost-effective, enabling businesses to address disruptions proactively and sustain continuity. Sustainable ICT solutions ensure that the environmental impact of digital operations is minimized, promoting responsible energy usage and waste management throughout the technology lifecycle. Scalable infrastructure empowers organizations to meet demand fluctuations without unnecessary resource expenditure while ensuring that their digital platforms remain interoperable and globally accessible. Investing in AI technologies provides businesses with strategic advantages, enabling them to enhance responsiveness, reduce operational costs, and maintain supply chain continuity amid uncertainty (Culot et al., 2024; Samuels & Singh, 2025). These technologies also support environmentally friendly practices by reducing energy consumption and improving logistics efficiency. Simultaneously, (Bharati & Podder,2022) highlight the necessity of sustainable ICT practices to ensure that technological advancements are in harmony with environmental priorities. Without these practices, companies risk amplifying their carbon emissions and worsening global climate challenges. The importance of scalable infrastructure is additionally highlighted by (Zhang et al., 2023), who claim that flexible digital systems enable organizations to effectively adjust their operations according to demand trends, thereby minimizing both expenses and complexity. (Ivanov et al., 2019) also emphasize that strong infrastructure supports coordinated efforts across supply chains, allowing for effective resource allocation, data-informed decision-making, and alignment with wider sustainability objectives. Collectively, these technological innovations create a formidable triad that bolsters supply chain resilience, enhances operational efficiency, and promotes environmental accountability.

The practical uses of this integrated approach are clear across various industries, including manufacturing and retail. For instance, global retailers utilize AI-driven forecasting systems to anticipate seasonal demand, thereby minimizing overstock and waste.In a similar vein, logistics companies have implemented AI-based routing algorithms to enhance delivery schedules and decrease fuel consumption (Caggiano, 2018). In the manufacturing sector, green ICT practices assist firms in monitoring energy consumption and optimizing production processes to reduce emissions.

Cloud-based platforms and modular data centers facilitate businesses in entering new markets while ensuring smooth operations and reducing downtime. As climate change and resource scarcity increasingly threaten global supply chains, the integration of AI, sustainable ICT, and scalable infrastructure is becoming an essential focus. Organizations that adopt these technologies are more capable of navigating disruptions, fulfilling regulatory obligations, and aligning with environmental stewardship initiatives. The combination of artificial intelligence, sustainable information and communication technology practices, and scalable infrastructure creates a holistic strategy for building resilient and environmentally responsible supply chains.

AI-driven optimization ensures that supply chain operations are more predictive, efficient, and cost-effective, allowing businesses to proactively manage disruptions and maintain continuity (Caggiano, 2018; Ivanov et al., 2019).

3. METHODOLOGY

This chapter employs a systematic review and content analysis based on the adaptation of Siripipatthanakul et al. (a-h); the systematic review process is presented as follows;

Step 1: Finding the Problem - Manual searches, literature databases, and other methods can find early entries. This investigation began with a random topic. A related article was selected, and duplicate papers were removed.

Step 2: Screening displays the remaining records. Recognised record titles and abstracts are assessed for relevance to the study's topic or goals. Google Scholar, Scopus, and Web of Science were used to sample scholarly papers in the second phase. This phase selected the most relevant articles.

Step 3: Eligibility describes the study possibilities following screening. This stage determines the eligibility of essential information articles. Phase Three involved deliberately sampling papers for rigorous examination. In 2021–2025, chosen papers, plagiarism, and AI similarity checking using Grammarly.

Step 4: Total systematic review analysis studies. These studies qualify for future research. The flow diagram explains why and how much research was stopped. Exclusions often result from inadequate study design, data shortages, or inclusion criteria. Content analysis verified the systematic review.

4. RESULTS

4.1 Conceptual Framework

Digital infrastructure is the essential facilitator for the transformation of healthcare, urban systems, and supply chains towards sustainability. It serves as the core enabler of the evolution of healthcare, urban systems, and supply chains toward sustainability (Ghobakhloo & Hong, 2015). These technologies empower organizations and cities to track the usage of energy, water, waste, and materials, produce predictive insights, and enhance operational decision-making to reduce environmental impacts and increase efficiency (Melville, 2010; Ivanov et al., 2019).In healthcare, digital infrastructure facilitates efficient hospital operations by enabling monitoring of medical equipment, energy consumption, and inventory through the Internet of Things (IoT). Additionally, telemedicine platforms help reduce travel and associated carbon emissions. The IoT-enabled monitoring of medical equipment, energy usage, and inventory management, along with telemedicine platforms, plays a crucial role in minimizing travel and its related carbon footprint. Hospitals can forecast demand, optimize the use of medical supplies, and reduce waste through data-driven insights, yielding both environmental and operational advantages (Zhang et al., 2023). This capability allows urban planners to model resource allocation scenarios, enhance infrastructure utilization, and engage citizens in sustainable development efforts, thereby improving environmental outcomes and overall quality of life (Batty et al., 2012; Albinus et al., 2015). Digital platforms empower citizens to engage in sustainable development initiatives, cultivating a sense of community and collective responsibility. Citizen involvement in sustainable development initiatives fosters a sense of community and shared accountability. Sustainability initiatives promote a sense of community and shared responsibility. Blockchain and other technologies offer traceability and transparency, encouraging the circular economy and minimizing resource inefficiency in global networks, thereby reducing resource inefficiencies across international networks (Brandenburg et al, 2014). Address disruptions, optimize energy and resource use, and enhance resilience to supply and demand uncertainties (Ivanov et al., 2019).

Figure 4. Conceptual framework

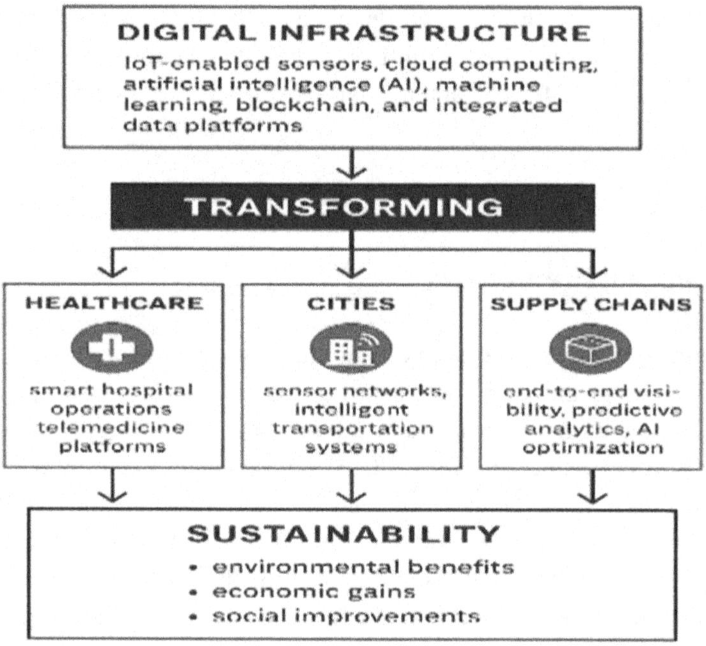

(Ghobakhloo & Hong, 2015; Melville, 2010; Ivanov et al., 2019; Zhang et al., 2023; Batty et al., 2012; Albino et al., 2015; Brandenburg et al., 2014)

Achieve sustainability results that encompass social enhancements such as improved well-being and fair access to services, economic benefits derived from heightened operational efficiency, and environmental advantages, including reduced emissions and waste. Adapt to evolving circumstances, enhance learning, and sustain long-term performance by connecting real-time monitoring and data analysis to both operational and strategic decision-making (Ghobakhloo & Hong, 2015; Melville, 2010).

4.2 HYPOTHESES

The proposed hypotheses collectively outline how digital infrastructure supports sustainable resource management in healthcare, smart cities, and supply chains. H1 establishes the foundational link by indicating that digital technologies such as IoT, AI, cloud platforms, and blockchain have a positive impact on resource management processes by facilitating real-time monitoring, automation, and data-driven decision-making.

Building on this, H2 asserts that enhanced resource management through better allocation, forecasting, and efficiency directly contributes to sustainability outcomes such as waste reduction, decreased energy consumption, and lower emissions.H3 acknowledges that digital infrastructure may also exert a direct influence on sustainability by enabling precision-based operations even before process redesign. H4 synthesizes these concepts through a mediating effect, suggesting that resource management processes act as the mechanism through which digital infrastructure achieves sustainability benefits. Additionally, H5 recognises that this relationship is not uniform across sectors; variations in regulatory environments, digital maturity, and governance structures affect the degree to which digital technologies improve resource management.

Finally, H6 and H7 propose that organisational readiness and data integration capabilities strengthen these relationships by enabling institutions to leverage digital tools for sustainable outcomes fully. Collectively, these hypotheses form a cohesive framework that links digital transformation to sustainability through enhanced resource management.

Figure 5. Hypotheses

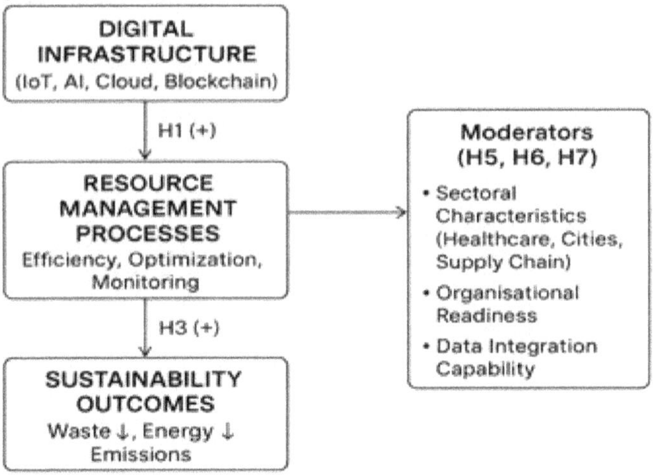

(Trist and Bamforth, 1951; Ghobakhloo & Hong, 2015; Albino et al., 2015; Brandenburg et al., 2014; Batty et al., 2012; Lee & Lee, 2015).

4.3 Theoretical Framework

The theoretical basis of digital infrastructure in sustainable resource management is grounded in sociotechnical systems theory, the resource-based view (RBV), and cyber-physical systems (CPS). These frameworks collectively elucidate the interactions among technology, human actors, and organizational processes to optimize resources and achieve sustainability outcomes (Madhani, 2010; Lee & Lee, 2015).

Sociotechnical systems theory emphasizes the interdependence of technology and human processes, asserting that for digital infrastructure to be effectively implemented, there must be a harmonious alignment among organizational structures, human behavior, and technological capabilities. This perspective is further supported by socio-technical systems theory, which similarly argues that effective deployment of digital infrastructure requires a cohesive relationship between organizational frameworks, human actions, and technological resources (Trist & Bamforth, 1951).

In the realms of healthcare, urban development, and supply chains, this viewpoint elucidates why simply implementing sensors or analytics systems is inadequate; achieving success hinges on the seamless integration of individuals, processes, and technology to effectively and sustainably manage resources (Barney, 1991). An additional theoretical framework is the resource-based view (RBV), which asserts that companies can secure a sustainable competitive advantage by leveraging unique, valuable, rare, and difficult-to-replicate resources and capabilities (Barney, 1991).

Within the scope of digital infrastructure, data, analytical capabilities, and cohesive digital platforms serve as strategic assets that empower healthcare providers, urban planners, and supply chain managers to enhance the efficiency of energy, water, waste, and material utilization. Furthermore, integrated digital platforms represent strategic resources that enable healthcare providers, city planners, and supply chain managers to optimize the consumption of energy, water, waste, and materials (Ghobakhloo & Hong, 2015). In a similar vein, supply chains that incorporate blockchain and IoT technologies can attain transparency, traceability, and operational efficiency (Brandenburg et al., 2014).

The theory of cyber-physical systems (CPS) also holds that digital infrastructure is transforming resource management. Sensors, actuators, and controllers integrated within CPS frameworks facilitate real-time monitoring, feedback, and optimization, which encompass the amalgamation of computing, networking, and physical processes (Lee & Lee, 2015). In the context of smart cities, CPS principles enable the dynamic management of energy, traffic, and water networks, ensuring efficient resource utilization while minimizing environmental impacts (Batty et al., 2012). Likewise, in supply chains, CPS fosters autonomous logistics and intelligent inventory management, where decisions are consistently guided by real-time data streams (Ivanov et al., 2019). Furthermore, sustainability-driven innovation theory enriches the theoretical framework by emphasizing how digital technologies can simultaneously create environmental, economic, and social value (Melville, 2010). By integrating predictive analytics, IoT, and cloud-based platforms, organizations and cities can implement proactive resource management strategies that align with the triple bottom line of profit, planet, and people (Albino et al., 2015).

Figure 6. Theoretical Framework

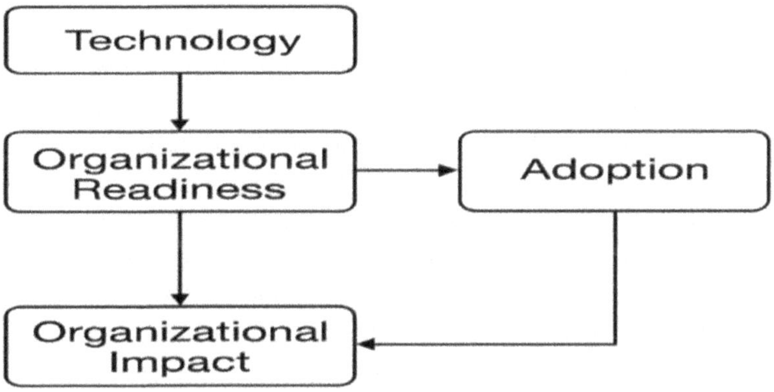

(Madhani, 2010; Lee & Lee, 2015; Trist & Bamforth, 1951; Barney, 1991; Ghobakhloo & Hong, 2015, Albino et al., 2015; Melville, 2010; Ivanov et al., 2019)

5. DISCUSSIONS

Digital infrastructure has emerged as a vital component in the sustainable management of resources across multiple sectors, such as healthcare, urban governance, and supply chain systems. As noted by Laurenza (2018), data-driven platforms significantly enhance the efficiency and transparency of processes that utilize substantial resources, enabling organizations to implement more sustainable and accountable practices.

In healthcare, integrating electronic health records (EHR), advanced diagnostics, and AI-powered decision support systems improves clinical accuracy, reduces resource waste, and enhances patient safety. These advancements highlight that digital infrastructure is not merely an additional technological aspect but a transformative element that reshapes how healthcare systems allocate resources, reduce redundancies, and ensure equitable access.

In urban environments, digital infrastructure similarly facilitates the transition from traditional administration to proactive, sustainability-oriented "smart city" frameworks. (Batty et al., 2012) observes that the deployment of sensors, integrated IoT platforms, and real-time data analytics results in enhanced management of energy, mobility, waste, and public services. Utilizing these tools, urban planners can forecast demand trends, optimize resource allocation, and formulate adaptive sustainability strategies. This viewpoint is consistent with the assertions of Albino et al. (2015), who contend that smart cities rely on digital innovation not only for operational efficiency but also for improving environmental resilience and quality of life. Such systems strengthen governance by enabling predictive, data-informed decision-making.

Supply chain systems have undergone significant transformations due to digital connectivity and automation. Winkelmann et al. (2024) argue that digital supply chains, characterized by blockchain technology, predictive analytics, and cloud-based logistics, improve resilience, reduce material waste, and foster transparency across the procurement process.

Digital infrastructure has emerged as a crucial component in the pursuit of sustainable development, offering transformative opportunities across various sectors. Its integration enables the combination of information and communication technologies with traditional infrastructure, leading to smarter and more efficient systems. For example, the advancement of digital infrastructure has been demonstrated to foster technological innovation, enhance the efficiency of resource allocation, and improve industrial frameworks, thus driving new levels of productivity (Ejiaku, 2014). It supports the creation of intelligent cities where data-driven decision-making enhances public services, reduces energy consumption, and raises the quality of life for residents. The intentional application of digital technologies in urban planning has led to more sustainable urban environments, highlighting the significant impact of digital infrastructure on urban development (Zhuo et al., 2025). In these smart cities, infrastructure such as interconnected sensors, advanced traffic management systems, and optimized energy grids plays a vital role in lowering urban carbon emissions while advancing social welfare and efficiency. To facilitate the growth of digital infrastructure, the formulation and execution of effective policies are critical. Governments ought to formulate comprehensive policies that encourage the development and enhancement of digital infrastructure while also addressing financial limitations. These policies should focus on reducing fiscal pressures, improving the quality of fiscal revenue, and reinforcing the regional financial ecosystem to alleviate the financing challenges faced by businesses, particularly those that are non-state-owned (Du & Wang, 2024).

Public-private partnerships have demonstrated their effectiveness in expediting the development of digital infrastructure. By combining resources, expertise, and strategic vision, collaborations between governmental bodies and private enterprises can tackle significant challenges such as funding deficits, technical deficiencies, and workforce constraints. These partnerships facilitate the execution of extensive digital initiatives that may otherwise be unfeasible due to resource limitations (Tolstolesova et al., 2021). Moreover, public-private partnerships guarantee that digital infrastructure projects are rooted in practical knowledge and can adapt responsively to changing sustainability objectives. Capacity building and skill enhancement are essential for the successful implementation and upkeep of digital infrastructure.

As infrastructure increasingly relies on sophisticated technologies, it is vital to equip individuals with the training needed to operate and manage these systems. Educational institutions and training centers should focus on equipping the workforce with essential skills to support digital initiatives, thereby ensuring that human capital evolves in tandem with technological progress (Melnyk et al., 2023). Urban planners are urged to incorporate digital tools into their development strategies, utilizing data analytics to inform decision-making and embracing technologies that minimize environmental impact. Initiatives for smart cities, including optimized transportation systems and intelligent energy management solutions, contribute to waste reduction and bolster urban resilience (Pereira et al., 2024). Regular evaluations enable stakeholders to assess the results of their actions, modify strategies, and ensure that infrastructure investments remain aligned with long-term objectives. Ongoing monitoring provides essential feedback on the performance and environmental effects of digital systems, facilitating adaptive governance and data-driven policymaking (Tolstolesova et al., 2021). The integration of digital infrastructure offers significant opportunities for advancing sustainable development. By enacting policies that promote investment, research and development, public-private collaborations, skill enhancement, sustainable urban planning, and effective monitoring, stakeholders can fully harness the potential of digital infrastructure (Agudelo-Vera et al., 2011).

While this chapter offers a comprehensive conceptual analysis, it is important to acknowledge its specific limitations. Firstly, the discourse relies heavily on secondary literature, which may limit the generalizability of empirical results. Given the significant variations in the adoption of digital infrastructure across countries and sectors, the insights provided may not fully capture the diverse realities of implementation, particularly in regions lacking adequate technological readiness or regulatory support (Zhang et al., 2023).

Secondly, this chapter highlights the transformative potential of digital infrastructure, while only briefly addressing challenges such as digital inequality, cybersecurity threats, and privacy concerns. Scholars such as Alshaflut (2024) argue that digital divides stemming from disparities in access, skills, and affordability pose significant barriers to achieving inclusive digital transformation. Similarly, (Supriya & Padaki, 2016) emphasize that improved data connectivity raises security and ethical data governance issues, which may undermine trust and hinder adoption. These risks indicate that sustainability outcomes are influenced not only by technological capabilities but also by strong governance and policy frameworks. Additionally, the rapid progress of digital technologies presents another limitation. The speed of technological evolution often exceeds the capacity of organizations and policies to adjust, suggesting that the digital solutions available today may soon become obsolete. Consequently, the insights presented in this chapter should be viewed as reflecting current trends rather than as long-term predictions. Moreover, comparisons across sectors such as healthcare, smart cities, and supply chains should be conducted with care, as each sector has unique structural, cultural, and regulatory characteristics that affect the adoption of digital technologies and associated sustainability outcomes (Garay-Rondero et al., 2020).

6. CONCLUSIONS

Digital infrastructure is transforming healthcare, urban environments, and supply chains by enabling more intelligent, more efficient systems, while green governance regulations ensure this transition is sustainable. In healthcare, digital technologies like AI diagnostics and telehealth increase efficiency and access; in cities, smart grids and IoT optimize energy use and traffic flow; and in supply chains, real-time tracking lowers waste and boosts resilience. This integration, guided by rules that demand resource efficiency and environmental responsibility, creates a unified strategy in which technical innovation and sustainable management are mutually reinforcing, ensuring that economic advantages coincide with long-term ecological health. By establishing conducive conditions for investment and innovation, governments can ensure that digital infrastructure plays a significant role in sustainable development while promoting inclusive growth. Furthermore, investment in research and development is essential for nurturing innovation within digital infrastructure. Both the public and private sectors should proactively endorse initiatives that investigate the use of digital technologies across industries such as culture, health, and education to boost productivity and operational efficiency. Supporting research that examines the effects of digital infrastructure helps create new technologies and frameworks aligned with sustainability goals. Such investments not only lead to technological progress but also lay the groundwork for resilient and adaptable infrastructure systems.

7. LIMITATION & RECOMMENDATIONS

This chapter is a systematic review and may not include respondents. Future research should build upon this foundation by incorporating longitudinal and sector-specific empirical studies, along with respondents, such as a questionnaire survey or an interview.

REFERENCES

Agudelo-Vera, C. M., Mels, A. R., Keesman, K. J., & Rijnaarts, H. H. (2011). Resource management as a key factor for sustainable urban planning. *Journal of Environmental Management, 92*(10), 2295–2303. DOI: 10.1016/j.jenvman.2011.05.016

Akhter, R., & Sofi, S. A. (2022). Precision agriculture using IoT data analytics and machine learning. *Journal of King Saud University. Computer and Information Sciences, 34*(8), 5602–5618. DOI: 10.1016/j.jksuci.2021.05.013

Al-kateeb, Z. N., & Abdullah, D. B. (2024). Unlocking the potential: Synergizing IoT, cloud computing, and big data for a bright future. *Iraqi Journal for Computer Science and Mathematics, 5*(3), 25. DOI: 10.52866/ijcsm.2024.05.03.001

Albino, V., Berardi, U., & Dangelico, R. M. (2015). Smart cities: Definitions, dimensions, performance, and initiatives. *Journal of Urban Technology, 22*(1), 3–21. DOI: 10.1080/10630732.2014.942092

Alshaflut, A. (2024). Enhancing smart city infrastructure: An IoT-integrated system approach for real-time urban management. *Indian Journal of Science and Technology, 17*, 3012–3017. DOI: 10.17485/IJST/v17i29.1070

Alzahrani, A. I., Chauhdary, S. H., & Alshdadi, A. A. (2023). Internet of Things (IoT)-based wastewater management in smart cities. *Electronics (Basel), 12*(12), 2590. DOI: 10.3390/electronics12122590

Bansal, N., & Choudhary, H. (2024). Fostering digital equity: Evaluating impact of digital literacy training on internet outcomes in rural marginalised communities in India. *International Journal of Lifelong Education, 43*(5), 473–493. DOI: 10.1080/02601370.2024.2347327

Barney, J. (1991). Firm resources and sustained competitive advantage. *Journal of Management, 17*(1), 99–120. DOI: 10.1177/014920639101700108

Batty, M., Axhausen, K. W., Giannotti, F., Pozdnoukhov, A., Bazzani, A., Wachowicz, M., Ouzounis, G., & Portugali, Y. (2012). Smart cities of the future. *The European Physical Journal. Special Topics, 214*(1), 481–518. DOI: 10.1140/epjst/e2012-01703-3

Beloglazov, A., Abawajy, J., & Buyya, R. (2012). Energy-aware resource allocation heuristics for efficient management of data centers for cloud computing. *Future Generation Computer Systems, 28*(5), 755–768. DOI: 10.1016/j.future.2011.04.017

Bestuzheva, K., Besançon, M., Chen, W. K., Chmiela, A., Donkiewicz, T., Van Doornmalen, J., & Witzig, J. (2023). Enabling research through the SCIP optimization suite 8.0. *ACM Transactions on Mathematical Software*, *49*(2), 1–21. https://dl.acm.org/doi/abs/10.1145/3585516

Bharati, S., & Podder, P. (2022). Machine and deep learning for iot security and privacy: Applications, challenges, and future directions. *Security and Communication Networks*, *2022*(1), 8951961. DOI: 10.1155/2022/8951961

Brandenburg, M., Govindan, K., Sarkis, J., & Seuring, S. (2014). Quantitative models for sustainable supply chain management: Developments and directions. *European Journal of Operational Research*, *233*(2), 299–312. DOI: 10.1016/j.ejor.2013.09.032

Caggiano, A. (2018). Cloud-based manufacturing process monitoring for smart diagnosis services. *International Journal of Computer Integrated Manufacturing*, *31*(7), 612–623. DOI: 10.1080/0951192X.2018.1425552

Couldry, N., MacDonald, R., Stephansen, H., Clark, W., Dickens, L., & Fotopoulou, A. (2015). Constructing a digital story circle: Digital infrastructure and mutual recognition. *International Journal of Cultural Studies*, *18*(5), 501–517. DOI: 10.1177/1367877913519313

Culot, G., Podrecca, M., & Nassimbeni, G. (2024). Artificial intelligence in supply chain management: A systematic literature review of empirical studies and research directions. *Computers in Industry*, *162*, 104132. DOI: 10.1016/j.compind.2024.104132

Du, Z. Y., & Wang, Q. (2024). Digital infrastructure and innovation: Digital divide or digital dividend? *Journal of Innovation & Knowledge*, *9*(3), 100542. DOI: 10.1016/j.jik.2024.100542

Ejiaku, S. A. (2014). Technology adoption: Issues and challenges in information technology adoption in emerging economies. *Journal of International Technology and Information Management*, *23*(2), 5. DOI: 10.58729/1941-6679.1071

Gamidullaeva, L., Shmeleva, N., Mityakov, E., Tolstykh, T., & Vasin, S. (2025). Strategic Tools for the Formation of Cluster Capital to Implement Technological Innovations. *Systems*, *13*(4), 270. DOI: 10.3390/systems13040270

Ganin, A. A., Mersky, A. C., Jin, A. S., Kitsak, M., Keisler, J. M., & Linkov, I. (2019). Resilience in intelligent transportation systems (ITS). *Transportation Research Part C, Emerging Technologies*, *100*, 318–329. DOI: 10.1016/j.trc.2019.01.014

Garay-Rondero, C. L., Martínez-Flores, J. L., Smith, N. R., Caballero Morales, S. O., & Aldrette-Malacara, A. (2020). Digital supply chain model in Industry 4.0. *Journal of Manufacturing Technology Management, 31*(5), 887–933. DOI: 10.1108/JMTM-08-2018-0280

Ghobakhloo, M., & Hong, T. S. (2015). The business value of information system-enabled e-collaboration capabilities. [IJeC]. *International Journal of e-Collaboration, 11*(1), 22–56. DOI: 10.4018/ijec.2015010103

Ghobakhloo, M., Iranmanesh, M., Grybauskas, A., Vilkas, M., & Petraitė, M. (2021). Industry 4.0, innovation, and sustainable development: A systematic review and a roadmap to sustainable innovation. *Business Strategy and the Environment, 30*(8), 4237–4257. DOI: 10.1002/bse.2867

Gleeson, N., & Walden, I. (2016). Placing the state in the cloud: Issues of data governance and public procurement. *Computer Law & Security Review, 32*(5), 683–695. DOI: 10.1016/j.clsr.2016.07.004

Huzzat, A., Anpalagan, A., Khwaja, A. S., Woungang, I., Alnoman, A. A., & Pillai, A. S. (2025). A comprehensive review of digital twin technologies in smart cities. *Digital Engineering, 100040*. Advance online publication. DOI: 10.1016/j.dte.2025.100040

Ivanov, D., Dolgui, A., & Sokolov, B. (2019). The impact of digital technology and Industry 4.0 on the ripple effect and supply chain risk analytics. *International Journal of Production Research, 57*(3), 829–846. DOI: 10.1080/00207543.2018.1488086

Jain, A. D. S., Mehta, I., Mitra, J., & Agrawal, S. (2017). Application of big data in supply chain management. *Materials Today: Proceedings, 4*(2), 1106–1115. DOI: 10.1016/j.matpr.2017.01.126

Kitchin, R., & Moore-Cherry, N. (2021). Fragmented governance, the urban data ecosystem, and smart city-regions: The case of Metropolitan Boston. *Regional Studies, 55*(12), 1913–1923. DOI: 10.1080/00343404.2020.1735627

Laurenza, E., Quintano, M., Schiavone, F., & Vrontis, D. (2018). The effect of digital technologies adoption in healthcare industry: A case based analysis. *Business Process Management Journal, 24*(5), 1124–1144. DOI: 10.1108/BPMJ-04-2017-0084

Lee, I., & Lee, K. (2015). The Internet of Things (IoT): Applications, investments, and challenges for enterprises. *Business Horizons, 58*(4), 431–440. DOI: 10.1016/j.bushor.2015.03.008

Liakos, K. G., Busato, P., Moshou, D., Pearson, S., & Bochtis, D. (2018). Machine learning in agriculture: A review. *Sensors (Basel), 18*(8), 2674. DOI: 10.3390/s18082674

Liew, E. F., Ong, D. E., & Peerun, M. I. (2025). Mapping the Future of Sustainable Construction: Review on Digitalisation and Life Cycle Assessment Trends in Asia-Pacific. *Energy and Building, 116547.* Advance online publication. DOI: 10.1016/j.enbuild.2025.116547

Lodha, L., Baghela, V. S., Bhuvana, J., & Bhatt, R. (2023). A blockchain-based secured system using the Internet of Medical Things (IOMT) network for e-healthcare monitoring. *Measurement. Sensors, 30*, 100904. DOI: 10.1016/j.measen.2023.100904

Lokmic-Tomkins, Z., Davies, S., Block, L. J., Cochrane, L., Dorin, A., Von Gerich, H., & Peltonen, L. M. (2022). Assessing the carbon footprint of digital health interventions: A scoping review. *Journal of the American Medical Informatics Association : JAMIA, 29*(12), 2128–2139. DOI: 10.1093/jamia/ocac196

Madhani, P. M. (2010). Resource based view (RBV) of competitive advantage: an overview. *Resource based view: concepts and practices, Pankaj Madhani, ed*, 3-22. https://doi.org/DOI: 10.54518/rh.2.6.2022.587-596

Maleki Varnosfaderani, S., & Forouzanfar, M. (2024). The role of AI in hospitals and clinics: Transforming healthcare in the 21st century. *Bioengineering (Basel, Switzerland), 11*(4), 337. DOI: 10.3390/bioengineering11040337

Melnyk, M., Blyznyukov, A., & Cieślik, J. (2023). The impact of digital education initiatives.http://doi.org/DOI: 10.61093/sec.7(3).1-9.2023

Melville, N. P. (2010). Information systems innovation for environmental sustainability. *Management Information Systems Quarterly*, •••, 1–21. DOI: 10.2307/20721412

Mishra, S., Jena, L., Tripathy, H. K., & Gaber, T. (2022). Prioritized and predictive intelligence of things enabled waste management model in smart and sustainable environment. *PLoS One, 17*(8), e0272383. DOI: 10.1371/journal.pone.0272383

Moutinho, A., Sousa, P., & Tereso, A. (2025). Applying the PRiSM™ methodology to raise awareness of the importance of using sustainable project management practices in organizations. *Systems, 13*(2), 69. DOI: 10.3390/systems13020069

Olatinwo, S. O., & Joubert, T. H. (2023). Resource allocation optimization in IoT-enabled water quality monitoring systems. *Sensors (Basel), 23*(21), 8963. DOI: 10.3390/s23218963

Osibanjo, O., & Nnorom, I. C. (2007). The challenge of electronic waste (e-waste) management in developing countries. *Waste Management & Research, 25*(6), 489–501. DOI: 10.1177/0734242X07082028

Pereira, G. V., Temple, L., & Klausner, L. D. (2024). Transferring Smart City Concepts to Smaller Urban and Rural Contexts. https://doi.org/DOI: 10.31219/osf.io/deuyc

Prange, C., Bruyaka, O., & Marmenout, K. (2018). Investigating the transformation and transition processes between dynamic capabilities: Evidence from DHL. *Organization Studies, 39*(11), 1547–1573. DOI: 10.1177/0170840617727775

Saberi, S., Kouhizadeh, M., Sarkis, J., & Shen, L. (2019). Blockchain technology and its relationships to sustainable supply chain management. *International Journal of Production Research, 57*(7), 2117–2135. DOI: 10.1080/00207543.2018.1533261

Samuels, A. B., & Singh, U. (2025). Education reimagined: South Africa's journey through the 4IR and beyond. *Transformation in Higher Education, 10*, 482. DOI: 10.3389/frai.2024.1477044

Satpathy, I., Chakravarty, D., Jain, V., & Patnaik, B. C. M. (2025). Towards Sustainable Intelligence: Enhancing Cloud Infrastructure Through Green Computing and AI-Driven Engineering. In *Advancements in Cloud-Based Intelligent Informative Engineering* (pp. 203–228). IGI Global Scientific Publishing. https://doi.org/DOI: 10.4018/979-8-3373-0781-7.ch010

Seng, J. K. P., Ang, K. L. M., Peter, E., & Mmonyi, A. (2022). Artificial intelligence (AI) and machine learning for multimedia and edge information processing. *Electronics (Basel), 11*(14), 2239. DOI: 10.3390/electronics11142239

Senthilkumar, R., Venkatakrishnan, P., & Balaji, N. (2020). Intelligent based novel embedded system based IoT enabled air pollution monitoring system. *Microprocessors and Microsystems, 77*, 103172. DOI: 10.1016/j.micpro.2020.103172

Siripipatthanakul, S., Chaiprakarn, S., Rungsiyanont, S., & Phuangsuwan, P. (2025, a). A Case Study of Sustainable Development in Hospitals in Thailand Using ChatGPT. In *Sustainable Development for Hospitals and Clinics* (pp. 1-32). IGI Global Scientific Publishing. DOI: DOI: 10.4018/979-8-3693-9800-5.ch001

Siripipatthanakul, S., Phuangsuwan, P., & Jaipong, P. (2025, e). Mobile Application Security: A Case Study of Bank of Ayudhya PCL.(KMA), Thailand. In *International Conference on Data-Processing and Networking* (pp. 479-493). Springer, Singapore. https://doi.org/DOI: 10.1007/978-981-96-3102-5_36

Siripipatthanakul, S., Phuangsuwan, P., & Jaipong, P., & Muthmainnha. (2025, f, September). Wi-Fi Networks and Security: A Systematic Review. In *International Conference on Artificial Intelligence and Networking* (pp. 81-92). Singapore: Springer Nature Singapore. https://doi.org/DOI: 10.1007/978-981-96-2015-9_6

Siripipatthanakul, S., Phuangsuwan, P., Siripipattanakul, S., & Jaipong, P. (2025, c). Sustainable Tourism Based on Challenges, Tourist Experiences, and Host Communities: A Case Study of Ayutthaya, Thailand. In *Best Practices for Human Resources Management in Tourism and Hospitality* (pp. 361-394). IGI Global Scientific Publishing. DOI: DOI: 10.4018/979-8-3693-8779-5.ch014

Siripipatthanakul, S., Phuangsuwan, P., Siripipattanakul, S., Limna, P., Jaipong, P., & Chaiprakarn, S. (2025, g, September). Sustainable Development Adopting Digital Intelligence Among University Students. In *International Conference on Artificial Intelligence and Networking* (pp. 211-220). Singapore: Springer Nature Singapore. https://doi.org/DOI: 10.1007/978-981-96-2015-9_14

Siripipatthanakul, S., Phuangsuwan, P., Siripipattanakul, S., Limna, P., Jaipong, P., & Chaiprakarn, S. (2025, h, September). Privacy and Information Security for Sustainable Community Among TikTok's Users. In *International Conference on Artificial Intelligence and Networking* (pp. 221-233). Singapore: Springer Nature Singapore. https://doi.org/DOI: 10.1007/978-981-96-2015-9_15

Siripipatthanakul, S., Rakpathum, W., Siripipattanakul, S., Kaewnaknaew, C., Phuangsuwan, P., & Jaipong, P. (2025, d). The Future Artificial Intelligence (AI) for the Hotel and Hospitality Industry: The Case of Hotel's Human Resource Management (HRM). In *Best Practices for Human Resources Management in Tourism and Hospitality* (pp. 395-422). IGI Global Scientific Publishing. DOI: DOI: 10.4018/979-8-3693-8779-5.ch015

Siripipatthanakul, S., Rungsiyanont, S., & Siripipattanakul, S. (2025, b). Sustainable Dentistry Using Artificial Intelligence (AI) for Digital Education and Business Among University Dental Hospitals. In *Sustainable Development for Hospitals and Clinics* (pp. 33-64). IGI Global Scientific Publishing. DOI: DOI: 10.4018/979-8-3693-9800-5.ch002

Supriya, S., & Padaki, S. (2016, December). Data security and privacy challenges in adopting solutions for IOT. In *2016 IEEE International Conference on Internet of Things (iThings) and IEEE Green Computing and Communications (GreenCom) and IEEE Cyber, Physical and Social Computing (CPSCom) and IEEE Smart Data (SmartData)* (pp. 410–415). IEEE. https://doi.org/DOI: 10.1109/i

Things-GreenCom-CPSCom-SmartData. (2016). 97.Tolstolesova, L., Glukhikh, I., Yumanova, N., & Arzikulov, O. (2021). Digital transformation of public-private partnership tools. *Journal of Risk and Financial Management, 14*(3), 121. DOI: 10.3390/jrfm14030121

Trist, E. L., & Bamforth, K. W. (1951). Some social and psychological consequences of the longwall method of coal-getting: An examination of the psychological situation and defences of a work group in relation to the social structure and technological content of the work system. *Human Relations*, *4*(1), 3–38. DOI: 10.1177/001872675100400101

Ullah, S., Awan, M. D., & Sikander Hayat Khiyal, M. (2018). Big data in cloud computing: A resource management perspective. *Scientific Programming*, *2018*(1), 5418679. DOI: 10.1155/2018/5418679

Valadas, C. (2024). Eurocities as "Collaboration Laboratories": Innovation and Policy Development within the European Union. *População e Sociedade*, (42), 90–105. DOI: 10.52224/21845263/rev42a7

Williams, P. A., Lovelock, B., Cabarrus, T., & Harvey, M. (2019). Improving digital hospital transformation: Development of an outcomes-based infrastructure maturity assessment framework. *JMIR Medical Informatics*, *7*(1), e12465. DOI: 10.2196/12465

Winkelmann, S., Guennoun, R., Moller, F., Schoormann, T., & van der Valk, H. (2024). Back to a resilient future: Digital technologies for a sustainable supply chain. *Information Systems and e-Business Management*, *22*(2), 315–350. DOI: 10.1007/s10257-024-00677-z

Wolfert, S., Ge, L., Verdouw, C., & Bogaardt, M. J. (2017). Big data in smart farming–a review. *Agricultural Systems*, *153*, 69–80. DOI: 10.1016/j.agsy.2017.01.023

Zhang, G., MacCarthy, B. L., & Ivanov, D. (2022). The cloud, platforms, and digital twins—Enablers of the digital supply chain. In *The digital supply chain* (pp. 77–91). Elsevier., DOI: 10.1016/B978-0-323-91614-1.00005-8

Zhang, W., Fan, H., & Zhao, Q. (2023). Seeing green: how does digital infrastructure affect carbon-emission intensity? *Energy Economics, 127*, 107085. https://doi.org/rights and contentDOI: 10.1016/j.eneco.2023.107085Get

Zhuo, Z., Ye, J., Wang, Y., Chen, H., & Liang, B. (2025). Smart cities, smarter land use? Unveiling the efficiency gains from China's digital urban transformation. *Ecological Indicators*, *171*, 113151. DOI: 10.1016/j.ecolind.2025.113151

KEY TERMS & Definitions

Artificial Intelligence (AI): AI helps forecast healthcare needs, optimize city services, and enhance supply chain operations.

Big Data Analytics: *enables predictive decision-making to manage resources efficiently, anticipate demand, and prevent* waste across sectors.

Cloud Computing: It enables scalability and reduces the need for energy-intensive local infrastructure.

Data Security and Privacy: Ensuring secure data exchange and maintaining patient confidentiality are vital when deploying digital infrastructure in healthcare and other sectors.

Digital Infrastructure: It serves as the foundation for deploying digital solutions across sectors such as healthcare, urban management, and supply chains.

Green Computing: The practice of designing, manufacturing, and using computing technologies in an energy-efficient and environmentally responsible manner. It includes optimizing data centers, using renewable energy sources, and reducing electronic waste.

Internet of Things (IoT): In healthcare, smart sensors monitor patient health conditions and track traffic and air quality; in supply chains, they monitor inventory and logistics.

Smart Cities: Urban areas that use digital technologies to enhance infrastructure, services, and governance.

Supply Chain Optimization: The usage of technical tools to enhance the efficiency and effectiveness of supply chain operations. It involves forecasting demand, reducing transportation costs, minimizing inventory levels, and ensuring timely delivery.

Sustainable Resource Management: Digital tools assist in monitoring and optimizing the use of these resources.

Chapter 5
Collaborative Ventures Between Public and Private Sectors in Technology and Sustainability

Ashka Prince
https://orcid.org/0009-0000-6228-8597
Christ University, Bangalore, India

Babita Chaudhary
https://orcid.org/0009-0003-2854-6670
Christ University, Bangalore, India

Charu Sehgal
https://orcid.org/0000-0002-4607-4625
Christ University, Bangalore, India

ABSTRACT

Governance models are needed in this age when climate change, resource depletion, and environmental degradation are now accelerating the pace of life. This chapter is a systematic review method using thematic-content analysis, in reality, critically analyzes public-private collaborative ventures as public-private partnership (PPP) matters, positioning them as the device to connect government policy frameworks with private sector technological expertise and investment capacity. With renewable energy and waste management, among other low-carbon technology applications, coupled with much public ground governance, one finds PPPs at the interface be-

DOI: 10.4018/979-8-3373-6821-4.ch005

tween technology policy and a green governance agenda. The chapter delves into the dynamics of government factors, barriers, and replicable strategies. It applies them here using the case-study methodology of the Rewa Ultra Mega Solar Project and the Indore Smart City waste management initiative.

1. INTRODUCTION

Governments around the world have prioritized sustainable development as a global challenge, seeking to address environmental degradation, rapid urbanization, energy shortages, and the shift to low-carbon systems. Traditional public sector mechanisms are not usually very effective at tackling these complex, tightly interconnected issues, especially in developing countries that face inadequate financial resources, technological gaps, and poorly organized institutions. In such a scenario, public-private partnerships (PPPs) have become a governance strategy that can attract private investment, introduce new technologies, and build institutional capacity for sustainable development (Adebayo et al., 2021; Alnaqbi & Alami, 2023). PPPs share burdens, risks, and resources between the public and private sectors, enabling more efficient infrastructure and service delivery than typical government-led models (Koppenjan & Enserink, 2009).

The current state of sustainability transitions highlights the increasing importance of PPPs since transitions demand the involvement of multiple parties, adaptive governance, and long-term policy commitment. Researchers point out that PPPs provide financial leverage as well as managerial skills and technological capacities required for modern environmental governance (Niesten et al., 2017; Mahardhani, 2023). In renewable energy, waste disposal, transport, and water systems, PPPs have supported the introduction of advanced technologies, including AI, IoT-based monitoring, digital platforms, waste-to-energy plants, and automated grid management (Zaman & Lehmann, 2011; Bibri et al., 2023). These technologies improve operational efficiency and contribute to governance by increasing transparency, enabling real-time performance monitoring, and supporting data-based decision-making, all of which are components of emerging smart governance models (Yigitcanlar & Cugurullo, 2021; Almulhim & Yigitcanlar, 2025).

India presents an especially pertinent scenario for the study of PPP-driven sustainability transitions. As an emerging economy, India faces the challenge of expanding renewable energy, improving urban areas, and reducing environmental impacts. Over the last decade, India has introduced PPP models across energy, waste management, transportation, and smart cities. The Rewa Ultra Mega Solar Project in Madhya Pradesh and the Indore Smart City Waste Management system underscore the impact of technology adoption, institutional design, and governance interaction on sustainability outcomes. Rewa gained global recognition for its transparent processes, competitive tariff structures, and innovative risk-sharing strategies (Marquardt, 2024). The use of digital monitoring, IoT devices, and public involvement has transformed Indore into India's cleanest city and demonstrated how these tools can jointly improve urban environmental performance (Pinilla-de la Cruz et al., 2022).

However, many countries still experience governance and capacity-related challenges despite PPP successes. Common issues include misaligned regulatory frameworks, fragmented institutions, ambiguity in risk allocation, limited technological readiness, and financial limitations that make private investment less appealing (Dutz & Sharma, 2012; Wang & Jin, 2007). Sustainability transitions require not only physical infrastructure but also behavioural change, institutional learning, and policy coherence across government levels (Shrivastava & Guimarães-Costa, 2017).

In light of this, the present chapter elucidates the role of Public-Private Partnerships (PPPs) in green governance and sustainable development, focusing on the use of technology and governance design. It integrates various academic fields, delineates the governance and institutional factors responsible for successful PPPs, and analyzes these factors in two major Indian cases: the Rewa Solar Project and the Indore Waste Management Initiative. The chapter's objectives are: first, to strengthen the theoretical foundation of PPPs, green governance, and smart technological systems; second, to study how governance, institutional capacity, and technology together affect PPP outcomes; third, to illustrate through Indian case studies the workings of PPPs during sustainability transitions; and fourth, to provide policy suggestions for enhancing governance capacity and integrating technology in PPPs to fast-track sustainable infrastructure development.

2. LITERATURE REVIEW

2.1 Theoretical Foundations of Green and Collaborative Governance

One of the most important perspectives for understanding PPPs is the collaborative governance theory, which considers the union of government offices, the private sector, community groups, and other relevant actors as the central point of the issue (Hofmann et al., 2022). The need for cooperation is acknowledged through examples of sustainability challenges, such as climate change adaptation, renewable energy development, and integrated waste management, that cannot be solved by public administrations alone. These issues need the integration of knowledge, sharing of risks, innovative technologies, and constant communication.

The main features of collaborative governance are the following:

- Co-operative decision-making, where the public and private sectors join their efforts in strategy design, planning, and implementation.
- Interdependent liability, where the partners involved in the project assess the progress and the outcomes jointly.
- Complementary competence, where the sectors involved share their technical, managerial, financial, and regulatory expertise.
- Trust, which is a necessary condition for establishing long-term relationships and reducing the quality of communication conflicts.

In the context of environmental governance, the use of collaboratives strengthens legitimacy, opens the process, and makes stakeholders more agreeable—all factors that lead to compliance and behavioral change. A case in point is the partnership between public and private sectors in renewable energy, where cooperative governance permits the government to set long-term energy targets. In contrast, the private sector contributes through technological and operational efficiency (Adebayo et al., 2021). The same holds in urban waste management, where cooperation among municipal workers, private waste-handling companies, and the public helps reduce waste and improve recycling (Pinilla-De La Cruz et al., 2022).

The theory of collaborative governance primarily posits that public-private partnerships (PPPs) are productive only if they share sustainability outcomes among the involved parties and thus enable the actors to address the problem together rather than in their individual corners. As illustrated in Figure 1, PPP governance operates through interconnected mechanisms involving institutional cooperation, policy networks, long-term contracting, community development, and urban renewal.

Figure 1. PPP Governance Structure

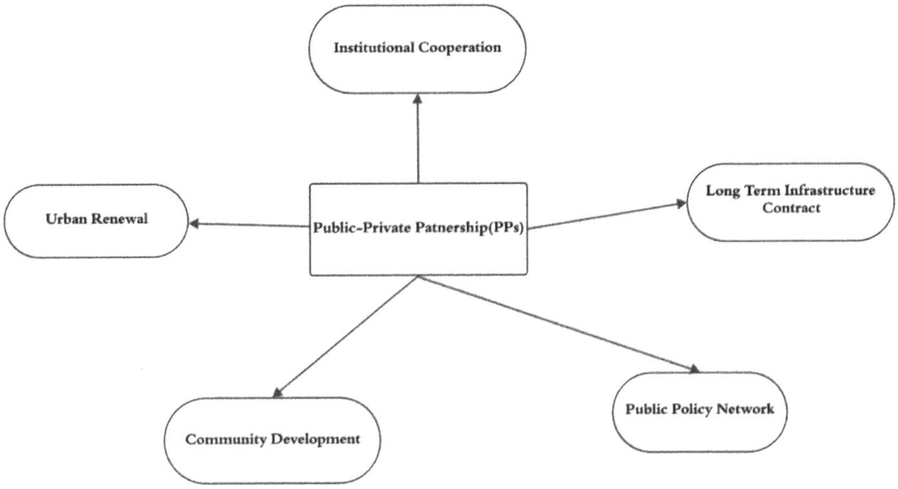

(adapted from Koppenjan & Enserink, 2009; Bovaird, 2004; Pierre, 2000).

2.1.1 Green Governance Frameworks

Green governance is a concept that not only extends the traditional governance model but also integrates environmental aims, such as the protection of ecosystems, the conservation of wildlife, and the development of low-carbon technologies, throughout the policymaking, institutional structuring, and infrastructure development processes. Along with this, it points out that the whole issue of sustainable development is not just a matter of technical or operational aims but rather a matter of governance involving coherent regulatory frameworks, long-term strategic planning, and cross-sector alignment (van Zanten & van Tulder, 2021).

The central tenets of green governance are:

- Ecological responsibility, which minimizes the negative impact on nature by developing suitable policies from the institutions.
- Low-carbon transition, which focuses on directing the use of energy that is efficient, harnessing the power of nature, and allowing environmentally friendly movement.
- Regulatory coherence, which guarantees that environmental policies are in sync with economic frameworks and technological standards.
- Resource circularity, which doubles the life of materials and products by reducing, reusing, and recycling.

- Multi-actor engagement, which is the process of inviting and involving all relevant stakeholders, including ordinary citizens, firms, NGOs, and government bodies, into the sustainability agenda.

Nexus-based governance models, which have emerged as the most significant and most promising segment of green governance, hold that different environmental systems (energy, water, waste, land, climate) are interrelated and should therefore be managed together rather than separately. This argument is most pertinent in the case of PPPs operating in smart cities and the renewable energy sector, where actions in one domain (e.g., energy use) have an immediate and profound impact on others (e.g., water demand, emissions).

Green governance theory, together with the concept of adaptive governance inferred from it, underscores the necessity of institutions to respond dynamically to environmental risks, technological innovations, and changes in sustainability targets (Stafford-Smith et al., 2017). Adaptive governance is particularly vital for the renewable energy and urban waste management industries, as they are undergoing rapid change.

2.1.2 Polycentric and Multi-Level Governance

Polycentric governance refers to a situation in which different authorities operate independently yet interact across different levels: local, regional, national, and global. Polycentric governance is not hierarchical but rather a distributed system of decision-making in which institutions with their distinct strengths and jurisdictions share authority.

Renewable energy governance in India is a typical case that illustrates the advantages of polycentric systems. Having established long-term solar energy capacity targets, the government not only sets national missions but also, through state-level regulatory frameworks and local agencies that handle land acquisition, grid connections, and project operations (Marquardt, 2024). The polycentric approach allows for more innovations of different kinds, since different jurisdictions can test different governance models, financial mechanisms, and technological solutions.

2.1.3 Smart Governance Models

Smart governance is characterized by the use of digital technologies in governance processes to enhance decision-making, operational efficiency, transparency, and citizen participation, among others. Unlike traditional governance, which relies heavily on manual processes and isolated systems, smart governance uses real-time data, automated systems, AI-driven analytics, and digital platforms to improve public service performance (Yigitcanlar & Cugurullo, 2021).

- Data-driven decision-making is one of the key features of smart governance. Digital dashboards, IoT sensors, and analytics tools facilitate this.
- Real-time monitoring is another crucial feature that supports authorities in tracking performance indicators and discovering inefficiencies.
- Participatory governance is yet another significant aspect where mobile apps and online platforms are employed to allow citizens to report issues, give feedback, and even take part in the planning.
- Interoperable systems that integrate across waste management, energy management, transport, and water systems are also critical.

In sustainability public-private partnerships (PPPs), smart governance is of exceptional value, as it is the technology that connects different systems—renewable energy platforms, automated waste management, and digital monitoring—to the overall environmental goals. For example, the city of Indore has set up a smart waste management PPP that relies on GPS tracking, digital route optimization, and real-time dashboards to enhance the efficiency and accountability of waste collection (Pinilla-De La Cruz et al., 2022). The combination of technology and integration leads to a situation in which operational goals and sustainability outcomes are perfectly aligned.

2.2 Key Variables Influencing PPP Performance

2.2.1 Governance Capacity and Accountability

The capacity to govern is the central pillar on which the success of PPP rests. It is the power of the public institutions to allow the partnership arrangements to be designed, implemented, regulated, and monitored most effectively. High governance capacity entails the existence and utilization of the following: open procurement processes, precise institutional and inter-institutional roles, performance monitoring systems, and mechanisms for dispute resolution (Koppenjan & Enserink, 2009).

Strong governance capacity leads to the following:

- Governance capacity that is transparent and efficient provides a precise distribution of responsibilities, thus minimizing the ambiguity of the operational duties.
- Government and private sector entities have a good and quick relationship, thus efficient coordination reduces the time taken for activities.

- The private partners monitor compliance to ensure that they meet the environmental and performance standards.
- There are fewer chances of corruption arising in the procurement and contract award processes if strict governance is in place.

2.2.2 Risk Allocation and Management

There is consensus that one of the most important factors in the success of PPP is the effective allocation of risk. The involvement of PPPs brings several risks with it—among these are financial, technical, regulatory, operational, social, and environmental risks. Mismanagement of these risks can lead to project failure and also to the withdrawal of the private sector (Li et al., 2018; Wang & Jin, 2007).

A well-structured PPP contract guarantees:

- The financial risk each party can handle is the basis for sharing it.
- The least competent party takes over the technological risk associated with renewable energy projects.
- Regulatory risk is at a minimum with a stable policy for the area.

The operational risk is the one handed over to the private operator, who can most likely provide the necessary technical expertise. Poorly designed risk allocation can lead to disputes, delays, cost overruns, and loss of enthusiasm for the project among partners. Countries with clearer risk allocation frameworks attract more private investment in green infrastructure (Bai et al., 2024). So, improving legal clarity and contract design is of utmost importance for the sustainable execution of PPPs.

2.2.3 Technological Integration

Technology is the driving force behind the transformation of sustainability projects based on PPPs. The different tech applications, such as renewable energy technologies, IoT sensors, AI-enabled analytics, automation systems, and digital governance platforms, are the main contributors to the project's efficiency, transparency, and environmental performance.

Research has established that the contribution of tech integration to the following areas is as follows:

- Real-time monitoring, thus allowing for the quick detection of inefficiencies or non-compliance.
- Optimizing operations like routing waste management trucks or balancing the load in solar energy (Zaman & Lehmann, 2011).

- Greater offsetting of environmental impacts, such as lesser emissions and better resource utilization.
- Increased institutional learning as data analytics provide insights for future improvements (Bibri et al., 2023).

2.2.4 Institutional and Policy Frameworks

The institutional and policy frameworks that constitute the rules, norms, and incentives guiding the design and implementation of PPPs. Strong institutions, those with regulatory coherence, well-defined mandates, and simpler administrative procedures, serve as the environment required for PPPs to be successful (Pandey et al., 2022).

Institutional strength has an impact on PPP performance in several ways:

- The stability of regulations draws in investment from the private sector that is willing to commit to the long run.
- The clarity of mandates prevents any overlap and fragmentation among the institutions.
- Supportive Policies, such as subsidization and tax credits, make the projects more viable.
- Quick and efficient approval processes lead to a reduction in delays and costs associated with the transaction.

On the contrary, fragmented institutions usually result in the issuance of conflicting mandates, the establishment of inconsistently applied regulations, and the occurrence of bureaucratic inefficiencies that together discourage the involvement of the private sector. According to Dutz and Sharma (2012), policy innovation, especially in green growth strategies, is crucial to institutional dynamics, as it helps overcome institutional barriers and promote technological advancement.

2.2.5 Financial and Economic Viability

The financial viability over the long term is critical to the PPPs that will be the most sustainable in the future, especially since the erection of green infrastructure entails very high initial costs, long payback periods, and uncertain revenues (Holmes et al., 2023). To attract the interest of the private sector, the government may, on the other hand, resort to green financing options, such as blended finance, concessional loans, tax incentives, project viability funds, and risk guarantees (Fu et al., 2023). The central part of financial viability relies primarily on predictable revenue models—such as renewable energy long-term power purchase agreements

(Adebayo et al., 2021)—and on clear cost-sharing arrangements, thereby alleviating pressure on public finances. Moreover, access to capital markets and reliable demand forecasts are essential, especially for projects in the waste management, transport, and energy sectors.

2.3 PPPs in Green Infrastructure and Renewable Energy

2.3.1 Global Perspectives on PPPs in Renewable Energy and Green Infrastructure

Numerous studies in the international literature provide evidence of the transformative role of PPPs in speeding up the transition to renewable energy and sustainable infrastructure. It is noted that PPPs allow governments to raise large amounts of private capital, attract global investors, and allocate risk in ways that stimulate long-term participation in renewable energy markets (Ma & Wang, 2023). All renewable energy sources—solar, wind, hydro, and waste-to-energy—demand significant up-front investments and depend on technological progress, which is why PPPs offer a more adaptable and dynamic model than traditional public procurement methods.

In all parts of the world, public-private partnerships (PPPs) have helped to increase the use of renewable energy:

1. Financial Leverage and Capital Mobilization

One advantage of PPPs is that they provide access to a wide range of financial instruments, including blended finance, concessional loans, private equity, and green bonds. These mechanisms make it feasible to create large-scale infrastructure such as solar parks and wind farms, which may be difficult for the government to finance on its own (Fu et al., 2023).

2. Technology Transfer and Modernization

PPPs often see the private sector transfer their renewable energy technology to public utilities, including photovoltaic systems, battery storage, grid-balancing software, and AI-enabled energy forecasting tools (Mahardhani, 2023). This way, the performance and dependability of the nation's energy systems are improved.

3. Risk Sharing and Improved Project Structuring

High risks, including technology performance, policy uncertainty, and long-term revenue unpredictability, accompany renewable energy investments. PPPs share these risks among public and private players, thereby allowing projects to proceed even in the face of uncertainty (Wang & Jin, 2007). Research indicates that proper risk-sharing mechanisms significantly increase not only the project's viability but also investors' confidence.

4. Strengthened Governance and Policy Alignment

Through their structured contracts, transparent bidding, and performance-based incentives, the PPPs foster regulatory innovation. These governance innovations make the projects more attractive and less corrupt, thus improving the overall good governance performance (Koppenjan & Enserink, 2009; Heldeweg et al., 2015).

5. Contribution to Global Climate Goals

The renewable energy infrastructure powered by PPPs is now regarded as a key factor in achieving the countries' climate compliance under the Paris Agreement. Studies confirm that renewable energy PPPs enable nations to reduce their fossil fuel consumption and improve their long-term energy security (Adebayo et al., 2021).

2.4 Key Findings From the Literature

The research on PPPs in the fields of renewable energy and green infrastructure presents a set of findings that are almost the same throughout:

1. The reduction of carbon emissions and the strengthening of the power grid are the outcomes of using PPPs. According to empirical evidence, the use of renewable energy PPPs leads to the closure of coal-fired power stations in the country and also to the large-scale introduction of solar, wind, and hybrid energy systems (Adebayo et al., 2021). In addition, the power grid is made more reliable by the use of renewable energy, which combines various power sources.
2. Legal and regulatory design is the strongest factor that determines the fate of a PPP. Heldeweg et al. (2015) point out that legal frameworks—including contract structures, regulatory guarantees, and compliance mechanisms—are the very building blocks of long-term stability. If done correctly, contracts will not only keep the private sector interested but also facilitate the introduction of new technologies.
3. Institutional capacity building is a prerequisite. Public-private partnerships are generally less successful when government institutions lack the power to carry out their responsibilities, have poor monitoring systems, and unclear mandates (Pandey et al., 2022). On the other hand, strong institutional capacity helps to avoid delays, regulatory uncertainty, and inefficiencies.
4. PPPs speed up sustainability by providing reforms in governance. PPPs, in addition to being an avenue for technological innovation, also lead to improve-

ments in procurement transparency, policy alignment, and accountability. The governance reforms resulting from this have positive effects on other areas and even upcoming sustainability projects.

5. Emerging economies benefit greatly from PPP models. Governments of developing countries, which lack financial resources, depend on private-public partnerships to fulfill their renewable energy goals, modernize and repair their infrastructure, and remain sustainable in the long run. Private-public partnerships are found to be a solution to the problems of limited funding, hard-to-use technology, and low operational efficiency (Alnaqbi & Alami, 2023).

3. METHODOLOGY

The current chapter employs a systematic literature review (SLR) methodology, enriched by two case studies from India as illustrations. By systematically reviewing the existing body of research, the authors provide a transparent and replicable strategy for identifying, selecting, and summarizing studies that offers a structured understanding of the interplay among public-private partnerships (PPPs), technological innovation, and green governance in the context of sustainability transitions. The method is grounded in the established SLR guidelines applied in sustainability and governance research, especially those highlighted by Hofmann, Choi, and Binder (2022), who emphasize meticulous selection, thematic synthesis, and clear reporting of the review process.

3.1. Research Design

The research design incorporates a comprehensive literature review concerning scholarly works on PPPs, green governance, sustainability policy, and technology-enabled environmental management, thematic synthesis of findings across different studies to pinpoint the common governance challenges, policy mechanisms, and technological enablers; and a descriptive case study of two Indian sustainability initiatives Rewa Ultra Mega Solar Project (Madhya Pradesh), and Indore's integrated solid waste management PPPs showcasing the collaborative models in practice.

The integrative design approach enhances analytical rigor, as the systematic review results provide widely applicable insights, while the cases demonstrate these insights within the specific local governance contexts.

3.2. Systematic Search Strategy

For conducting the literature search, four principal academic databases were used as a basis for comprehensive coverage:

- Scopus
- Web of Science
- Google Scholar (peer-reviewed results filtered)
- ScienceDirect

The search terms were created iteratively and grouped into four categories:

Public–Private Partnerships (PPPs)

- "public–private partnerships."
- "PPP governance."
- "PPP sustainability."

Green Governance & Policy

- "green governance."
- "environmental governance."
- "sustainability governance."
- "SDG governance."

Technology-Enabled Governance

- "smart governance."
- "AI for sustainability."
- "IoT environmental monitoring."
- "data-driven governance."

Sector-Specific Domains

- "renewable energy PPPs
- "Solar Governance India."
- "waste management PPPs."
- "smart cities sustainability."
-

3.3. Inclusion and Exclusion Criteria

In order to ensure quality and relevance, the following criteria were set:

Inclusion Criteria

- Peer-reviewed journal articles, books, and book chapters
- Studies published in English
- Research addressing PPPs in environmental, sustainability, or technological governance
- Studies examining smart cities, renewable energy, waste management, or sustainability policy
- Empirical, conceptual, and review papers

Exclusion Criteria

- Non-peer-reviewed publication
- Articles unrelated to governance or sustainability
- PPP studies in unrelated sectors (e.g., defense, healthcare, unless green governance was involved)
- Reports lacking methodological transparency

3.4. Data Extraction and Coding

The extraction of data was done step by step through a structured coding sheet that noted down the following:

- the study's aim
- the governance model (collaborative, polycentric, smart governance)
- the sustainability domain (renewable energy, waste, urban systems)
- the PPP characteristics (risk-sharing, contract type, institutional actor roles)
- the technological components (AI, IoT, data platforms, monitoring tools)
- the outcomes (environmental, economic, institutional)

Later, the extracted data were grouped into thematic clusters, which were used to write the Results section.

3.5. Synthesis Method

According to the studies by Siripipatthanakul et al. (2025; a-h), content analysis and thematic synthesis are used in a systematic review because these flexible qualitative research methods enable the detection, evaluation, and interpretation of patterns (themes) within data, identifying recurring meanings, ideas, or concepts relevant to the study.

A thematic analysis and synthesis were used in this chapter to identify recurring ideas, governance difficulties, enabling conditions, and insights that cross-cut.

Three main analytical themes were obtained:

1. Governance Conditions for Effective PPPs: Risk distribution, institutional capability, regulatory consistency, and stakeholder support were all included.
2. Technological Enablers of Green Governance: AI-, IoT-, and data-enabled systems supporting monitoring, transparency, and operational efficiency technologies.
3. Policy and Institutional Mechanisms: Various, such as SDG nexus governance, smart governance frameworks, environmental policy instruments, and the polycentric federalism of India.

These themes are linked to the concept framework presented later in the chapter.

3.6. Case Selection Rationale

The selection of the Rewa Solar and Indore Waste Management cases was based on their overall recognition as general conceptions of PPP success in India; they involve the use of technology, policy innovation, and the governance of various stakeholders. They are also very close opposites in the areas of sustainability, energy, and waste. Moreover, both cases have strong documentation and measurable outcomes, making their inclusion even more interesting as it brings real-world insights closer to the theoretical principles established through the review of literature.

4. RESULTS

4.1 Case 1: The Rewa Ultra Mega Solar Project (Renewable Energy PPP)

The Rewa Ultra Mega Solar Project (RUMS) is a yardstick for other renewable energy-based PPPs in the state of Madhya Pradesh, India, where the project symbolizes India's journey toward low-carbon growth. One of Asia's largest solar installations on one site, the project was commissioned in 2018 with an installed capacity of 750 MW. A joint venture between the Madhya Pradesh Urja Vikas Nigam (MPUVN) and the Solar Energy Corporation of India (SECI), the Rewa Ultra Mega Solar Limited (RUMSL), and private-sector developers, including Mahindra Renewables, ACME Solar, and Solenergi Power, jointly developed the project.

4.1.1 Policy and Institutional Framework

The success of RUMS is grounded in a well-structured institutional framework reflecting an effective PPP governance model. The Government of India had put in place such enabling policies through the Solar Park Scheme-2014 for providing Technical and Financial assistance through concessional loans from the World Bank. About the Regulation and mitigating risks for private investors, the primary functions of RUMSL were as a nodal agency for land acquisition, transmission infrastructure, and power evacuation facilities. The Power Purchase Agreement (PPA) for the project includes a three-tier payment security structure. Once payment was made to DMRC (Delhi Metro Rail Corporation), one of the major buyers, revenue uncertainty was mitigated, and investor confidence increased.

4.1.2 Technological and Financial Innovation

A genuinely independent trademark and hallmark characteristic of Russia's USSI was its emphasis on technological and financial innovation. A competitive bidding mechanism could make it simple to offer a tariff of only 2.97 per kWh for the initial periods of operation, when such a rate would be lower than the national average for solar power. For this project, advanced photovoltaic (PV) technology and optimal land use were adopted through modular layouts. In addition to transparent tendering and innovative risk allocation approaches, the project allocated obligations in a balanced manner among all stakeholders.

It was a perfect illustration of how the technology policy, together with good governance, fosters market efficiency and sustainability in one go.

4.1.3 Sustainability Outcomes and Challenges

The Rewa project helps India fulfil its climate action targets by displacing nearly 1.5 million tons of CO_2 per year. It delivers clean power to the Delhi Metro and then onto the Madhya Pradesh grid, further proof of how urban transport and renewables can be well-integrated. Nevertheless, challenges persist, especially in managing unresolved long-term maintenance costs and in reconciling land use with policy consistency. Notwithstanding these obstacles, RUMS is a good example of how such a PPP could lead to super-sized renewable infrastructure that combines economic viability, social acceptance, and environmental soundness.

4.2 Case 2: Indore Smart City Waste Management Initiative (Urban Sustainability PPP)

Indore Smart City Waste Management initiative stands as a good model for urban governance driven by effective PPP and technological innovation in waste management. The IMC formed an integrated waste segregation, collection, and processing system with private entities, including Eco-vision Environmental Services, A2Z Infrastructure Ltd, and EPR-compliant recyclers under the Indore Smart City Mission.

4.2.1 Institutional Mechanism and Governance Framework

The project runs under the general umbrella of the Smart Cities Mission established in 2015, which advocates citizen-centric, data-driven, and technology-enabled solutions. IMC chose decentralized governance and built collaboration with private partners for collection, logistics, material recovery, and waste-to-energy conversion. The public domain performed regulatory oversight, gave infrastructure, and accounted for performance through digital monitoring platforms like Swachh Indore. The project was financed through a hybrid model of municipal budget allocations, central government grants, and private capital investments, thereby ensuring its sustainability and scalability.

4.2.2 Technological and Operational Components

The entire waste management system introduced several technological innovations, including RFID-enabled bins, GPS-tracked collection vehicles, and IoT-enabled monitoring to optimize routes and improve operational efficiency. The segregation of waste at the source, driven by a long series of public awareness campaigns, helped the city reach 100% door-to-door waste collection and zero landfill status. The setup of a bio-methanation plant and composting units helped convert organic waste into biogas and fertilizer.

4.2.3 Outcomes and Impact

The Indore Smart City initiative has brought about drastic changes in the city's environmental profile. Indore has been ranked the cleanest city in India for six consecutive years (2017-2023) in the Swachh Survekshan rankings by the Ministry of Housing and Urban Affairs. Nevertheless, the PPP model has helped combat urban pollution, promote sanitation, and create green employment opportunities. Citizen engagement-building was an important component of green governance because the initiative helped embed sustainability in the concrete day-to-day practices of citizens through behavioral change and awareness campaigns.

4.3 Comparative Insights

A comparison of the Rewa Solar Project with Indore highlights the nature of PPPs in the context of technology-based sustainability. Rewa and Indore lie in different sectors-energy and urban management-though they share quite similar collaborative governance structures, tech integration, and shared accountability.

4.3.1 .Governance & Institutional Design.

Both Rewa and Indore emphasized the need for stronger governance and institutional clarity. The building of synergies between central and state agencies greatly aided the Rewa project, enabling it to operate in a streamlined regulatory environment with minimal bureaucratic friction. The same efforts exhibited decentralized modalities: close interaction of municipal authorities with private firm partners and direct engagement with local communities. These are examples that show institutional design-smoothly working institutions with clearly defined roles, accountability mechanisms, and inter-agency coordination-is a key determinant in the functionality of PPP. This comparative study also stressed that both types of governance could work, provided that their policy frameworks are consistent, flexible, and well-aligned with the project objectives.

4.3.2. Technology as an Enabler

For both projects, technology acted as an enabler of systemic changes rather than a standalone fix. In fact, this was evident in Rewa, with large-scale solar photovoltaic systems and grid management technologies to enable the integration of renewable energy sources and rapid deployment. In Indore, the model integrated IoT-enabled waste tracking, automated collection, and data-capturing analytics to reform urban waste governance. Importantly, technology would realize its promise only when embedded in governance that allows for transparency, accountability, and adaptive decision-making. The two examples also show how PPPs can leverage technological novelty to deliver scalable, measurable solutions for sustainability outcomes.

4.3.3. Financial and Risk-Sharing Mechanisms

In both cases, technology served as a facilitator of systemic transformation rather than an isolated solution. Rewa took advantage of extensive photovoltaic infrastructure integrated with grid management technologies, enabling the swift deployment of renewable energy. Indore undertook IoT-enabled waste tracking, automated collection systems, and data-powered analytics to drive reform in urban waste management. The salient point is that technology will fulfill its promise only when it is subject to governance philosophies that embrace transparency, accountability, and adaptive decision-making. The twinning of the two projects demonstrates how PPPs can enable technology splendor to deliver measurable, sizable sustainability impacts.

4.3.4. Stakeholder participation and social inclusion

The shoal was significant for both initiatives through stakeholder engagement. Here, behavioral change campaigns and citizen participation programs in Indore increased public compliance and the project's widespread adoption. On the other hand, Rewa illustrates how inter-organizational collaboration, along with corporate responsibility, has worked for land acquisition, workforce mobilization, and societal legitimation. These cases show that inclusive governance, transparency, and genuine participation in the governance process are paramount to improving social acceptance and the sustainability of PPP projects.

4.4 Synthesis

These also serve as case studies in which PPPs can be used as one of the instruments for powerful green transformation, provided that all of these are packaged under coherent policy frameworks, equitable governance structures, and robust technological ecosystems. Such that when sustainability is studied, it is shown that one does not depend on technological adoption alone, but rather collaborative governance around innovation, regulation, and social participation. Between the two capitals, that is, Rewa and Indore, lies the intrinsic strategic value of the PPP in dual purposes: technological advancement and environmental care; thus, theoretically and practically, it enlightens policymakers, scholars, and practitioners and draws a future for sustainable development.

5. DISCUSSIONS

5.1.Governance Structures Impact on PPP Performance

The quality of governance has been established as a primary factor in the literature, determining a project's success or failure. It was argued that collaborative governance models would lead to better coordination, more aligned policies, and greater stakeholder trust, ultimately allowing PPP to function more efficiently (Hofmann et al., 2022). One of the main arguments was that the sustainability issues are "multi-actor and multi-institutional," meaning they are not the responsibility of a single actor, but rather that sharing among them is the way to go, eliminating the hierarchical control model.

The perspectives on polycentric governance placed a significant emphasis on the need for multi-level alignment—national, state, local, and private actors must operate with well-defined responsibilities. In the case of renewable energy projects, the disparity between federal incentives and state-level regulations often led to uncertainty, resulting in delayed implementation (Marquardt, 2024). The PPPs that were most effective relied on regulatory alignment and a clear strategic direction.

5.2. Technology is a Critical Enabler of Green Governance

Digital technology is an environmental management tool that changes the way modern societies treat nature. The implementation of AI, IoT sensors, predictive analytics, and digital monitoring systems has increased transparency in operations, accountability, and operational efficiency. Research has indicated that technology-enhanced governance will bridge the information gap between public and private partners (Bibri et al., 2023).

Smart city literature has supported this finding by demonstrating that the provision of timely environmental data enables government agencies to monitor compliance indicators, assess performance, and impose standards more effectively (Zaman & Lehmann, 2011). Moreover, technology-integrated governance models are more resilient, adaptive, and responsive to environmental changes (Yigitcanlar & Cugurullo, 2021).

5.3. Nature-Based and Hybrid Solutions Strengthen Environmental Outcomes

Along with digital solutions, ecological and nature-based approaches offer another way to advance urban sustainability. Mahmoud et al. (2022) demonstrate that nature-based solutions-green corridors, permeable surfaces, and urban vegetation-are made more effective through a governing model that embraces citizens' participation.

The combination of ecological interventions with the deployment of digital monitoring systems was a hybrid approach that led to stronger environmental performance and long-term resilience.

5.4.Green Finance and Institutional Support Enable PPP Sustainability

Among the financing mechanisms of pulpit capital, green bonds, and public guarantees, the above are significant to the sustainability and durability of PPP projects. According to Fu, Lu, and Pirabi (2023), green finance is the driver of private investment in renewable energy, waste management, and smart urban systems.

Researchers pointed out that highly institutionalized partnerships, regulatory frameworks that provide certainty, and financial risk-sharing practices were the main factors behind the enormous sustainability compared to conventional projects with low institutional support.

5.5.Policy Alignment and SDG Nexus Governance: Improve Outcomes

Governance of environmental issues should not only be sectoral but also multidisciplinary, and hence, Van Zanten and van Tulder (2021) proposed that sustainable development goals (SDGs) should be treated as one and managed as a 'nexus'. Policy measures in energy, water, waste, land use, and transportation should be designed with consideration of each other.

The review found that PPP projects aligned with broader sustainability frameworks (e.g., SDGs, national renewable energy targets, or smart city missions) benefited from greater policy consistency, broader stakeholder engagement, and social acceptance.

5.6. Institutional Capacity and Accountability Determine Implementation Quality

The institutional competence, with its components of skills, staffing, regulatory oversight, and data capabilities, directly impacted the project's success. The studies concluded that weak institutions would lead to:

- delays in getting approvals
- poor monitoring
- disputes over contracts

- inconsistent application of the law

Data-driven decision-making, complemented by the Smart governance literature, has been shown to increase institutional accountability and performance assessment (Almulhim & Yigitcanlar, 2025).

5.7. Stakeholder Engagement Strengthens Legitimacy and Compliance

Research across various sectors found that community involvement in the project increased its legitimacy, reduced resistance, and led to better environmental outcomes. In the waste management and urban services areas, the combination of behavioral interventions and digital incentives boosted public participation.

The literature on Nature-based solutions indicates that when communities are allowed to co-design and co-manage, sustainability outcomes are more long-lasting (Mahmoud et al., 2022).

6. CONCLUSIONS

6.1 Governance Capacity as the Foundational Determinant of PPP Effectiveness

The literature stating the obvious identifies governance capacity as the most important factor governing how different organizations and sectors work together through a public-private partnership (PPP) arrangement. Institutional clarity, policy coherence, transparent procurement, and monitoring systems, along with strong inter-organizational coordination, are key elements of successful PPPs (Koppenjan & Enserink, 2009; Pandey et al., 2022).

The Rewa Ultra Mega Solar Project is a perfect example of this. The project's governance framework consisted of a transparent competitive-bidding process, bankable power purchase agreements, stringent compliance monitoring, and co-ordinated actions among national and state governmental authorities and private developers. Such features reduced investment risk, thereby widening participation and, in turn, achieving record-low solar prices (Marquardt, 2024). Rewa confirms that governance is not just an administrative function but also a strategy that can accelerate the deployment of renewable energy.

Similarly, at the urban level, Indore's waste management PPP shows how governance innovations can benefit people and the environment, even beyond the urban level. The city adopted formal PPP contracts, daily digital monitoring, performance-based evaluation, transparent public reporting, and sustained citizen engagement campaigns. These governance practices collectively reshaped institutional behavior and public participation, thus contributing to Indore's repeated ranking as the cleanest city in India. The case shows that strong local governance capacity can significantly improve PPP outcomes in urban sustainability, especially for complex initiatives, as in Indore.

6.2 Technology as a Transformational Catalyst in PPP-Driven Sustainability

There is no doubt that technology is the backbone of modern PPP systems. According to the literature and practical examples, technology is the primary driver of improvements in transparency, operational efficiency, environmental performance, and citizen engagement (Bibri et al., 2023; Zaman & Lehmann, 2011).

Indore's municipal solid waste management PPP is a prime example of tech-enabled smart governance. Indore's waste collection and disposal system was made more efficient, less corrupt, and data-driven through the use of IoT-enabled monitoring, an RFID system, GPS tracking, automated route optimization, and digital dashboards. It was a significant shift from the manual, opaque system to a completely different world where everything was known, and the operation was highly efficient.

The technological systems made it possible to have:

- service delivery with real-time transparency
- analysis of resource allocation based on predictions
- community participation through citizen reporting apps
- evaluation of private contractors based on performance

Rewa is not as digital-intensive as the previous case, but it is still using advanced technologies in:

- solar photovoltaic system optimization
- automated monitoring of generation
- real-time energy forecasting
- improved grid integration

These technologies help stabilize the energy supply and improve grid reliability, thereby highlighting the complementary roles of technology, governance, policy, and finance in renewable energy PPPs.

Technology + Governance = Enhanced Sustainability

In the two cases, technology is definitely not only an operational tool, but it is also a governance instrument that brings about the following improvements:

- compliance
- monitoring
- decision-making
- stakeholder coordination
- public trust

6.3 Risk Allocation, Financial Structuring, and the Dynamics of Investor Confidence

Risk allocation is recognized as one of the most important predictors of the success of public-private partnerships (PPPs) (Li et al., 2018; Wang & Jin, 2007). In the case of green infrastructure and renewable energy projects, the risks are as follows:

- market risks
- policy and regulatory uncertainty
- collapse of the technology
- difficulties in land acquisition
- exceeding the budget
- unpredictable revenues

6.4 Policy Coherence, Institutional Alignment, and the Importance of Enabling Environments

PPPs are highly dependent on the institutional and policy environment in which they operate. Studies have shown that the presence of disconnected policies, regulatory inconsistencies, and weak institutional structures negatively affects the performance of PPPs (Dutz & Sharma, 2012).

The Rewa project received the following benefits:

- potent leadership from the state government
- explicit policies regarding land
- the government's commitment to renewable energy
- measures aimed at reducing the financial risks involved
- Aligning institutions at all federal levels created a supportive ecosystem.
- Urban Policy Integration in Indore

Indore's waste PPP was reinforced through alignment among the following:

- Swachh Bharat Mission
- Smart Cities Mission
- VB municipal regulations
- behavioral campaigns
- The coherence of institutions resulted in a single framework being created for sustainability.

6.5 The Often-Underestimated Role of Stakeholder Participation and Social Acceptance

One key observation is that the transition to sustainability is strongly influenced by the behavior of the local community, not only by the infrastructure or financial support provided (Mahmoud et al., 2022; Shrivastava & Guimarães-Costa, 2017).

Indore has unique levels of:

- citizen participation
- public awareness
- community-level initiatives
- reporting systems
- instant feedback loops

The transformation of Indore shows that environmental success is dependent on public involvement.

In the case of Rewa, stakeholder engagement manifests in:

- negotiations with the landowners
- processes for environmental clearances
- compensation programs for farmers
- creation of jobs for locals

Acceptance by society leads to less conflict, quicker project implementations, and greater public trust.

6.6 Integrated Interpretation: PPPs as Governance-Based Sustainability Systems

By bringing together the evidence, the public-private partnerships (PPPs) are presented not only as means of financing but as governance systems that integrate the following sectors:

- institutions
- policies
- technology
- finance
- citizen behavior
- environmental objectives

Three integrated insights are as follows:

1. Governance + Technology = Stronger Institutional Capacity

Technology enhances governance capacity through monitoring, transparency, and efficiency.

2. Policy Coherence + Risk Allocation = Investor Confidence

Stable policies and well-designed contracts reduce uncertainty and, thus, attract investors.

3. Stakeholder Engagement + Behavioral Change = Long-Term Sustainability. Environmental improvements demand social participation, not only infrastructure.

These interactions have shown that PPPs are dynamic systems that allow transitions in sustainability across multiple dimensions. As shown in Figure 2, PPP sustainability depends on the combined inputs of the public and private sectors, citizen satisfaction, and the strength of governance oversight.

Figure 2. Sustainability Outcomes Pathway Model

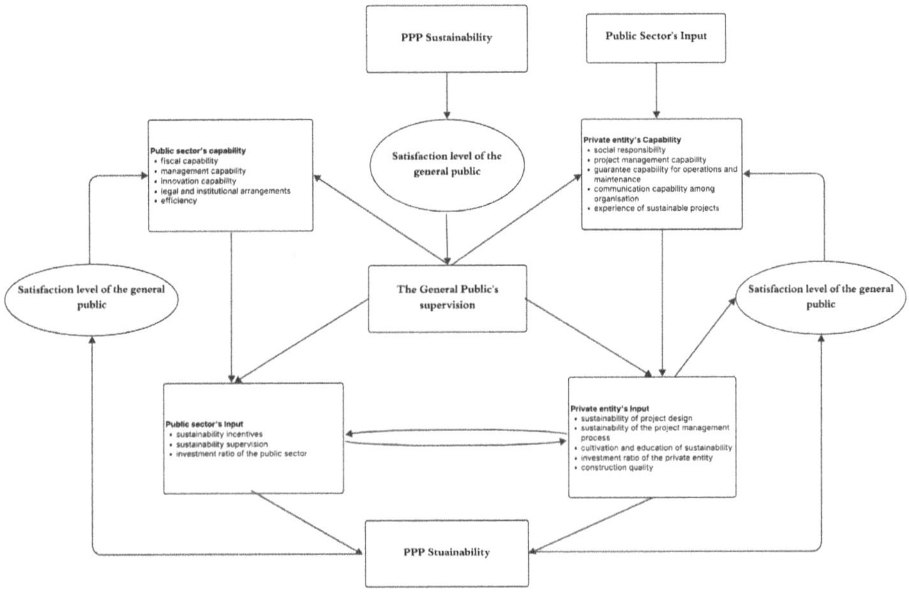

(adapted from Hodge & Greve, 2017; Yong & Wang, 2021; Yigitcanlar & Cugurullo, 2021).

6.7 Future Trends in PPP-Driven Sustainability

- Increasing reliance on digital governance systems, advanced analytics, and automation to manage environmental challenges.
- Growth of AI-enabled monitoring, blockchain-based contracting, and digital twins to enhance transparency, predictive planning, and regulatory oversight.
- Shift toward polycentric and networked governance models, replacing traditional hierarchical structures.
- Greater collaboration across public agencies, private firms, and community stakeholders.
- Expansion of financial innovation, including green bonds, blended finance, and climate-risk instruments, to improve PPP feasibility.
- Movement toward PPPs that are more technologically integrated, participatory, and resilience-oriented.

7. IMPLICATIONS

In the 21st century, transitions towards sustainability will not be possible without new governance models that can coordinate diverse groups of stakeholders, mobilize resources, and lead the long-term transformation of nature. The present chapter evaluated the contribution of public/private partnerships (PPPs) to green governance with respect to technology integration, institutional design, and collaborative governance frameworks. The analysis, based on a systematic literature review and two major Indian case studies—the Rewa Ultra Mega Solar Project and the Indore Smart City Waste Management Initiative—reveals that PPPs are not only indispensable tools for building green infrastructure but also the enablers of the governance ecosystems that are necessary for the realization of sustainable development.

Perhaps one of the most significant discoveries that makes this research relevant is that governance capacity is by far the most critical factor in determining whether a PPP will succeed or fail. Transparent procurement, strong contractual frameworks, regulatory stability, and comprehensive monitoring systems are important ingredients for successful PPPs. The Rewa project, for example, has national-state coordination, bankable contracts, and well-defined risk-sharing arrangements, which can be attractive to large-scale private investment and thus accelerate the rollout of renewable energy. The same holds in Indore, where city-level governance reforms-digitalization, citizen engagement, and performance-based contracts-can transform not only urban environmental systems but institutional practices. These two cases show that PPPs are most successful in environments where governance institutions are transparent, credible, and able to change.

Technology appears to be a significant factor in today's green governance. In both studied examples, technology was not just an auxiliary but also a primary function of governance by making everything more transparent, constantly monitoring, and facilitating decision-making. Digital platforms and IoT devices, together with automated processes, have brought not only accountability but also ease to authorities' operations, which were previously complicated. Such results are in line with international research showing the need for smart governance systems to accelerate sustainability transitions.

Another point is the significance of financial structuring and risk allocation.

PPPs are only profitable when the risks—the financial, regulatory, operational, or technological—are assigned to actors who are best trained to manage them. The payment security mechanisms and the competitive tariff structure of the Rewa project illustrate how proper risk allocation can increase investors' trust and reduce costs. Similarly, Indore's revenue diversification model underscores the need for robust, multi-source financing to ensure the long-term survival of a PPP. Thus, well-thought-out and appropriately structured financial arrangements not only guarantee the viability of the projects but also attract investors.

The chapter underscores the necessity of essential policy coherence and institutional alignment. Even the best technically designed partnerships might collapse in the fragmented policy realms. As much as possible is achieved, the effectiveness of PPPs increases enormously through extensive coordination, including multi-level coordination, predictable institutional mandates, and transparent regulatory frameworks. The convergence of India's renewable energy policies with urban governance programs appears to create a policy ecosystem that is more conducive to embedding technology, thereby making sustainability even more permanent.

The chapter asserts that the fates of green infrastructure and renewable energy - their development - will hinge on the intergovernmental technology, capacity, finance, and community participation model. When public-private partnerships are nested within clear policy frameworks and robust governance systems, they can yield outcomes that are environmentally responsible, economically viable, and socially inclusive. The results add to existing knowledge about the impact of PPPs on global sustainability goals and provide recommendations for policymakers, practitioners, and researchers seeking to create effective, technology-led, and governance-driven PPP frameworks that address roadblocks.

8. Limitations and Recommendations

8.1 Limitations

One limitation of this method is that systematic reviews may also be affected by excluding literature that, although non-indexed, is still relevant. The method also has the drawback of being confined to English-language publications. Furthermore, it does not have access to proprietary reports or government documents, and there is insufficient detailed PPP contract data. The limitations have been recognized, but they do not lessen the validity of the findings.

8.1.1. Scope of the Case Studies

The chapter focuses on two Indian cases: the Rewa Ultra Mega Solar Project and the Indore Smart City Waste Management Initiative. Although these cases surely aid in understanding governance innovation and technological integration, they may not represent the full spectrum of PPP engagements across India or many developing countries. Different sectors, such as water supply, transportation, and sanitation, may carry various institutional architectures, risk characteristics, and levels of community involvement.

8.1.2. Reliance on Secondary Data

The chapter is primarily based on secondary research, including peer-reviewed publications, project reports, governmental documents, and policy analyses. While these sources are credible, they may not accurately reflect the operational challenges, contractor opinions, politics, or informal governance practices on the ground. The absence of primary interviews with stakeholders or field data limits how much can be made empirical.

8.1.3. Changes in the Policy and Technological Landscape

These realms are in contention, basically under green governance and renewable energy. They are continuously updating the various policies, institutional structures, and technical methods in response to new environmental issues and global market trends. Therefore, specific results may be affected by the emergence of new rules, breakthroughs, or public-private partnership models—especially in areas such as smart cities, digital governance, and the like.

8.1.4. Generalizability of Findings

The findings are primarily based on Indian contexts influenced by a mix of socio-political, economic, and institutional conditions. Although most insights are relevant worldwide, direct generalization to high-income or fragile-state contexts should still be treated cautiously. The different countries involved have varying governance capacities, legal frameworks, and technological readiness.

8.1.5. Limited Sectoral and Longitudinal Comparison

PPPs are not the same in their operation across sectors, and sustainability outcomes are sometimes visible only in the long term. The chapter, however, does not conduct a longitudinal study of PPPs or profile the inter-sectoral differences in such depth due to the limited space. Such comparisons might uncover governance mechanisms that are particular to the sector or the bottleneck locations of institutional evolution.

8.2 Recommendations

8.2.1. Strengthen Governance Capacity at All Levels

Investments in institutional fortification, which include:

- transparent procurement systems
- contract management training
- performance monitoring tools
- independent regulatory oversight

It should be policymakers' priority. Stronger governance power means greater accountability, reduced risks, and higher-quality private investment.

8.2.2. Integrate Technology as a Governance Tool, Not Only an Operational Tool

The use of digital technologies such as IoT, data analytics, digital dashboards, and automation should be integrated by both governments and private companies, not only for the purpose of making the operations better, but also for the enhancement of technology, which should be a part of the governance structure, as the Indore example illustrates.

8.2.3. Improve Risk Allocation and Financial Structuring

The risk-sharing arrangements must be devised adequately so that the risks are distributed according to each party's capability to control them. Among the proposals are actions that can increase investor trust and extend the project's life.

8.2.4. Enhance Policy Coherence Across Government Levels

It is compulsory to have proper coordination of governance at different levels. The different layers of government should ensure a common understanding that leads to no fragmentation and faster project delivery, as seen in the case of Rewa, where different agencies worked closely together.

8.2.5. Promote Cross-Sectoral and Longitudinal PPP Research

The research in the future should investigate:

- performance metrics across different sectors (water, mobility, energy, waste)
- the long-term impacts of sustainability
- changes in governance over time
- The outcome of such research will be theoretical frameworks enriched and more robust guidance for policymakers provided.

REFERENCES

Adebayo, T. S., Genç, S. Y., Castanho, R. A., & Kirikkaleli, D. (2021). Do public–private partnership investment in energy and technological innovation matter for environmental sustainability in the East Asia and Pacific region? An application of a frequency domain causality test. *Sustainability, 13*(6), 3039. Advance online publication. DOI: 10.3390/su13063039

Almulhim, A. I., & Yigitcanlar, T. (2025). Understanding smart governance of sustainable cities: A review and multidimensional framework. *Smart Cities, 8*(4), 113. DOI: 10.3390/smartcities8040113

Alnaqbi, M., & Alami, A. (2023). Sustainability and Renewable Energy in the UAE: A Case Study of Sharjah. *Energies, 16*(20), 7034. Advance online publication. DOI: 10.3390/en16207034

Bai, X., Lu, C., Dong, X., & Li, Y. (2024). Role of collaborative governance in unlocking private investment in sustainable projects. *Humanities & Social Sciences Communications, 11*, 699. Advance online publication. DOI: 10.1057/s41599-024-03175-2

Bibri, S. E., Alexandre, A., Sharifi, A., & Krogstie, J. (2023). Environmentally sustainable smart cities and their converging AI, IoT, and big data technologies and solutions: An integrated approach to an extensive literature review. *Energy Informatics, 6*, 9. Advance online publication. DOI: 10.1186/s42162-023-00259-2 PMID: 37032812

Dutz, M. A., & Sharma, S. (2012). *Green growth, technology, and innovation* (World Bank Policy Research Working Paper No. 5932). https://papers.ssrn.com/sol3/papers.cfm?abstract_id=1980586

Fu, C., Lu, L., & Pirabi, M. (2023). Advancing green finance: A review of sustainable development. *Digital Economy and Sustainable Development, 1*, 20. Advance online publication. DOI: 10.1007/s44265-023-00020-3

Heldeweg, M. A., Sanders, M., & Harmsen, M. (2015). Public-private or private-private energy partnerships? Toward good energy governance in regional and local green gas projects. *Energy, Sustainability and Society, 5*(1), 9.

Hofmann, M., Choi, S., & Binder, C. (2022). Collaborative governance for sustainability transitions: A systematic literature review. *Journal of Cleaner Production, 338*, 130606. Advance online publication. DOI: 10.1016/j.jclepro.2022.130606

Koppenjan, J., & Enserink, B. (2009). Public–private partnerships in urban infrastructure: Reconciling private sector efficiency and public accountability. *Public Administration Review*, *69*(2), 284–296. DOI: 10.1111/j.1540-6210.2008.01974.x

Li, Y., Yang, H., & Zhang, L. (2018). Green Governance: New Perspective from Open Innovation. *International Journal of Project Management*, *36*(5), 763–778. DOI: 10.3390/su10113845

Ma, W., & Wang, M. (2023). Discussion on the relationship between environmental regulation and green technology innovation from the perspective of innovation external cooperation: Evidence from Chinese private enterprises. *Sustainability*, *15*(23), 16333.

Mahardhani, A. J. (2023). The role of public policy in fostering technological innovation and sustainability. [ADMAN]. *Journal of Contemporary Administration and Management*, *1*(2), 47–53.

Mahmoud, I. H., Morello, E., Salvia, G., & Puerari, E. (2022). Greening cities, shaping cities: Pinpointing nature-based solutions in cities between shared governance and citizen participation. *Sustainability*, *14*(12), 7011. Advance online publication. DOI: 10.3390/su14127011

Marquardt, J. (2024). Promises and pitfalls of polycentric federalism: The case of solar power in India. *Global Environmental Politics*, *24*(3), 75–96. DOI: 10.1162/glep_a_01315

Niesten, E., Jolink, A., de Sousa Jabbour, A. B. L., Chappin, M., & Lozano, R. (2017). Sustainable collaboration: The impact of governance and institutions on sustainable performance. *Journal of Cleaner Production*, *155*, 1–6.

Pandey, R., Bhattacharya, M., & Gupta, S. (2022). Institutional frameworks for sustainable PPP implementation: Challenges and opportunities in India. *Journal of Infrastructure Development*, *14*(3), 211–229. DOI: 10.1002/wene.422

Pinilla-De La Cruz, G. A., Rabetino, R., & Kantola, J. (2022). Unveiling the shades of partnerships for the energy transition and sustainable development: Connecting public–private partnerships and emerging hybrid schemes. *Sustainable Development*, *30*(5), 1370–1386.

Shrivastava, P., & Guimarães-Costa, N. (2017). Achieving environmental sustainability: The case for multi-layered collaboration across disciplines and players. *Technological Forecasting and Social Change*, *116*, 340–346.

Siripipatthanakul, S., Chaiprakarn, S., Rungsiyanont, S., & Phuangsuwan, P. (2025a). A case study of sustainable development in hospitals in Thailand using ChatGPT. In *Sustainable development for hospitals and clinics* (pp. 1–32). IGI Global Scientific Publishing., DOI: 10.4018/979-8-3693-9800-5.ch001

Siripipatthanakul, S., Phuangsuwan, P., & Jaipong, P. (2025e). Mobile application security: A case study of Bank of Ayudhya PCL. (KMA), Thailand. In *International conference on data-processing and networking* (pp. 479–493). Springer. https://doi.org/DOI: 10.1007/978-981-96-3102-5_36

Siripipatthanakul, S., Phuangsuwan, P., & Jaipong, P., & Muthmainnha. (2025f). Wi-Fi networks and security: A systematic review. In *International conference on artificial intelligence and networking* (pp. 81–92). Springer Nature Singapore. https://doi.org/DOI: 10.1007/978-981-96-2015-9_6

Siripipatthanakul, S., Phuangsuwan, P., Siripipattanakul, S., & Jaipong, P. (2025c). Sustainable tourism based on challenges, tourist experiences, and host communities: A case study of Ayutthaya, Thailand. In *Best practices for human resources management in tourism and hospitality* (pp. 361–394). IGI Global Scientific Publishing., DOI: 10.4018/979-8-3693-8779-5.ch014

Siripipatthanakul, S., Phuangsuwan, P., Siripipattanakul, S., Limna, P., Jaipong, P., & Chaiprakarn, S. (2025g). Sustainable development adopting digital intelligence among university students. In *International conference on artificial intelligence and networking* (pp. 211–220). Springer Nature Singapore. https://doi.org/DOI: 10.1007/978-981-96-2015-9_14

Siripipatthanakul, S., Phuangsuwan, P., Siripipattanakul, S., Limna, P., Jaipong, P., & Chaiprakarn, S. (2025h). Privacy and information security for sustainable community among TikTok's users. In *International conference on artificial intelligence and networking* (pp. 221–233). Springer Nature Singapore. https://doi.org/DOI: 10.1007/978-981-96-2015-9_15

Siripipatthanakul, S., Rakpathum, W., Siripipattanakul, S., Kaewnaknaew, C., Phuangsuwan, P., & Jaipong, P. (2025d). The future artificial intelligence (AI) for the hotel and hospitality industry: The case of hotel's human resource management (HRM). In *Best practices for human resources management in tourism and hospitality* (pp. 395–422). IGI Global Scientific Publishing. https://doi.org/DOI: 10.4018/979-8-3693-8779-5.ch015

Siripipatthanakul, S., Rungsiyanont, S., & Siripipattanakul, S. (2025b). Sustainable dentistry using artificial intelligence (AI) for digital education and business among university dental hospitals. In *Sustainable development for hospitals and clinics* (pp. 33–64). IGI Global Scientific Publishing., DOI: 10.4018/979-8-3693-9800-5.ch002

Stafford-Smith, M., Griggs, D., Gaffney, O., Ullah, I., Reyers, B., Kanie, N., Stange, B., Shrivastava, P., Leach, M., & O'Connell, D. (2017). Integration and implementation for sustainable development goals: Addressing trade-offs and building coherence. *Sustainability Science*, *12*(6), 911–919. DOI: 10.1007/s11625-016-0383-3 PMID: 30147763

van Zanten, J. A., & van Tulder, R. (2021). Towards nexus-based governance: Defining interactions between economic activities and Sustainable Development Goals (SDGs). *International Journal of Sustainable Development and World Ecology*, *28*(3), 210–226.

Wang, N., & Jin, X. (2007). PPP project evaluation: A framework for sustainability assessment. *International Journal of Project Management*, *25*(2), 212–219.

Zaman, A. U., & Lehmann, S. (2011). Urban growth and waste management optimization towards 'zero waste city'. *City, Culture and Society*, *2*(4), 177–187.

KEY TERMS AND DEFINITIONS

Collaborative Governance: This model of decision-making involves government agencies, private companies, and stakeholders throughout the entire process of policy or project planning, implementation, and management, with shared responsibility, openness, and learning from one another as the main principles.

Green Governance: Such a governance model is based on environmental sustainability and encompasses the whole process of decision-making, from low-carbon development, resource efficiency, ecological protection, and long-term ecological responsibility perspectives across all institutions, policies, and partnerships.

Public–Private Partnerships (PPPs): They are collaborative arrangements in which governmental bodies and private firms share responsibilities, risks, and resources, making the production of public infrastructure, services, or sustainability projects more effective than under traditional public-sector methods.

Renewable Energy Public-Private Partnerships: Model partnerships that are dedicated to the establishment and operation of renewable energy infrastructure through joint investment, technological cooperation, and risk sharing between the public and private sectors; for example, solar, wind, or biomass projects.

Smart Governance: The system of governance applies data analytics and integrates it with an IoT platform, smart infrastructure, and automation to pave the way for highly transparent, responsive, and coordinated public services and sustainability initiatives.

Sustainability Transitions: The long-term, gradual systemic transformation from traditionally borrowed technologies to environmentally friendly systems and technologies meant to mitigate climate change effects, resource depletion, pollution, and socio-economic vulnerability.

Technological Innovation in Governance: Governance structures include the new technologies such as AI, IoT, digital monitoring, and automation to ensure that they are resource-efficient and accountable at the same time, enhancing their environmental performance.

Urban Sustainability Public-Private Partnerships: Public-private partnerships are directed toward enhancing the urban environment's performance in waste management, smart city technology, recycling, pollution control, and the provision of resource-efficient services.

Chapter 6
AI-Driven Policy Frameworks for Net-Zero Economies:
Digital Pathways to Decarbonization

Sitesh Kumar Singh
https://orcid.org/0000-0002-7108-0808
College of Engineering, National University of Science and Technology, Muscat, Oman

Jaishri Gothania
https://orcid.org/0000-0002-2656-642X
Lingaya's Vidyapeeth, Faridabad, India

ABSTRACT

This chapter discusses the disruptive role of AI in governance and sustainable policymaking regarding climate issues. It examines the role of AI in both defending and making evidence-based decisions, carbon accounting, optimization of renewable energy, decreasing and decarbonizing transportation, and strengthening urban resilience, as well as informing international and regional collaboration. This chapter employed a systematic review. The findings show examples of AI implementation across different countries reveal its potential to enhance emissions reporting, policy modeling, and net-zero transitions. Some of the opportunities and issues that they point out include ethical governance, fair access, and data sovereignty. The chapter concludes that AI is a keystone of second-generation climate solutions, able to transform finance, energy, and cities. Moving forward, it recommends inclusive

DOI: 10.4018/979-8-3373-6821-4.ch006

governance systems in place that can enact AI-driven policies that are transparent, participatory, and internationally equitable in the pace towards resilient and net-zero futures.

1. INTRODUCTION

The integration of AI into governance systems will allow cities to gain faster access to pathways to net-zero and to cultivate resilience and inclusive sustainability.

1.1 Smart City Governance Frameworks

The concept of innovative city governance frameworks can be considered an essential route to align digitalization with climate neutrality objectives. It improves urban management by supporting co-integrated decisions needed in urban settings, including energy, transportation, waste treatment, and housing. The effective frameworks have achieved this by providing a mechanism that integrates technological innovation with robust mechanisms that enforce sustainability and accountability.

Urban data platforms powered by AI permit urban authorities to aggregate data on sensors, the IoT, and civic amenities. These platforms enable predictive modelling to estimate demand patterns, optimize energy consumption, and locate high-emission hotspots. In another example, Barcelona has already deployed the "CityOS" platform, which uses AI and combines data on energy, water, and transportation to help define decarbonization plans (Noori, 2021).

By equipping policymakers, AI will support the design of evidence-based zoning and land-use policies. Machine learning algorithms are used to model the effects of densification, green infrastructure implementation, and the introduction of renewable sources on emissions and resistance. These types of frameworks reduce the spread-out nature of city planning while facilitating energy-efficient, compact city planning.

On the governance side, transparency and inclusiveness are the keys. Decision systems that involve AI should be complemented by participatory mechanisms of governance that incorporate citizens, communities, and civil society organizations. This will achieve legitimacy, credibility, and compatibility with climate justice.

There are still challenges in data privacy, cross-platform interoperability, and unequal access to digitalization tools. To fill these gaps, we need adaptive policies in which AI innovation is placed on the broader sustainability contexts.

1.2 Urban Resilience and Adaptation Policies

In addition to decarbonization, AI can significantly enhance urban resilience in the face of climate-driven hazards such as heatwaves, flooding, and sea-level rise. Policymakers are gradually using AI-powered tools in climate risk modelling, infrastructure adaptation, and emergency response preparedness.

The AI-powered predictive analysis enables cities to anticipate weather spikes and assess critical infrastructure weaknesses. In another example, deep learning networks can be used to identify flood-prone areas in satellite images and other hydrological data, enabling policymakers to prioritize green infrastructure and adaptive drainage investments (Shirzadi et al., 2025).

The AI can assist with adaptive building code for cooling and heating loads, optimizing them in extreme climatic conditions. In the same manner, AI-enabled smart grids offer on-demand demand-response capabilities, helping keep the energy resource steady even during a heatwave or high-power draw.

Policymakers also have an opportunity to develop aid policies that seriously consider vulnerable communities through AI-enabled adaptation. For example, machine learning would enable plotting of socio-economic exposure to climate risks and directing resources and policies accordingly. These frameworks put resilience planning into the context of climate justice.

Nonetheless, the fact that resilience policies are challenged by fragmented data ownership and low local government capacity suggests that these policies can be difficult to govern. Joint efforts by governments, the private sector, and academia are required to scale AI-enabled resilience solutions.

2. LITERATURE REVIEW

2.1 Foundations of AI-Driven Climate Governance

Climate change governance is at a stage where conventional policy instruments are no longer sufficient and must be supplanted by cutting-edge technologies capable of handling complex data and uncertainty. AI has become a game-changing capability in this environment, promising new options for formulating, executing, and tracking climate policies. Compared with the traditional methodology, which typically uses a static model and differentiated data, AI offers multi-dimensional, real-time insights that help bolster the responsive governance system. Through the integration of data across sectors, AI enables evidence-based policymaking to be predictive rather than reactive, so that policy interventions must be geared toward setting short-term priorities as well as enduring sustainability interests (Son et al.,

2023). Figure 1 shows the theoretical picture of AI in the climate regime. It demonstrates how machine learning, predictive modelling, and decision-support systems can be applied to the design, implementation, and evaluation of policies. The figure highlights the feedback circuits that can enable monitoring and responsive policy to achieve net-zero goals.

Figure 1. AI-Driven Climate Policy Framework (authors, 2026).

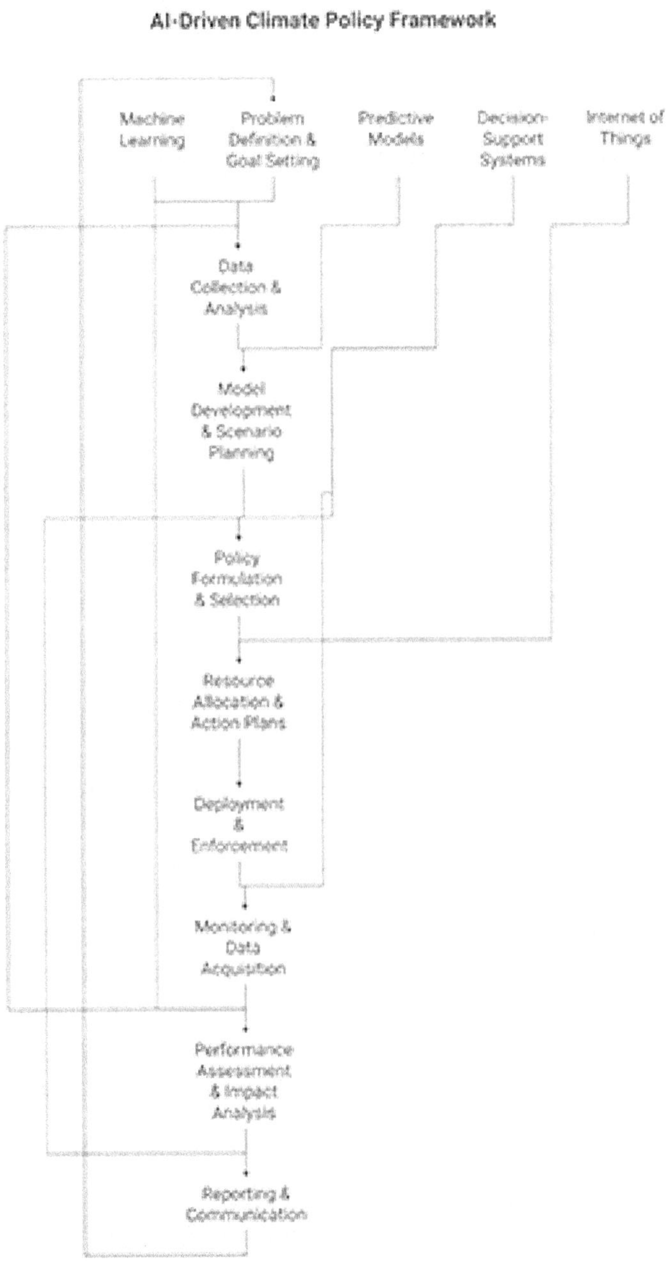

The other role will be as the spine of digital ecosystems through which collaborative decisions can be made. AI integrates technologies such as the Internet of Things (IoT), blockchain, and cloud computing, resulting in interconnected governance platforms that promote transparency, accountability, and resilience (Marengo, 2024). These ecosystems can facilitate coordination of actions across different levels of governance, from local to national governments and international organizations, while remaining inclusive and equitable. Nevertheless, the introduction of AI raises ethical issues related to data privacy, algorithmic bias, and the digital divide. In this way, the very pillars of AI-based climate governance lie in the balance between technological innovation and the values of justice, trust, and sustainability.

2.2 AI as a Tool for Evidence-Based Policymaking

Policy formulation in the climate area has relied on hindcasts, post hoc positioning, and imaginative estimates by specialists. Such approaches are valuable, but they tend to lack the pace and dynamism needed to keep pace with the rapidity of climate change. Artificial Intelligence (AI) is a paradigm shift, as it enables policymaking in an evidence-based way that combines multiple data sources, applies predictive modelling, and provides action-oriented insights in real time (Son et al., 2023). AI applications can analyze climate models, socioeconomic indicators, energy demand trends, and city development data in parallel using machine learning and natural language processing. This multidimensional capacity will ensure that the holistic picture informs policies of environmental, economic, and social factors.

One way AI shapes policymaking is in urban mobility development. Neural networks and deep learning algorithms have been effectively used to identify transport demand patterns, evaluate the progress of emissions-reduction measures through proposed interventions, and optimize infrastructure investment (Wang et al., 2024). In contrast to traditional models of transport, artificial intelligence systems have access to dynamic data from sensors, satellite imagery, and citizen feedback to adjust policies in real time. Adopting this approach in agriculture, AI-based models can predict crop production under different climate conditions, enabling governments to create adaptive policies that balance risks without compromising food security (Aijaz et al., 2025).

The important role of AI in policymaking extends beyond efficiency and can be applied to anticipatory governance. Predictive analytics enable policymakers to gauge the long-term outcomes of policies before introducing them. As an example, governments can model the economic and social impacts of introducing a carbon tax or reducing renewable energy subsidies to reduce unintended impacts. Furthermore, AI supports inclusivity by visualizing complex data in understandable formats, enabling citizens and stakeholders in a local community to engage in deliberations about a specific policy (OECD, 2023).

Nevertheless, the problematic nature of AI use in policymaking persists. Algorithmic opacity and biases in training data can be used to discriminate based on existing inequalities in society, and this is what machine learning can address. To address these risks, clear data governance policies and participatory validation procedures must be built. Policymakers must ensure that technology using artificial intelligence does not replace intellectual judgment but instead supplements it to maintain accountability and democratic legitimacy.

2.3 Digital Ecosystems for Sustainability Transitions

The dimension of climate change demands governance mechanisms that go beyond sectoral fragmentation and beyond national borders. One solution that can provide this efficiency is digital ecosystems that leverage AI, combining technologies such as the Internet of Things (IoT), blockchain, big data analytics, and cloud computing into collaborative governance platforms. Such ecosystems enable real-time observatories, transparent governance, and inter-sectoral coordination, thereby amplifying the sustainability transitions of scale.

At the heart of such ecosystems, IoT devices are transforming into continuous flows of environmental information: air-quality gauges in cities and satellite tracking of deforestation in nature. All this data is analyzed by AI algorithms that can detect patterns and anomalies and, accordingly, send the required signal to carry out interventions. Blockchain technology can increase trust in these systems by ensuring data integrity and transparency in transaction processes, including carbon credit trading schemes. Wholesomely, these technologies will make a digital infrastructure essential in mitigation and adaptation strategies in climate governance (UNDP, 2021).

AI-driven digital ecosystems have a clear role in urban planning and can promote sustainability in this city. Recent developments integrate innovative cyber-physical systems into governance procedures, helping municipalities minimize energy and emissions and remain resilient to climate risks (Jain et al., 2025). For example, an AI-powered platform can integrate renewable energy into an urban grid and, at the same time, manage the water supply and waste. Digital ecosystems can achieve coherence among policies by connecting traditionally unrelated spheres.

The essential advantage of digital ecosystems is that they encourage multi-stakeholder and multi-level interactions. These data platforms provide access to shared data by national governments, private-sector actors, civil society, and international bodies, thereby increasing transparency and accountability. Global interdependence is necessary for problem-solving regarding global commons, such as global warming, which require cooperation across jurisdictional boundaries.

However, management of the digital ecosystem also raises some governance issues. Disparity in the availability of digital infrastructure across locations poses a threat to marginalization, whereas the inherent opacity of algorithms undermines the trustworthiness of AI-based decisions. The use of data in various areas necessitates addressing ethically sensitive matters such as data privacy, ownership, and data surveillance, which are making it difficult to take up (OECD, 2023). To meet this challenge, the design of digital ecosystems should be inclusive, equitable, and democratically controlled.

2.4 Green Governance in the Artificial Intelligence Age

Green governance can describe institutional contexts, control systems, and engagement efforts aimed at integrating environmental sustainability goals into population decision-making. When viewed through the prism of climate action, it focuses on transparency, accountability, equity, and inclusiveness, and coordinates economic development with ecological protection. The development of Artificial Intelligence (AI) is transforming green governance by introducing data-driven regulatory controls, more adaptive policy implementation, and improved stakeholder engagement.

AI enhances green governance by advancing evidence-based regulation through continuous environmental monitoring, compliance verification, and outcome evaluation. For example, real-time emission tracking systems enable regulators to shift the focus from static reporting to dynamic enforcement mechanisms. Predictive analytics also plays a role in identifying non-compliance risks, enabling early interventions. These abilities improve institutional responsibility and maximize the effectiveness of the regulatory responses.

The significance of AI in fostering participatory governance is also important. Complicated environmental data can be turned into meaningful information and insights with the help of sophisticated visualization applications and open-data platforms, enabling communities, civil society organizations, and local authorities to participate more effectively in policymaking. This information democratization on the environment supports procedural justice and confidence in the climate governance processes.

Nonetheless, AI-facilitated green governance has its own difficulties. Transparency can be compromised by algorithmic obscurity, data biases, and unequal digital infrastructure, which exacerbate inequalities in governance. These risks require strong ethical provisions, non-discriminatory data governance systems, and transparent, just, and open regulatory control systems. AI can be a pillar of green governance, enabling responsibility and enhancing institutional capacity alongside democratic accountability.

2.5 AI For Emissions Tracking and Carbon Accounting

Precise measurement and monitoring of greenhouse gas (GHG) emissions enable the achievement of international climate targets, including those set in the Paris Agreement. Conventional methods of emissions accounting rely on self-reporting by industries, surveys, or periodic inventories, which may be insufficient, inconsistent, or delayed. Artificial Intelligence (AI) provides a paradigm shift by enabling real-time, high-resolution, and verifiable emissions-tracking systems. By incorporating data on satellite images, IoT sensors, and industrial surveillance systems, AI will provide governments and organizations with a clear view of emissions at different levels – local, regional, and global.

Additional functions of AI include supporting effective carbon accounting systems that underpin climate policies, such as carbon taxes, emission trading schemes, and climate reporting. With machine learning, AI can identify anomalies within emissions reports, decreasing fraudulent reporting and making reporting more accountable. Not only do these capabilities increase adherence to environmental laws, but they also build more investor confidence in the green markets. Still, the introduction of AI into emissions governance is also associated with the following issues: data ownership, cybersecurity challenges, and technological gaps between countries. These issues will need to be addressed to use AI as an effective and even-handed tool of carbon governance.

2.6 Real-Time Data Monitoring Using IoT and AI

The combined use of the Internet of Things (IoT) devices and Artificial Intelligence (AI) is transforming the way and means of managing and monitoring emissions. The IoT devices, such as air quality sensors, industrial meters, and satellite-based observation platforms, produce sky-high amounts of real-time information about energy use, the transportation stream, and industrial production. AI algorithms compute these data streams to detect emission patterns, identify hotspots, and predict the future. This synergy creates a dynamic system of emissions governance that addresses the weaknesses of fixed inventories and self-reporting.

Industrial emissions tracking is one of the most potent applications of IoT-AI integration. For example, implementations of AI in monitoring systems at power plants or manufacturing facilities can continually analyze flue gas composition and detect variations, in accordance with legal requirements (Pourrahmani et al., 2025). These systems can reduce reporting times and provide the regulator with almost instant alerts for intervention. AI models applied to satellite imagery have also been used to monitor methane emissions from oil and gas facilities with unprecedented precision, enabling rapid action against their sources (Mohammadimanesh et al., 2025).

In the transportation industry, IoT systems supported by AI help analyze on-road traffic flows, emissions, and fuel utilization in real time. Other cities are using AI-powered digital twins of the urban mobility system to simulate the effects of measures to reduce emissions and advise on policy interventions (Koirala et al., 2024). In addition to reducing carbon footprints, these systems have been shown to improve air quality and public health significantly.

The transparency and accountability in climate governance is also boosted by AI-driven in-the-moment surveillance. With blockchain integration, all data on the control and measurement of emissions can be stored, verified, and transferred without the possibility of manipulation by all stakeholders. This is more than necessary for facilitating international agreements, where trust and verification are major priorities.

Challenges stay even though their promise is present. The infrastructure of an IoT is costly to install, and low-income countries cannot afford to set it up. Also, AI models based on incomplete or biased data cannot provide accurate policies, which would negatively affect policy credibility. These gaps need to be resolved through international collaboration, capacity, and the formulation of inclusive digital governance structures.

2.7 Smart Carbon Tax and Trading Systems

Carbon taxes and emissions trading systems (ETSs) are among the most effective policy strategies for reducing greenhouse gas emissions. They work, however, depending on good emissions measurement, a well-functioning market, and good compliance measures. Artificial Intelligence (AI) provides excellent features to make these systems smarter and more transparent by making them adaptive.

In *carbon taxation*, AI can support the government in setting tax levels based on real-time emissions, economic conditions, and sectoral performance. For example, machine learning models can be used to illustrate the economic and environmental consequences of different tax rates, enabling policymakers to create tax frameworks that reconcile emissions reduction and economic development (Lilliestam et al., 2025). Its scheme can also help detect anomalies in declared emissions, reduce tax evasion, and enforce fair acts.

In *emissions trading systems*, AI can be used to optimize the trading of carbon credits and the allocation of credits to different entities. AI-enhanced trading platforms have the potential to forecast demand, gauge market volatility, and advise participants on how best to implement their pricing strategy. Moreover, AI systems can be integrated with blockchain to verify carbon credits, eliminating risks such as double-counting and fraud that have led to the decline in the credibility of ETS (Boumaiza & Maher, 2024).

It also gives added strength in verifying carbon markets. Conventional verification is characterized by effort-consuming auditing and third-party evaluation. In comparison, cross-validation between emissions reports and satellite data, as well as that between IoT data, can be performed automatically by AI, and compliance verification can be achieved near-instantaneously. This saves the transaction costs and adds trust and efficiency to carbon markets.

Practical applications are developing worldwide; for example, the European Union is experimenting with AI-based digital platforms to supplement monitoring and verification activities for ETSs. Elsewhere, pilot schemes in China and Canada are trialing the use of AI-powered predictive analytics to stabilize carbon markets and prevent price volatility (Wei et al., 2024). The innovations above depict how AI could help create more robust and functional carbon markets, which, in turn, would drive decarbonization.

However, the incorporation of AI in carbon pricing has its problems. Markets have become highly dependent on digital infrastructure, which creates cybersecurity vulnerabilities, and a lack of understanding of how algorithms work can lead to issues of fairness and inclusivity. Furthermore, the data and national capabilities to implement AI can exacerbate global disparities in climate governance. Global collaboration and standardization are required to have fair results.

3. METHODOLOGY

The chapter follows a systematic narrative review approach, which entails summarizing the recent scholarly literature on the use of artificial intelligence (AI) in climate governance and decarbonization policy. Instead of a meta-analytical research design, a systematic narrative review was selected because the chapter aims to combine and synthesize theoretical, policy, and applied research from multiple disciplines, including environmental governance, energy systems, transportation planning, urban studies, and data science.

The methodology of quantitative meta-analysis was not appropriate due to the heterogeneity in study designs, data types, and outcome measures across this body of research. Instead, systematic narrative synthesis offers the flexibility needed to synthesize findings while remaining analytically transparent and rigorous.

A comprehensive literature search was conducted in key academic databases, including Scopus, Web of Science, Google Scholar, and ScienceDirect. The mixed keywords included 'Artificial Intelligence', 'Machine Learning', 'Climate Governance', 'Decarbonization Policy', 'Carbon Accounting', 'Smart Cities', 'Transportation Decarbonization', and 'Energy Policy'. The search targeted peer-reviewed journal articles and extensive institutional reports from 2017 to 2025 for discussion as relevant to recent AI technologies and climate policy frameworks.

According to the study by Siripipatthanakul et al. (2025; a-c), a systematic thematic content analysis is used to synthesize the content. Systematic thematic content analysis is a meticulous qualitative approach for identifying, analyzing, and reporting patterns (themes) within data, transitioning from specific data points to overarching themes, thereby ensuring comprehensiveness and yielding profound insights into intricate subjects.

4. RESULTS

4.1 Machine Learning for Policy Simulation and Impact Forecasting

Climate policy design is typically associated with ignorance of the future course of emissions, economic impacts, and social consequences. Conventional models of policymaking are based on static assumptions that may not fully represent the true nature of real-world interactions. ML offers a dynamic approach to policy simulation and forecasting by identifying obscured trends in large datasets, enabling more reliable future predictions.

Policymakers, using ML, can model the likely effects of carbon taxes, renewable energy subsidies, or car policies even before they are implemented. These simulations can provide decision-makers with evidence-based information on policy risk reduction and efficient solutions. For example, ML models could be used to assess the impact of various demographic cohorts under changes in energy prices and to enable more equitable climate policy measures (Danish & Senjyu, 2023).

Furthermore, ML can draw on numerous data types, including climate models and economic forecasts, as well as social media trends, to provide a comprehensive projection of policy effectiveness. This ability is especially Germane to predicting the achievement of net-zero targets, where such measures need to be delivered against changing economic and social contexts. Though encouraging, obstacles to tackling the problem of ML-driven policy simulations include algorithmic bias, interpretability, and data governance that need to be overcome to make the use of ML in policy-related simulations transparent, inclusive, and trustworthy. Figure 2 presents the AI-enabled decision-making process for climate policy. It charts the iterative path of data aggregation to AI processing, policy recommendation, execution, and feedback-driven adaptation. The figure highlights the role of AI in supporting evidence-based, dynamic policy formulation to accelerate the drive towards net-zero ambitions.

Figure 2. AI-Enabled Decision-Making Process for Climate Policy (authors, 2026).

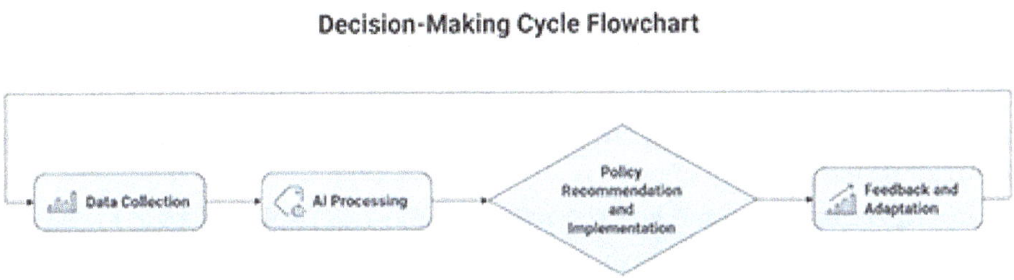

4.2 Predictive Models for Policy Outcomes

One of the strongest tools for applying machine learning in the context of climate governance is predictive modelling. Predictive models can help policymakers gauge the implications of a policy initiative by leveraging historical and real-time data to assess how an intervention will play out. The models perform well in modeling nonlinear relationships and multidimensional interactions that econometric models cannot easily capture.

For example, predictive ML models can be used to emulate the effects of changes in carbon tax rates on industrial competitiveness, energy demand, and household welfare. A paper by Bulut and Samuel (2025) showed that ML-based models could better predict the distributional effects of carbon taxes across income groups than conventional regression models, providing policymakers with more nuanced, valuable information. Similarly, reinforcement learning methods can be applied to continuously adapt the policy by adjusting predictions in response to new information.

In the energy industry, predictive models aid in evaluating renewable energy subsidies. Evaluation of weather trends, energy use fluctuations, and grid data using ML predicts the effects of subsidies on renewable penetration, system reliability, and emissions reduction (Shahverdi et al., 2025). Through such foresight, policymakers can optimize subsidies to be as effective as possible without trying to achieve the goal of subsidizing fuels by causing energy prices to soar.

The other emerging use is transportation policy. Predictive models with traffic data, fuel consumption, and urban demographics can be used to model the actions of congestion pricing, electric vehicle incentives, or fuel efficiency standards. As an example, an ML simulation study in European cities showed that dynamic congestion pricing has the potential to reduce emissions from traffic by up to a quarter and to make traffic more efficient (Su et al., 2024).

Despite the advantages mentioned, predictive models have limitations. The excessive use of historical information may re-entrench historical injustices in future forecasts, leading to unfairness in policy decisions. Besides, complex ML models are frequently not very interpretable, and policymakers cannot elucidate the causal processes underlying their forecasts. To address this, explainable AI methods are under development to present model outputs in a more understandable, more likely-to-be-understood format.

Altogether, predictive models are among the most valuable tools for developing climate policies that are adaptive and data-driven on the one hand and socially inclusive on the other. With the ability to test policy options and predict outcomes in advance, they can significantly strengthen the supporting evidence for achieving climate and sustainability objectives.

4.3 Scenario Planning for Net-Zero Transitions

Net-zero emissions require long-term planning amid uncertainty in technology, markets, and social behavior. The use of machine learning has brought a new dimension to the already existing scenario planning commonly applied in energy and climate policy. ML is more inclusive of self-guided scenario planning, as it can produce probabilistic forecasts, stress-test a policy under a variety of conditions and identify an optimal pathway towards achieving net-zero transitions.

Machine learning software is also capable of crunching through enormous volumes of climate, economic, and social data to draw up adaptive scenarios. As an example, deep learning methods can emulate multiple stages of renewable energy implementation, transport electrification, or carbon capture implementation and their impact on the emissions path (Q. Wang et al., 2025). Such models aid policymakers in making relevant comparisons and determining strategies that can strike a balance among emissions reduction, economic development, and social equity.

One significant benefit of the ML-enhanced scenario planning is its ability to deal with uncertainty. With traditional models, fixed assumptions about technological prices or behavior do not change over time, whereas ML is flexible and adapts accordingly. This flexibility must be implemented in the changing spheres of renewable energy and electric mobility, where cost reductions or consumer uptake rates may vary widely across decarbonization trajectories (Bataille et al., 2020).

Real-world applications are being realized worldwide; for example, the UK Climate Change Committee has used ML-based models to investigate net-zero pathways across various carbon-pricing and renewable-investment scenarios. On a similar note, ML has also been used in India to predict electricity demand under different electrification policies and to analyze how renewable growth can integrate with future consumption patterns (Taloba & Rayan, 2025).

The stakeholders also perform scenario planning with the body of ML. ML can estimate the political liveness of various climate policies by incorporating information on how people form opinions, based on survey data, consumer behavior, and social media. This ensures that scenarios are not only technically and economically feasible but also socially acceptable.

4.4 AI and Energy Policy Innovation

Allowing more revision of the energy sector, the relationships of world climate policy are mainly concentrating on this EU sector because it is a leading contributor to greenhouse gas emissions (Eurostat, 2025). Innovation in the energy sector has been enabled and fostered by Artificial Intelligence (AI), leading to higher efficiency, greater reliability, and sustainability throughout the energy value chain. AI systems analyze large amounts of data on the generation, use, and delivery of energy to support more intelligent decision-making and the incorporation of renewable energy sources into national energy grids.

Policymakers and regulators use AI to design dynamic policies through predicting energy demand, optimizing renewable integration, and determining risk exposures to intermittency or volatile prices. As a result, an increased load during peak times could be predicted, and corresponding demand-side interventions could be suggested to reduce the use of fossil-fuel-based backup systems. Equally, machine learning models can evaluate the effects of the subsidization schemes on the process of renewable adoption and inform the upscale of clean technologies.

The regulation of energy markets is another critical area. The use of AI tools offers real-time insights into pricing, storage, and trading and helps align policy frameworks with dynamic energy transitions (Antonopoulos et al., 2020). Data governance and cybersecurity, as well as algorithmic transparency issues, still exist, however. Despite all such obstacles, AI-based solutions are crucial to reach more resilient, equitable, and net-zero-aligned energy systems.

4.5 Optimizing Renewable Energy Systems

The move toward renewable energy is demanding *innovative policy mechanisms* to span the challenge of variability and intermittency. Figuring out renewable energy systems using AI is critical for optimizing supply and demand forecasting, enhancing storage management, and integrating systems. Predictive analytics enables policymakers to formulate policies that maximize renewable integration while maintaining grid stability through AI.

One of the most important applications is *forecasting renewable generation.* Machine learning algorithms use meteorological and satellite data, along with historical production patterns, to model the output of solar and wind generation with the greatest possible accuracy. Better forecasts help to facilitate the policy incentives to incentivize grid balancing and storage to minimize renewables curtailment. As an example, AI-assisted prediction has decreased the need to curtail solar power in India by 20 percent and, likewise, in China (Hussein et al., 2025).

AI informs *policy innovation* in another area: energy storage optimization. Storage systems play a great role in providing reliability when the supply of renewable resources is scarce. The AI models can identify optimal charging and discharging cycles, increase battery longevity, and inform grid-scale subsidy plans. Policymakers can therefore offer incentives following the investments in energy storage that meet the needs of system reliability.

The modelling of *complex grid interactions* also aids the integration of renewable energy through IAI. Reinforcement learning can be used to enable energy management systems to adapt as supply and demand change. This flexibility facilitates regulatory systems that encourage the creation of smart energy markets and the pricing mechanisms in a dynamic environment. Another example is that regulatory authorities in Europe are implementing AI-overlaid demand response, in which utility providers can design flexible tariffs to incentivize households to shift consumption to renewable off-peak periods (Hofmann & Lindberg, 2023).

Through a policy lens, such AI-enabled insights help governments calibrate their renewable energy subsidy plans, craft expansion plans, and ensure equitable clean energy access. Technical optimization, paired with regulatory actions, makes AI a tool that can ensure the scalability of renewable transitions and their fairness.

4.6 Smart Grids and Demand-Side Management

Smart grids are the infrastructure of modern energy transitions that facilitate the transition to decentralized generation and bi-directional energy flow and provide consumer involvement in the energy system. AI will enhance smart grid capability by providing real-time monitoring, predictive maintenance, and demand-side management. These applications can be directly used to design policy in the areas of energy efficiency, affordability, and emissions reduction.

Investing in modern, real-time monitoring and anomaly detection is what is required to enhance grid resilience. With the help of AI algorithms, information from sensors, smart meters, and the Internet of Things can be analyzed to identify faults in systems and equipment, and blackouts, ahead of time. The implementation of policies that embrace AI-enabled predictive maintenance can help minimize the time a system is in a down state and boost energy security (Ucar et al., 2024).

Another policy area where AI is helpful in innovative policy design is demand-side *management (DSM)*. With consumption patterns, AI helps utilities predict demand

and suggest dynamic pricing schemes. This will enable consumers to shift electricity use to hours when renewable supplies are most abundant, thereby minimizing fossil fuel use. According to a WEF study (2024), AI-assisted DSM policies would reduce peak demand by up to 15 percent, thereby reducing emissions.

Distributed energy resources (DERs), including rooftop solar and community batteries, also benefit from AI applications. The ML-based models can be used to optimize power flows and manage DERs at the neighborhood scale to ensure a match between supply and demand (Singh et al., 2024). Policymakers can facilitate this by regulations that encourage decentralized energy trading and community energy platforms.

In addition, AI may be used to *supply energy equity policies*, that is, to provide vulnerable households with subsidies to mitigate the risk of energy poverty. As an example, AI models in the UK have been effectively applied to map energy-poverty hotspots, making it clear where targeted interventions are needed (Spandagos et al., 2023).

Although beneficial, the use of AI in smart grids raises governance issues, including privacy concerns related to smart meter data, cybersecurity threats, and algorithmic bias. To improve these, policies need to be implemented that balance innovation and safeguards to ensure the transparent, ethical, and secure deployment of AI.

4.7 AI Transportation Decarbonization Policies

The transportation sector is a high priority for decarbonizing the planet, as it accounts for approximately one-fourth of carbon dioxide emissions from the energy sector. Artificial Intelligence (AI) is a game-changer in the field of transportation policy, as its use would ensure that policies are planned intelligently and operated efficiently, and that green technologies for clean mobility are swiftly incorporated. With the help of large mobility datasets, AI can help policymakers determine measures that minimize emissions while increasing accessibility and affordability. According to Figure 3, the sphere of transportation and energy is to be decarbonized, with AI-based apps mentioned. The schematic shows how smart grids using AI, renewable energy forecasting, intelligent traffic, and autonomous vehicles can reduce emissions and improve system performance, working together. It provides a clear visualization of sector-specific interventions that can lead the country to net zero.

Figure 3. AI Applications in Energy and Transportation Decarbonization (authors, 2026).

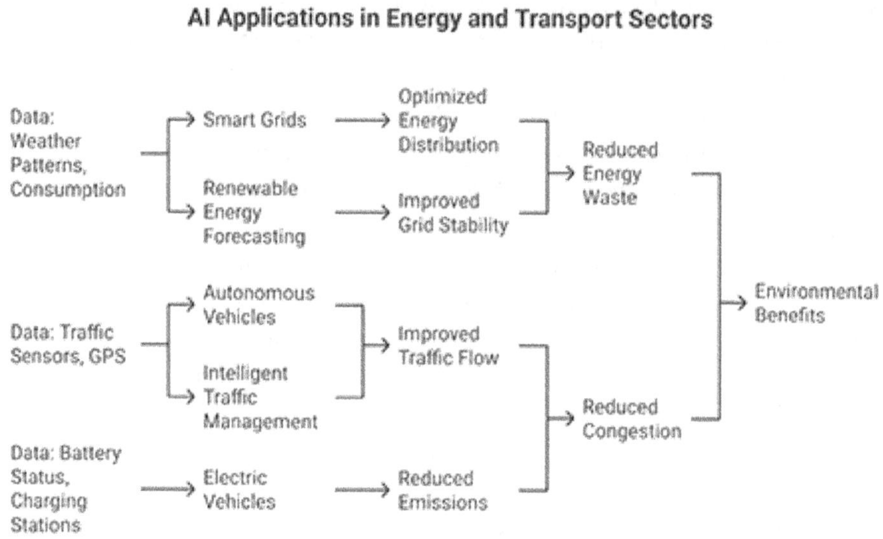

Automation will also help streamline public transport routes, minimize traffic congestion, and integrate low-emission modes of transport, in line with climate and sustainable development agendas (Sobczuk & Borucka, 2024). For example, predictive modelling is applied to dynamic traffic control and urban mobility planning to reduce fuel consumption and greenhouse gas emissions.

Furthermore, AI enhances regulatory systems for electric and autonomous cars by enabling models to analyze the life cycle of photovoltaics, infrastructure requirements, and consumer patterns. This assists governments in being innovative and balancing equity, safety, and sustainability. Despite this, obstacles remain, including data privacy, system interoperability, and unequal access to AI-enabled solutions across regions.

Despite these challenges, AI-driven transportation policies are now essential to advancing climate action. They offer governments tools grounded in accurate evidence to accelerate decarbonization and pursue a more sustainable, future-ready mobility system.

4.8 Intelligent Transport Systems

Intelligent Transport Systems (ITS) can serve as a paradigm for AI-enabled transportation policies. ITS uses real-time data, connectivity, and machine learning to optimize traffic, decrease congestion, and establish sustainable mobility. Uncalibrated cameras can also be used for traffic surveillance (Gupta et al., 2022). When AI is applied to transportation planning, policymakers will be able to develop a more efficient system with fewer emissions and increased accessibility.

The most effective applications are traffic management and reducing traffic congestion. Traffic models analyze data gathered by traffic sensors, GPS, and cars connected to the Internet to infer traffic congestion and suggest adaptive control of the traffic signal. Bahamazava (2025) states that AI-enabled adaptive traffic systems can lower emissions associated with traffic congestion by 10-20 percent. Policymakers may contribute to such systems by investing in intelligent infrastructure and policies to homogenize data sharing across cities.

Another policy-relevant use is that of public transportation optimization. The AI tools help find underused routes, optimize schedules, and integrate real-time data on demand to increase effectiveness. As an example, Singapore has set up the AI bus routing systems that reduce fuel consumption and overall customer satisfaction (Avagyan, 2025). Governments may expand those policies to include shared and micro-mobility services to shift transportation away from cars.

Freight and logistics decarbonization also largely depend on IA use. Machine learning apps can perform highly efficient delivery routing, freight consolidation, and minimization of empty-truck movements, directly reducing emissions from freight transportation. The concept of AI-enabled logistics can support green policies, as exemplified by the European Union's digital freight corridors initiative (ALICE, 2024).

As a governance construct, the ITS policies should discuss equity and inclusivity. Although AI increases efficiency, it can unintentionally discriminate against disenfranchised people when the data is biased. Policymakers require sound governance mechanisms to create similar opportunities to use AI-based mobility solutions equitably.

Lastly, AI-supported ITS can help integrate climate policies across all aspects of city planning and transportation. The pairing of real-time emissions monitoring with adaptive traffic management is another potential area where governments can dynamically assess and expand decarbonization efforts.

4.9 Policy Support for Electric and Autonomous Vehicles

They made reducing carbon emissions in the transport industry a key to the development of electric vehicles (EVs) and autonomous vehicles (AVs). AI could reinforce policymaking in this field by providing evidence on best practices in implementation, consumer adoption, and life cycle emissions, among others.

A significant proportion of policy problems to address is infrastructure planning. AI tools analyze traffic flow, consumer charging behavior, and grid capacity to suggest ideal locations for charging stations. These insights can guide governments on whether to invest and ensure equal access to EV infrastructure.

Predictive models maximize charging cycles to increase battery life and thereby minimize emissions of energy output over the entire lifecycle (Paneru et al., 2024). These pieces of knowledge can be integrated into government policies that encourage the use of batteries in their second life and recycling.

AVs' AI supports safety and regulatory control systems. Autonomous driving relies on AI perception systems and decision-making processes, and policymakers must impose regulations to ensure safety and accountability.

Another important element is incentive design. AI models could be used to analyze the success of subsidies and tax credits, as well as the ways congestion charges can promote the adoption of EVs and reduce unintended outcomes. As an example, policy simulations of AI-based EV policies in China have informed reforms to subsidies, making them cost-efficient and equalizing EV adoption (Sheldon & Dua, 2020).

Nevertheless, the use of AI to aid EV and AV policy requires solutions for data control, cybersecurity, and international compatibility. Policymakers should create international requirements for data exchange and the ethical use of AI to ensure categories of trust and fairness are achieved.

4.10 Urban Policy and Net-Zero Cities Through AI

However, cities are leaders on the frontline of climate action because they account for over 70 percent of global greenhouse gas emissions and are home to the majority of the world's population (UN-Habitat, 2024). It takes transformative urban policies to support net-zero cities that use Artificial Intelligence (AI) to harness sustainable planning, governance, and resilience. Using AI, it will be possible to analyze multidimensional city data, allowing policymakers to develop interventions that maximize decarbonization without being exclusionary to the singular economic or social unit in the city.

The applications of AI in urban governance are energy optimization, adaptive mobility management and assessment of climate risks. As an example, machine learning can simulate the effects of building retrofits or the installation of renewable energy systems, and city officials can have a scientifically supported plan on how to achieve carbon neutrality (Xu et al., 2021). Equally, using AI-based decision support systems, mobility, energy, and land-use data could all be incorporated into comprehensive solutions to urban planning.

AI-assisted urban transformations need to be responsive to the issues of data equity, algorithmic discrimination and digital gaps. Policymakers must make sure that the technological solutions in the smart city are in line with the principles of climate justice and can be used by everyone in the metropolitan population. Figure 4 shows an AI-enhanced urban system of net-zero cities. It shows how smart buildings and energy-efficient infrastructure, intelligent transport and waste management systems can interact with each other through the integration of IoT and AI-based algorithms. The figure shows how urban AI technologies interrelate to enhance sustainability, resilience, and climate adaptation on a city level.

Figure 4. Urban AI Ecosystem for Net-Zero Cities (authors, 2026).

5. DISCUSSIONS

5.1 Global Case Studies of AI-Driven Climate Policies

Applying AI to climate policy is gaining traction across the globe, and different countries are becoming more innovative in how they use AI to decarbonize faster and increase resilience and sustainable development. These case studies demonstrate how AI can not only supplement climate governance but also help achieve national and international commitments under the Paris Agreement and the Sustainable Development Goals (SDGs).

The flexibility of AI in policy contexts is also reflected in national experiences. For example, countries with advanced digital infrastructure are using AI to streamline energy, mobility, and carbon accounting. In contrast, emerging economies are using AI to manage resource efficiency and disaster resilience. The application of AI at global and regional levels enables the establishment of carbon markets, the management of emissions, and international cooperation on climate outcomes.

These cases of study offer important insights into opportunities, challenges, and replicable models. It also raises some important issues, including ethical governance, equity in access to technology, and sovereign rights over data, that climate policies worldwide based on AI need to address.

5.2 National Policy Innovations

The governments of some countries are leading with AI in policies to intensify climate action.

China: It has linked AI to its carbon monitoring and trading system. Under Chinese leadership, heavy industries and power plants are monitored using satellite imagery and machine learning to contribute to the world's largest carbon market (Chen et al., 2024). This system facilitates the enforcement of emissions caps and enhances corporate accountability.

India: It has used AI for renewable energy optimization, especially for solar and wind integration. With the help of AI models, the National Institute of Wind Energy can forecast wind patterns, thereby stabilizing the grid and promoting India's renewable energy ambitions (Algburi et al., 2025). Its implementation is also in the field of water resources management, where it aids climate adaptation in drought-prone regions.

European Union: Members of the European Union (EU) are integrating AI into smart city strategies. As an example, the "AI4Cities" initiative in Finland uses AI to reduce energy demand in transport and buildings across six European cities

(Jørgensen & Grace, 2025). Such innovations relate to the city's decarbonization and the national climate framework.

United States: The Department of Energy is one of the United States agencies that utilizes AI to upgrade its grid and forecast extreme weather. AI expands resilience planning (of hurricanes and wildfires) in line with the net-zero commitments of the Biden government (U.S. Department of Energy et al., 2024).

The national innovations demonstrate the role of AI in improving both adaptation and mitigation, but also raise concerns about scalability, interoperability, and technology equity. Table 1 summarizes key AI tools and their roles in national climate policies. It points out that technologies such as machine learning, decision support systems, IoT, and natural language processing can support evidence-based decision-making, energy systems optimization, and urban resilience. Clearly identifying the tool, sector, and function, the table can easily demonstrate the areas where AI can empower national programs toward net-zero goals.

Table 1. AI Tools and Their Roles in National Climate Policy

AI Tool	Policy Role	Sector/Application	Key Function
Machine Learning	Predictive analysis	Energy systems	Forecast renewable energy output and optimize grid management
Decision Support Systems	Policy evaluation	Urban planning	Simulate outcomes of city-level decarbonization strategies
IoT & Sensors	Monitoring	Emissions & air quality	Track real-time emissions and resource consumption
NLP & Data Analytics	Policy review	Climate regulations	Analyze policy documents and monitor compliance trends
Simulation Models	Scenario planning	Transportation & mobility	Model policy impacts electric and autonomous vehicles

5.3 Regional and International Best Practices

Climate governance beyond national boundaries: As a result of the international focus on climate governance, national borders cannot be avoided when considering institutions and instruments. Some best practices illustrate how collaborative frameworks use AI to achieve common climate objectives.

Regional Cooperation

In *Europe*, the European Green Deal focuses on digital ecology linkages in accordance with which to ensure integration of renewables, carbon trading, and carbon emissions monitoring; AI would support these processes. The European Union uses the Copernicus Climate Change Service, an AI-powered system of satellite analysis that monitors emissions in all member states and ensures that they meet EU-wide carbon goals (Jørgensen & Grace, 2025).

In *Africa*, regional efforts, such as the African Climate Policy Centre, are also looking into AI to manage climate risk. An AI-based early-warning system is used in East Africa to manage droughts and floods, and cross-border collaboration is necessary in the management of shared water basin and agricultural resilience (Mienye et al., 2024).

In the *Asia-Pacific*, ASEAN states are collaborating on smart energy grids, AI, and cross-regional greenhouse-gas tracking. These projects illustrate the promise of common digital infrastructure both as a source of resiliency and decarbonization.

World Best Practices

On the *international stage*, the United Nations (UN) is experimenting with AI in climate international relations. As another example, AI-powered language problem solvers are used to analyze texts in the context of negotiations at COP summits, enabling them to trace commitments and gaps in policy wording (UNFCCC, 2021).

The *World Bank* has introduced a set of climate finance projects aided by artificial intelligence, which is used to forecast and assess risks in infrastructure projects and to direct investments toward a low-carbon future (World Bank, 2022).

The *WEF* (World Economic Forum) uses AI to generate real-time energy statistics worldwide, which are used by governments and other researchers for emissions forecasting (Kuhlmann et al., 2021).

Table 2. AI Applications for Regional and Global Climate Governance

AI Application	Scale	Function	Policy Outcome
Smart Energy Grid Management	Regional	Optimize renewable energy distribution	Improved energy efficiency across regions
Early-Warning Climate Systems	Regional	Predict extreme weather events	Enhanced preparedness and resilience
AI-Based Policy Simulation	International	Model climate policy scenarios	Supports negotiation and commitment tracking
Data Analytics for Emissions	Global	Monitor and report emissions	Transparency and accountability in climate action
AI-Driven Resource Planning	Regional & Global	Allocate water, energy, and other resources	Optimized sustainability and adaptation strategies

Table 2 depicts AI tools that can address regional and global climate governance. It demonstrates how AI systems, such as smart energy grids, policy simulations, and early-warning climate systems, enable coordination, monitoring, and optimization of resources across borders. The table is a reminder of the chapter discourse on how digitization is critical to increasing transparency and resilience, as well as sustainability at larger levels.

Replicability and Learning Lessons

These practices indicate the possibility of AI to increase transparency, collaboration, and accountability in climate governance. Nonetheless, they also indicate the digital divide since the developing countries are usually ill-equipped to use AI. The first should instead aim to build capacity, ensure the availability of open platforms, and ensure equitable access to AI innovations.

Besides, regulatory systems should establish ethical AI in climate policy usage. The problems of algorithmic discrimination, monitoring, and the right to data remain unsolved globally and require regulatory frameworks that involve everyone.

6. CONCLUSION

Artificial intelligence has become an innovative engine in climate management, reinventing the design, delivery, and assessment of environmental policies. AI can help governments and institutions to achieve data-driven, adaptive, and precision policymaking through emissions monitoring and carbon accounting, resilient city structures, and efficient energy systems, among others. These examples, presented at the national and regional levels in this chapter, demonstrate that AI is not just an extension of technology but also a core facilitator of evidence-based governance, cross-sector coordination, and accelerated decarbonization.

In addition to technological efficiency, the effectiveness of AI in climate action is increasingly determined by the strength of green policies. To achieve sustainable success, regulatory frameworks need to incorporate ethical benchmarks, transparency, community engagement, and digital innovation. Policymaking should ensure that the implementation of AI supports the values of accountability, equity, and environmental justice, and protects the rights of data and inclusiveness – especially for vulnerable and technologically disadvantaged communities.

However, considerable obstacles still exist. Issues of algorithmic bias, digital divides, data sovereignty, and institutional capacity asymmetries remain a threat to equitable implementation, particularly in the developing world. To overcome these obstacles, it is important to resolve them through comprehensive governance regimes to avoid inequality in the benefits of AI-driven sustainability transitions.

To conclude, the opportunities for AI to enhance the efficacy of climate policies are enormous, but its disruptive effect will depend on how well it is incorporated into effective green governance frameworks aligned with the long-term objectives of sustainability. The AI can only play a fair role in global decarbonization and climate resilience with the help of a transparent regulation process, participatory governance, and ethically driven innovation.

7. RECOMMENDATION: FUTURE PROSPECTS

Its future in climate policy scales to the stars. In the future, it can be expected that machine learning models will become increasingly integrated into national climate strategies, where emissions will be monitored in real time, and trends predicted relative to climate impacts, as well as automated policy assessments based on machine learning models. Governments will turn to AI not only to undertake mitigation – optimize renewable energy and decarbonize transportation – but also to undertake adaptation – forecast disasters and manage resources using AI.

The use of AI at the global level may help transform the course of climate diplomacy by enabling the creation of transparent, shared datasets and negotiation support tools, reducing ambiguity and encouraging nations to become more accountable. AI-based risk analysis and green investment models are set to transform the landscape of climate finance, opening a new avenue for green development.

Nevertheless, to shape AI as an agent in climate governance, its risks should be addressed. It will be necessary to build fair access, relevant algorithmic transparency, and inclusive digital ecosystems. Policymakers will face the challenge of maximizing AI's potential and diffusing its benefits.

There are possibilities for AI to serve as one of the building blocks in overall efforts to go-to-zero and build climate resilience, but only when its application is ethically managed.

REFERENCES

Aijaz, N., Lan, H., Raza, T., Yaqub, M., Iqbal, R., & Pathan, M. S. (2025). Artificial intelligence in agriculture: Advancing crop productivity and sustainability. *Journal of Agriculture and Food Research, 101762*. Advance online publication. DOI: 10.1016/j.jafr.2025.101762

Algburi, S., Kareem, S. S. A., Sapaev, I., Mukhitdinov, O., Hassan, Q., Khalaf, D. H., & Jabbar, F. I. (2025). The role of artificial intelligence in accelerating renewable energy adoption for global energy transformation. *Unconventional Resources, 100229*. Advance online publication. DOI: 10.1016/j.uncres.2025.100229

ALICE. (2024). Leveraging the digital transition for efficient and zero-emissions logistics – ALICE Alliance for Logistics Innovation through Collaboration in Europe. https://www.etp-logistics.eu/leveraging-the-digital-transition-for-efficient-and-zero-emissions-logistics/

Antonopoulos, I., Robu, V., Couraud, B., Kirli, D., Norbu, S., Kiprakis, A., Flynn, D., Elizondo-Gonzalez, S., & Wattam, S. (2020). Artificial intelligence and machine learning approaches to energy demand-side response: A systematic review. *Renewable & Sustainable Energy Reviews, 130*, 109899. DOI: 10.1016/j.rser.2020.109899

Avagyan, A. (2025). AI-Integrated smart traffic systems for Carbon-Neutral cities. Trends Research & Advisory. https://trendsresearch.org/insight/ai-integrated-smart-traffic-systems-for-carbon-neutral-cities/?srsltid=AfmBOoqtne30QOrZYW5fgC34fGubBBMSspz8h-ppLJXBPtvXBWBIseqj

Bahamazava, K. (2025). *AI-driven scenarios for urban mobility: Quantifying the role of ODE models and scenario planning in reducing traffic congestion*. Transport Economics and Management., DOI: 10.1016/j.team.2025.02.002

Boumaiza, A., & Maher, K. (2024). Leveraging blockchain technology to enhance transparency and efficiency in carbon trading markets. *International Journal of Electrical Power & Energy Systems, 162*, 110225. DOI: 10.1016/j.ijepes.2024.110225

Bulut, H., & Samuel, R. (2025). Public support for carbon taxes varies across countries and policy design must consider the national context. *Communications Earth & Environment, 6*(1). Advance online publication. DOI: 10.1038/s43247-025-02562-0

Chen, Y., Zhou, J., Ge, Y., & Dong, J. (2024). Uncovering the rapid expansion of photovoltaic power plants in China from 2010 to 2022 using satellite data and deep learning. *Remote Sensing of Environment, 305*, 114100. DOI: 10.1016/j.rse.2024.114100

Danish, M. S. S., & Senjyu, T. (2023). Shaping the future of sustainable energy through AI-enabled circular economy policies. *Circular Economy*, *2*(2), 100040. DOI: 10.1016/j.cec.2023.100040

Eurostat. (2025). Quarterly greenhouse gas emissions in the EU. Eurostat. https://ec.europa.eu/eurostat/statistics-explained/index.php?title=Quarterly_greenhouse_gas_emissions_in_the_EU

Gupta, S., Kartha, R. S., Ramirez-Asis, E., Flores-Albornoz, J., Vílchez-Vásquez, R., & Singh, S. K. (2022). A novel traffic surveillance system using an uncalibrated camera. *Journal of Advanced Transportation*, *2022*, 1–10. DOI: 10.1155/2022/5867524

Hofmann, M., & Lindberg, K. B. (2023). Evidence of households' demand flexibility in response to variable hourly electricity prices – Results from a comprehensive field experiment in Norway. *Energy Policy*, *184*, 113821. DOI: 10.1016/j.enpol.2023.113821

Hussein, J. B., Workneh, T. S., Kassim, A., Ntsowe, K., Melesse, S. F., El-Mesery, H. S., & Zixheng, H. (2025). A review of the use of artificial intelligence in renewable energy for food processing and preservation process optimisation, challenges, and future prospects. *Renewable & Sustainable Energy Reviews*, *223*, 116076. DOI: 10.1016/j.rser.2025.116076

Jain, R., Singh, S. K., Palaniappan, D., Parmar, K., & Premavathi, T. (2025). Integrating smart cyber-physical systems into urban planning. In Poonia, R. C., Upreti, K., & Khan, M. S. (Eds.), *Smart cyber physical systems: Innovations and real-world implications* (pp. 82–106). CRC Press., DOI: 10.1201/9781003542513-5

Jørgensen, B. N., & Grace, Z. M. A. (2025). Impact of EU regulations on AI adoption in smart city Solutions: A review of regulatory barriers, technological challenges, and societal benefits. *Information (Basel)*, *16*(7), 568. DOI: 10.3390/info16070568

Koirala, B., Cai, H., Khayatian, F., Munoz, E., An, J., Mutschler, R., Sulzer, M., De Wolf, C., & Orehounig, K. (2024). Digitalization of Urban Multi-Energy Systems – Advances in Digital Twin Applications across Life-Cycle Phases. *Advances in Applied Energy*, *100196*. Advance online publication. DOI: 10.1016/j.adapen.2024.100196

Kuhlmann, A., Moore, J., & Mehlum, E. (2021). Harnessing artificial intelligence to accelerate the energy transition. In World Economic Forum, BloombergNEF, & Deutsche Energie-Agentur (dena), W H I T E P a P E R [Report]. https://www3.weforum.org/docs/WEF_Harnessing_AI_to_accelerate_the_Energy_Transition_2021.pdf

Lilliestam, J., Eckardt, J., & Bersalli, G. (2025). Sequencing, spending, and symbolism: Low carbon taxes primarily serve purposes other than emissions reduction. *One Earth, 101390*. Advance online publication. DOI: 10.1016/j.oneear.2025.101390

Marengo, A. (2024). Navigating the nexus of AI and IoT: A comprehensive review of data analytics and privacy paradigms. *Internet of Things : Engineering Cyber Physical Human Systems*, *27*, 101318. DOI: 10.1016/j.iot.2024.101318

Mienye, I. D., Sun, Y., & Ileberi, E. (2024). Artificial intelligence and sustainable development in Africa: A comprehensive review. *Machine Learning with Applications, 100591*. Advance online publication. DOI: 10.1016/j.mlwa.2024.100591

Mohammadimanesh, F., Mahdianpari, M., Radman, A., Varon, D., Hemati, M., & Marjani, M. (2025). Advancements in satellite-based methane point source monitoring: A systematic review. *ISPRS Journal of Photogrammetry and Remote Sensing*, *224*, 94–112. DOI: 10.1016/j.isprsjprs.2025.03.020

Noori, N. (2021). Policy transplantation for smart city initiatives. In Erasmus University of Rotterdam [Thesis]. https://repub.eur.nl/pub/135367/thesis-draftnegarversion-0044rev-est.pdf

OECD. (2023). *AI in the climate and environmental domain: Policy opportunities and challenges*. OECD Publishing. https://www.oecd.org/en/topics/policy-issues/artificial-intelligence.html

Paneru, B., Mainalil, D. S., Paneru, B., & Sapkota, S. C. (2024). Advancing Sustainable Mobility: Dynamic predictive modeling of charging cycles in electric vehicles using machine learning techniques and predictive application development. *Systems and Soft Computing, 200157*. Advance online publication. DOI: 10.1016/j.sasc.2024.200157

Pourrahmani, H., Amiri, M. T., Madi, H., & Owusu, J. P. (2025). Revolutionizing carbon sequestration: Integrating IoT, AI, and blockchain technologies in the fight against climate change. *Energy Reports*, *13*, 5952–5967. DOI: 10.1016/j.egyr.2025.05.042

Shahverdi, N., Saffari, A., & Amiri, B. (2025). A systematic review of artificial intelligence and machine learning in energy sustainability: Research topics and trends. *Energy Reports*, *13*, 5551–5578. DOI: 10.1016/j.egyr.2025.05.021

Sheldon, T. L., & Dua, R. (2020). Effectiveness of China's plug-in electric vehicle subsidy. *Energy Economics*, *88*, 104773. DOI: 10.1016/j.eneco.2020.104773

Shirzadi, A., Salvati, A., Tahan, M. H., Shahabi, H., Nodoushan, E. J., Ramezani, M., Hashim, M., & Clague, J. J. (2025). Urban flood susceptibility mapping using deep and machine learning algorithms as a management tool: A case study of Sanandaj City, Iran. *Ecological Indicators, 178*, 113886. DOI: 10.1016/j.ecolind.2025.113886

Singh, A. R., Kumar, R. S., Bajaj, M., Khadse, C. B., & Zaitsev, I. (2024). Machine learning-based energy management and power forecasting in grid-connected microgrids with multiple distributed energy sources. *Scientific Reports, 14*(1). Advance online publication. DOI: 10.1038/s41598-024-70336-3 PMID: 39160194

Siripipatthanakul, S., Chaiprakarn, S., Rungsiyanont, S., & Phuangsuwan, P. (2025, a). A Case Study of Sustainable Development in Hospitals in Thailand Using ChatGPT. In *Sustainable Development for Hospitals and Clinics* (pp. 1-32). IGI Global Scientific Publishing. DOI: DOI: 10.4018/979-8-3693-9800-5.ch001

Siripipatthanakul, S., Phuangsuwan, P., Siripipattanakul, S., Limna, P., Jaipong, P., & Chaiprakarn, S. (2025, c). Privacy and Information Security for Sustainable Community Among TikTok's Users. In *International Conference on Artificial Intelligence and Networking* (pp. 221-233). Singapore: Springer Nature Singapore. https://doi.org/DOI: 10.1007/978-981-96-2015-9_15

Siripipatthanakul, S., Rakpathum, W., Siripipattanakul, S., Kaewnaknaew, C., Phuangsuwan, P., & Jaipong, P. (2025, b). The Future Artificial Intelligence (AI) for the Hotel and Hospitality Industry: The Case of Hotel's Human Resource Management (HRM). In *Best Practices for Human Resources Management in Tourism and Hospitality* (pp. 395-422). IGI Global Scientific Publishing. DOI: DOI: 10.4018/979-8-3693-8779-5.ch015

Sobczuk, S., & Borucka, A. (2024). Recent advances for the development of sustainable transport and their importance in case of global crises: A literature review. *Applied Sciences (Basel, Switzerland), 14*(22), 10653. DOI: 10.3390/app142210653

Son, T. H., Weedon, Z., Yigitcanlar, T., Sanchez, T., Corchado, J. M., & Mehmood, R. (2023). Algorithmic urban planning for smart and sustainable development: Systematic review of literature. *Sustainable Cities and Society, 94*, 104562. DOI: 10.1016/j.scs.2023.104562

Spandagos, C., Reaños, M. A. T., & Lÿnch, M. Á. (2023). Energy poverty prediction and effective targeting for just transitions with machine learning. *Energy Economics, 128*, 107131.

Su, Y., Ghaderi, H., & Dia, H. (2024). The role of traffic simulation in shaping effective and sustainable innovative urban delivery interventions. *EURO Journal on Transportation and Logistics, 13*, 100130. DOI: 10.1016/j.ejtl.2024.100130

Taloba, A. I., & Rayan, A. (2025). Machine learning based on reliable and sustainable electricity supply from renewable energy sources in the agriculture sector. *Journal of Radiation Research and Applied Sciences, 18*(1), 101282. DOI: 10.1016/j.jrras.2024.101282

Ucar, A., Karakose, M., & Kırımça, N. (2024). Artificial intelligence for Predictive maintenance Applications: Key components, trustworthiness, and future trends. *Applied Sciences (Basel, Switzerland), 14*(2), 898. DOI: 10.3390/app14020898

UN-Habitat. (2024). *World Cities Report 2024: Climate and Climate* https://unhabitat.org/sites/default/files/2024/11/wcr2024_-_full_report.pdf

UNDP. (2021). More than just cryptocurrencies - using blockchain for climate action in agriculture | UNDP Climate Change Adaptation. (2021, April 21). https://www.adaptation-undp.org/more-just-cryptocurrencies-using-blockchain-climate-action-agriculture

United Nations Framework Convention on Climate Change (UNFCCC). (2021). *Digital innovation in climate negotiations.* UNFCCC. https://sdgs.un.org/un-system-sdg-implementation/united-nations-framework-convention-climate-change-unfcc-44210#:~:text=The%20most%20recent%20COP%2026,the%20necessary%20finance%20for%20both

U.S. Department of Energy. Benes, K. J., Porterfield, J. E., Yang, C., Lawrence Livermore National Laboratory, Brownell, E., Helms, J., McGrath, J., Moughan, R., Ponce, C., Trinh, W., Wert, J., National Energy Technology Laboratory, Rose, K., National Renewable Energy Laboratory, Choi, S. L., Jain, R., Emami, P., Pacific Northwest National Laboratory, . . . Freyermuth, V. (2024). AI for Energy: Opportunities for a Modern Grid and Clean Energy Economy. https://www.energy.gov/sites/default/files/2024-04/AI%20EO%20Report%20Section%205.2g%28i%29_043024.pdf

Wang, Q., Li, Y., & Li, R. (2025). Integrating artificial intelligence in energy transition: A comprehensive review. *Energy Strategy Reviews, 57*, 101600. DOI: 10.1016/j.esr.2024.101600

Wang, X., Dong, X., Zhang, Z., & Wang, Y. (2024). Transportation carbon reduction technologies: A review of fundamentals, application, and performance. [English Edition]. *Journal of Traffic and Transportation Engineering*. Advance online publication. DOI: 10.1016/j.jtte.2024.11.001

WEF (World Economic Forum). (2024). AI and energy: Will AI help reduce emissions or increase power demand? Here's what to know. Energy Transition. World Economic Forum. https://www.weforum.org/stories/2024/07/generative-ai-energy-emissions/

Wei, Y., Li, Y., Chevallier, J., & Wojewodzki, M. (2024). The efficiency of China's carbon trading schemes: A tale of seven pilot markets. *Commodities*, *3*(3), 355–375. DOI: 10.3390/commodities3030020

World Bank. (2022). *Artificial intelligence for climate-smart investments*. World Bank. https://documents1.worldbank.org/curated/en/099455012062224218/pdf/P1725690cd27300c30b5f101fd728276fe7.pdf

Xu, Y., Loftness, V., & Severnini, E. (2021). Using machine learning to predict retrofit effects for a commercial building portfolio. *Energies*, *14*(14), 4334. DOI: 10.3390/en14144334

KEY TERMS AND DEFINITIONS

Artificial Intelligence (AI): Artificial Intelligence can be defined as computational systems that perform computations that normally involve human intelligence, including learning, reasoning, pattern recognition, and decision-making. AI is useful in climate governance by contributing to evidence-based policymaking, surveillance, and forecasting related to decarbonization plans.

Carbon Accounting: Carbon accounting is the process of systematically quantifying, disclosing, and confirming greenhouse gas emissions related to activities, organizations, or regions. Carbon accounting with AI will enhance accuracy, transparency, and real-time monitoring, thereby improving compliance with climate policy and the credibility of carbon markets.

Climate Governance: Climate governance refers to institutional structures, policy tools, and decision-making mechanisms used by governments and stakeholders to address climate change mitigation and adaptation. It includes regulatory systems, global collaboration, and multilateral governance systems that can be used to meet climate targets.

Machine Learning (ML): Machine learning is a subfield of AI that enables systems to learn on their own and react to data. ML is a common tool in climate policy, including emissions projections, policy modelling, policy design, and the optimization of renewable energy systems.

Net-Zero Economy: A net-zero economy is an economy where emissions of greenhouse gases are decreased to the lowest possible amount, and the rest of the emissions are offset by carbon removal or an offset program. To reach net-zero, the energy, transportation, and industrial systems, as well as urban systems, will have to transform systemically.

Smart Grids: Smart grids are digitally enabled electricity systems that use AI, IoT, and other advanced analytics to efficiently manage energy generation, distribution, and consumption. They help incorporate renewable energy, demand-side control, and real-time grid resilience to support the net-zero targets.

Urban Resilience: Urban resilience is the ability of urban areas to predict, absorb, adapt, and respond to climate-related disasters such as heat waves, floods, and power outages. AI improves urban resilience by enhancing predictive risk modeling, adaptive infrastructure modeling, and data-driven emergency response systems.

Chapter 7
Revisiting Environmental Security and Climate Justice:
A Human Security Perspective

Jipson Joseph
https://orcid.org/0009-0005-1795-5051
Christ University, India

Jobin Joseph
https://orcid.org/0009-0004-3228-2572
Independent Researcher, India

S. Srinivasan
https://orcid.org/0009-0002-0179-9849
Graphic Era University, Dehradun, India

ABSTRACT

A healthy and secure environment is essential to the survival of all living beings. Human interactions as individuals, social beings, and economic agents depend on the ecosystem. However, humans brutally exploited the ecosystem beyond its regenerating capacity, causing severe damage to environmental sustainability. This chapter employs a systematic review and thematic content analysis. The finding shows that humans' profit-oriented and egoistic interests caused unparalleled damage to nature, resulting in global warming, biodiversity loss, and anthropogenic climate change. Such a precarious situation compelled global leaders to prioritize environmental security primarily through the 1987 sustainable development paradigm. In this regard, the United Nations Development Program (UNDP) introduced the human security paradigm in 1994 as a broader framework, encompassing environmental

DOI: 10.4018/979-8-3373-6821-4.ch007

security as its core category. Both Millennium and Sustainable Development Goals echo the spirit of human security's people-centered approach. From this perspective, this chapter critically revisits environmental security and climate justice through the lens of the human security paradigm.

1. INTRODUCTION

Human survival largely depends on environmental security (Huo & Peng, 2023). From the dawn of civilization, humans have prioritized a harmonious lifestyle with the surrounding nature (Wada et al., 2017). However, the Industrial Revolution and the growth of the profit-oriented private sector compromised environmental concerns in favor of economic motives (Kabir et al., 2021). Realizing the human-generated complexities in the ecosystem, the world focused on sustainability in the second half of the 20th century. The Brundtland Report of 1987 further reinforced this principle by proposing the inspirational concept of sustainable development, which aimed to balance the developmental and environmental needs of both present and future generations (WCED, 1987). The declarations of the Millennium Development Goals (MDGs) and the Sustainable Development Goals (SDGs) in 2000 and 2015, respectively, broadened the scope of sustainable development and affirmed the importance of environmental protection for a global sustainable future. During this evolving process, the United Nations Development Program (UNDP) proposed the human security paradigm in 1994, which comprises seven categories, including environmental security, as a people-centered approach to development.

As stated above, international efforts to protect the environment are increasing. However, environmental degradation, pollution, and unpredictable climate change are causing serious concerns worldwide (Sultana, 2022). The global carbon footprint is closer to 4 tons. The world must reduce its emissions to 2 tons by 2050 to address rising global temperatures (The Nature Conservancy, 2025). There are also imbalances between developed and developing countries. All these necessitate that the globe prioritize environmental security to ensure climate justice for all people, regardless of socioeconomic, cultural, and regional differences (Sengupta & Sengupta, 2022; Stoett et al., 2024).

1.1 Significance of the Study

Environmental security remains a primary concern for the global population, despite the world being nearly four decades into the proposal of the concept of sustainable development (WCED, 1987). Global environmental security is currently unsatisfactory, as environmental degradation, anthropogenic climate change, and biodiversity loss are increasing significantly (Nguyen et al., 2023). Robinson and Igini (2025) outline several of the most significant environmental security issues of 2025: 2024 has been confirmed as the hottest year of human history, in which the global warming reached 1.60°C above pre-industrial levels; the concentration of greenhouse gas (GHG) has reached new heights in 2023; the continuing increase in dependency on fossil fuels; and almost 1.3 billion tons of food is wasted every year, causing unpredictable rise in GHG emissions. The issues, such as deforestation (Nor & Yusof, 2025), air pollution (Chen X.H. et al., 2023), plastic pollution (Stoett et al., 2024), ocean acidification (Allen & Bergin, 2009), melting of ice caps (Wang et al., 2025), and sea level rise (Das & Swain, 2024), further cause adverse insecurities to the environment. Environmental degradation and anthropogenic climate change are complementary (Chen X.M. et al., 2023; Kolawole & Iyiola, 2023). The global aspirations for climate justice will become irrelevant in the context of environmental insecurity (Porter et al., 2020). Both environmental degradation and climate change have become more complex with the advancement of Artificial Intelligence (AI). AI is like a double-edged sword, with benefits and drawbacks concerning environmental sustainability.

On the one hand, the world is widely promoting the benefits of environmental security and sustainable development for present and future generations. On the other hand, it fails to implement international policies and programs, such as the Paris Agreement and Carbon taxes, by influencing all global actors to prioritize environmental security over profit-centered motives and interests of industrialists, businesspeople, and others (King & van den Bergh, 2021). The trade-off between developed and developing countries concerning environmental security and climate-resilient activities is also a significant concern in contemporary times. Based on the sustainable development paradigm, the UN has proposed the MDGs and SDGs to address the issues of environmental insecurity and climate change. They have made significant progress in mitigating the drawbacks of pollution, deforestation, and the exploitation of natural resources. However, as mentioned above, the world continues to experience the threatening effects of climate change and environmental insecurity, demanding a more concentrated policy perspective.

In 1993, the UNDP introduced the concept of human security. In 1994, the UNDP expanded it to include seven categories, such as environmental security, to address various threats to human lives. Since it adopts the principles of universalism, sustainable human development, and early prevention strategies, it has the potential to systematically address various insecurities, such as environmental degradation and climate change. From this perspective, this chapter explores the possibility of adopting a human security lens to reexamine environmental security and climate change in the contemporary era.

2. LITERATURE REVIEW

Environmental security and climate justice gained momentum with the global acceptance of sustainable development (WCED, 1987). Various ecological movements and individual initiatives also contributed significantly to spreading environmental awareness among people. In response to the growing awareness of environmental protection, numerous international and national policies and programs have emerged over the past few decades, marking a progressive commitment to a sustainable future (Khagram et al., 2003). The UNDP's human security paradigm also belongs to this inspirational global action program. Both environmental security and climate justice ultimately aim to ensure human security.

2.1 Environmental Security

Environmental security has a broader scope, derived from two fundamental and broader concepts: environment and security. The environment refers to the biological, physical, and chemical elements and systems essential for sustaining life and ecosystems. Security refers to the required protection for referent objects (Barnett, 2009). The referent object varies depending on the circumstances, such as state, health, people, environment, and job. Conventionally, society has given primary place to state security, affirming the importance of national security. Security can also refer to various risks, such as territorial invasion, underdevelopment, unemployment, poverty, and environmental degradation. Security can also be concerned about survival and emergency matters (Trombetta, 2008). The survival of the ecosystem and the urgency of addressing various environmental challenges necessitate integrating environmental concerns within the security framework (Zurlini & Müller, 2008). When the referent object is the ecosystem/environment and the connected risks include environmental degradation, pollution, and depletion of natural resources, it can be referred to as environmental security. The notion of security must be expanded to include the growing insecurities emerging from environmental stress (WCED, 1987). Incorporating the environment into the security framework benefits global sustainability efforts, as it assigns additional responsibilities to governments and the people (Chen X.M. et al., 2023).

Although many thinkers, such as realists, argued that the security perspective cannot be attributed to environmental issues, as they are considered part of low politics (Trombetta, 2008), environmental security received global attention after the Cold War. Environmental sustainability and security are essential to achieving peace, national security, and human rights. No national security can be realized without promoting environmental protection (Zurlini & Müller, 2008). Environmental security emerged as a global discussion topic, inspiring traditional security analysts to focus on issues that truly matter for a sustainable global future (de Wilde, 2008). Environmental problems gradually became essential concerns on the political agenda (Trombetta, 2008), accompanied by a new set of values, including ecology, globality, and governance (Dyer, 2001). Global warming and ozone depletion were the primary concerns that prompted the world to shift its attention towards environmental protection. In 1993, Gorbachev affirmed in the UN General Assembly that the threat from the sky for the present generation is not missiles but global warming (Trombetta, 2008). Such a progressive approach shifted the discussions on security from the state borders to other vital concerns, such as the environment.

2.2 Climate Justice

Climate change has become a global problem, endangering the lives of millions of people and threatening biodiversity and ecosystem security (King & van den Bergh, 2021). Unpredictable and rapid climate change occurred due to the increased emission of GHGs, such as carbon dioxide (CO_2), methane (CH_4), and nitrogen monoxide (NO_2), released from various human activities, such as urbanization, industrialization, population explosion, transportation, fossil fuel consumption, and unscientific agricultural activities (Kabir et al., 2021). The specific focus on anthropogenic climate change started with the realization of a hole in the ozone layer. Many progressive thinkers and scientists affirmed the unimaginable drawbacks of climate change for ecosystems, human health, and the planet's survival. This led to the establishment of the Intergovernmental Panel on Climate Change (IPCC), which assesses and evaluates human activities contributing to climate change. Thereafter, the United Nations Framework Convention on Climate Change (UNFCCC) was signed in 1992 in Rio de Janeiro, marking the beginning of a systematic global approach towards climate change. However, the world slowly realized that discussions on climate change require a shift in present economic engagements and interactions, as well as people's way of life, which was not easy. This led to the marginalization of climate change discussion and the call for climate justice (Trombetta, 2008). However, the world began to refocus its activities on addressing climate change with the enhanced vision of SDG 13, which underscores the importance of taking urgent action against climate change and its impacts.

Climate change is occurring rapidly and raising alarm bells globally (Sultana, 2022). Since it is a global problem, it requires an international climate justice effort in which all national and international actors participate, intending to ensure environmental sustainability. The Paris Agreement (UNFCCC 2015, Art. 2(1a)) has provided the option for such a global effort with the inspirational formula that all international and national actors should strive for "holding the increase in the global average temperature to well below 2°C above pre-industrial levels and pursuing efforts to limit the temperature increase to 1.5°C above pre-industrial levels." Although many countries have ratified the Paris Agreement, with notable exceptions including Iran and Libya, the US withdrawal from the agreement has hindered its global implementation. When the leading economies, such as the US, fail to contribute effectively to global endeavors for climate justice, it hinders the fundamental objective of environmental sustainability (King & van den Bergh, 2021).

The global initiatives to tackle COVID-19 demonstrate that the world has the potential to address issues, such as climate change, systematically (Sultana, 2022). However, biased attitudes and motives of global actors are notable obstacles to achieving this objective. Previous attempts to address climate change have focused on scientific and technical aspects, whereas contemporary discussions emphasize climate justice (Lamb et al., 2020). It "helps to reframe mainstream debates to usher in critical attention to social impacts, outcomes, and justice concerns" (Sultana, 2022, p. 118). Climate justice discussions focus on presenting climate change as a moral and justice issue, rather than a restrictive scientific, technical, or managerial issue (Shue, 2014). It calls for focusing "how climate change impacts people differently, unevenly, and disproportionately, as well as redressing the resultant injustices in fair and equitable ways" (Sultana, 2022, p. 118). From this perspective, climate justice affirms that proportionate and fair treatment is needed to address the drawbacks of climate change.

2.3 Human Security Paradigm

Security revolves around the question, "What or who is to be secured?" If the central focus is on state boundaries, it is national security; if it focuses on individuals, it is human security. Besides, if security concerns focus solely on the survival of the state, it is national security, whereas if they focus on the survival and dignity of an individual, it is human security. Similarly, national security aims at securing freedom from fear. However, human security prioritizes freedom from fear, freedom from want, and freedom from indignity perspectives (Khagram et al., 2003). Threats to state security evolve from outside its boundaries. However, threats to human security are primarily due to domestic policy failures, human rights violations, and developmental issues.

The concept of human security evolved from the *Human Development Reports*. The UNDP (1993) introduced the human security paradigm within its developmental perspective, affirming that a comprehensive, inclusive, and people-centered approach will ensure justice in global efforts towards sustainability. It stated that a people-centered world order revolves around five pillars, with human security being the most significant. The UNDP (1994) explained that the other four pillars — sustainable human development, partnerships between states and markets, national and global governance patterns, and forms of international cooperation — are subsidiary to the broader human security framework. Underlying the importance of people-centered security, the UNDP (1993) argues that the world primarily requires well-designed policies, not soldiers in uniform. It thus shifted the focus of security discussions from territorial boundaries to people and their concerns, including environmental issues.

Figure 1. Categories of Human Security

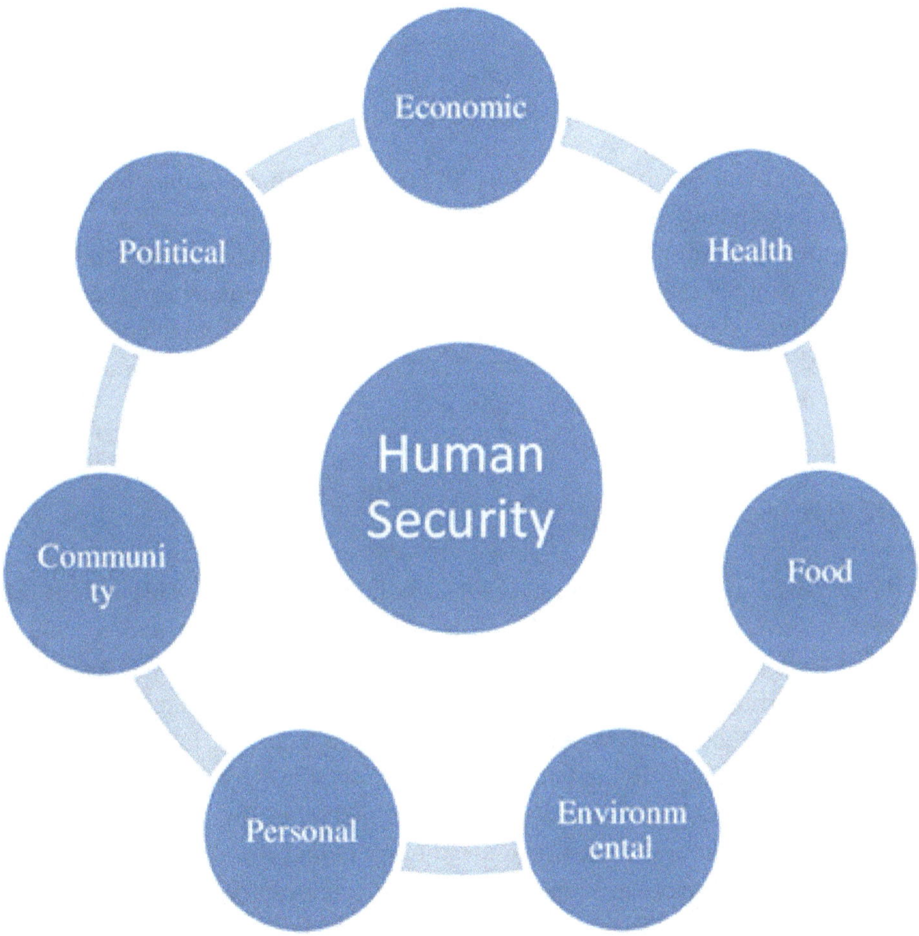

The concept of human security was expanded in the 1994 UNDP Report to encompass seven categories and many characteristics. It emerged as a significant policy concern, prioritizing individual well-being over national security. It does not undermine national security. However, it argues that states should prioritize national security during a warlike situation or armed rebellion that poses a threat to people's lives. When there is no such situation, but millions of people experience the drawbacks of underdevelopment, environmental degradation, climate change, health issues, and hunger, states must focus on ensuring people-centered development, rather than making huge investments in the military. Its ultimate objective is to address the severe and widespread threats that adversely affect people's lives (UNCHS, 2003). Since it is rooted in human rights, it has the potential to ensure better security for people across all vital areas (Popovski & Stanchev, 2021), including environmental and climate justice perspectives.

3. METHODOLOGY

According to Siripipatthanakul et al. (2025; a & b), the book chapter is suggested to employ a systematic review and thematic content analysis.

Thus, this chapter employs the analytical, exploratory, descriptive, and critical methods of the qualitative methodology to examine the contemporary scenario of environmental security and climate justice. Based on this critical analysis, it also aims to explore the feasibility of the human security policy perspective in addressing such insecurities.

4. RESULTS

Environmental security points to the responsible use of ecosystems and natural resources. Since the Industrial Revolution, the world has witnessed an unprecedented rise in pollution, GHG emissions, and the exploitation of natural resources. Industrialized countries primarily contribute to environmental issues that have plagued humanity throughout history (King & van den Bergh, 2021). Ecological issues complicate the health of living beings and the survival of the ecosystem. Humans are both primary causes and victims of environmental problems (Kabir et al., 2021). Ecological issues cause insecurities among human beings that cannot be perceived through the lens of national security. It is significantly diverse, as its immediate focus is on individual well-being and societal welfare. For instance, water scarcity may not immediately pose a threat to national security. However, it can cause dehydration and endanger human lives (Khagram et al., 2003).

The impact of environmental insecurity on human lives is multifaceted, encompassing health, food, economic, personal, community, and political security. It can also affect diverse subjects, ranging from individuals to the international community. While many environmental threats are context-specific (localized), many are beyond specific boundaries (Khagram et al., 2003). The UN enlisted the contemporary environmental issues and insecurities in its 2015 declaration of the SDGs: "Natural resource depletion and adverse impacts of environmental degradation, including desertification, drought, land degradation, freshwater scarcity, and loss of biodiversity, add to and exacerbate the list of challenges which humanity faces" (UNGA, 2015, p. 5). It affirms that humans are primarily at risk due to the uncontrolled exploitation of natural resources and degradation of environmental quality.

4.1 Pollution

Pollution has expanded its impact across various fields of human interaction, including water, air, and soil. These pollutions are caused by burning fossil fuels, industrial emissions, unsystematic waste disposal, and the increased use of plastic. Plastic pollution has become increasingly conspicuous in recent years, with its prevalence in water, landfills, and even human bodies posing serious security issues (Stoett et al., 2024). Gautam et al. (2024) thoroughly examine plastic pollution and its potential adverse impacts on ecosystems and human health. The World Bank (2025) notes that pollution is a significant barrier to economic development, as it contaminates natural resources, damages ecosystems, deepens poverty and inequality, and contributes to anthropogenic climate change. The most vulnerable sections of society experience the worst burden of environmental pollution. It also states that almost 2.3 billion people are exposed to outdoor levels of fine particulate matter (PM2.5), resulting in approximately 5.7 million deaths annually and an economic loss of nearly 5% of global GDP.

4.2 Carbon Footprint

Discussions of carbon footprint originated with the idea of ecological footprint in the early 1990s. In the first decade of the 21st century, British Petroleum popularized the concept of individual carbon footprint. A carbon footprint is the total amount of GHGs, including CO_2 and CH_4, that are emitted to the atmosphere by human activities. Since the carbon footprint can lead to numerous insecurities for people and the ecosystem, such as climate change, measuring GHG emissions is essential to address this issue. The carbon footprint is an effective tool for scientifically and systematically measuring carbon emissions (Shi & Yin, 2021). Primarily, CO_2 emissions result from the use of fossil fuels. Reducing the dependency on fossil fuels is a prerequisite for environmental sustainability.

The world's leading CO_2-emitting countries are China, the USA, India, Russia, and Japan. China released approximately 11.9 metric tons of CO_2 in 2023. However, the USA has a high per capita emission rate, about 16 tons per person (Blokhin, 2025). The USA is also responsible for 20% of the global historical CO_2 emissions

(Evans, 2021). The global average carbon footprint is approximately 4 tons (The Nature Conservancy, 2025). When the carbon footprint increases, it contributes to global warming and unpredictable climate change (Evans, 2021). The world must strive to reduce the average global carbon footprint to at least 2 tons by 2050 to avoid a 2°C rise in global temperature (The Nature Conservancy, 2025).

4.3 Biodiversity Loss

Biodiversity "includes diversity within species, between species and across ecosystems, representing the genetic makeup of plants, animals, microorganisms, and the complexity of ecosystems" (WHO, 2025). Biodiversity loss is another conspicuous environmental concern, originating from the imprudent and unsystematic use of ecosystems. The preservation of biodiversity is an essential component for environmental sustainability and human security. Forest preservation, biodiversity sustainability, and aquatic conservation are integral to promoting environmental protection. The profit-oriented agrarian and industrial activities result in the inconceivable exploitation of natural resources and the direct or indirect destruction of many species, leading to biodiversity loss in both land and sea (Das & Swain, 2024). Rising sea levels can cause insecurities for marine beings. The pollutants deposited in the sea can harm fish, other aquatic species, and the coastal ecosystem, leading to declines in fisheries, biodiversity loss, and the disruption of the ecosystem's balance (Burns, 2023). It is estimated that forests absorb around 2.6 billion tons of CO_2 annually. The increased absorption of CO_2 is crucial for regulating climate change and mitigating the effects of pollution. However, the world is witnessing an unparalleled increase in deforestation (Maza-Maza et al., 2025). It causes biodiversity loss at an accelerating rate. Almost 1 million species are on the verge of extinction, causing serious concerns for environmental sustainability (WHO, 2025).

4.4 Depletion of Natural Resources

Natural resources are vital for the survival and well-being of all living beings. The environment provides clean air, water, energy, soil, and other resources crucial for sustainable development. However, the natural resources are being depleted at an alarming rate. The use and exploitation of natural resources are occurring at a rate that exceeds their regenerative capacity. Population growth, climate change, and unsustainable human activities are the primary reasons for the increased use of natural resources (Huo & Peng, 2023). Scientific and technological advancements increased life expectancy, improved quality of life, and increased energy needs (Liu et al., 2022). All these factors significantly contribute to the depletion of natural resources and pose a challenge to environmental security. The unsustainable use and exploitation of natural resources, such as forests and minerals, have led to pollution, biodiversity loss, soil erosion, and climate change (Tawiah et al., 2021). It becomes more complicated within the context of human activities, such as deforestation, mining, and industrial agriculture, which largely pollute and contaminate air, water, and soil. The population explosion is the biggest challenge to natural resources. As the global population increases, so do its demands, resulting in severe pressure on the environment. The ongoing and increasing extraction of natural resources is causing alarming environmental insecurity. Furthermore, the increased population will lead to overcrowding and poor sanitation facilities, which will also adversely affect water, air, and soil. When these are severely compromised through the unsystematic exploitation of natural resources and human activities that pollute the environment, it poses health risks to humans and challenges the survival of other living beings (Huo & Peng, 2023).

4.5 Climate Change

Climate change fosters meaningful human interactions, enhances agricultural productivity, and promotes socioeconomic growth. However, anthropogenic climate change is "a pervasive existential civilizational challenge" (Morgan & Patomäki, 2021). Several human activities, such as the emission of large quantities of carbon and other GHGs, contribute to environmental degradation and unpredictable climate change. Ulrich Beck's sociological theory, "Risk Society," which is closely connected to climate change, affirms that "modern societies are characterized by a shift from industrial to post-industrial forms of organization." This shift brings forth many benefits and risks to society. Environmental risks are the most notable among them, which ultimately lead to anthropogenic climate change (Das & Swain, 2024, p. 1010). That is, in addition to contributing to economic growth, industrial activities cause environmental harm. When a society transitions from an industrial to a post-industrial status, it carries forward the drawbacks of industrialization, including pollution, waste accumulation, and emissions, which adversely affect people, society, and the ecosystem.

Climate change causes an increase in sea level, which will persist for centuries, even if the world stabilizes its GHG emissions (Church et al., 2013). The latest studies suggest that the sea level will rise and reach 1m or more by 2100 (Lindsey, 2023). Human activities have adversely affected the movement of water on land. They increasingly use water for homes, farms, and industries, decreasing freshwater sources and contributing to rising sea levels (Wada et al., 2017). Many coastal areas have faced permanent flooding and increased flood threats over the last few decades. Rising sea levels will place unimaginable pressure on coastal areas, affecting the economic aspects of these regions and the people living near them, and harming marine resources (Das & Swain, 2024). They also pollute underground freshwater sources, causing water scarcity and insecurity for people living in towns and cities, and harming the natural environment (Lindsey, 2023).

5 DISCUSSIONS

This section discusses the current state of environmental insecurity and climate injustice through the lens of various global efforts with a specific focus on the human security paradigm.

5.1 Global Efforts for Environmental Security: A Critique

Environmental issues cannot be considered from a restrictive perspective. They have both domestic and international impact. In other words, ecological problems know no boundaries. They occur globally and locally in various ways and to different degrees, underscoring the depth and breadth of the problem. They have become prominent economic, political, social, and cultural problems that challenge sustainability due to their adverse impacts, including decreasing forest cover, biodiversity loss, water pollution and scarcity, land contamination, sea-level rise, and climate change (Barnett, 2009). Environmental sustainability and security can be achieved by focusing on immediate challenges and long-term goals.

The political leadership of almost all countries tends to overlook environmental issues in favor of industrialization and economic growth. Since most international institutions and national governments evaluate a country's well-being based on its GDP growth rate, state governments prioritize infrastructure development and other measures to promote economic growth, regardless of their environmental impact. Many governments believe they become safer by adopting a neutral stance in international affairs regarding environmental security and by specifically promoting measures to promote economic development within their own boundaries (Khagram et al., 2003). Although sustainable development is possible, most countries fail to follow it due to its increased financial burden. This necessitates international cooperation, in which developed countries progressively support the positive and systematic initiatives of developing and underdeveloped countries. However, it rarely takes place. Even when they are willing to help, numerous conditions and regulations accompany them, burdening developing countries.

The global efforts for environmental protection originated from the increased environmental consciousness among various actors. The last four decades of the 20th century have significantly developed such a progressive vision (Barnett, 2009). The WCED's framing of the concept of sustainable development marked a historic moment in this process. Based on this, the UN established the MDGs and SDGs to continue the spirit of sustainable development and ensure a global sustainable future. They have made a crucial impact on environmental protection and promoted an eco-friendly culture. However, the desired objective has not yet been fully achieved.

5.1.1 Sustainable Development

There are no military solutions to environmental issues (WCED, 1987). Instead, the world requires a developmental paradigm that does not compromise environmental concerns. The concept of sustainable development has been foundational in establishing the increased interconnection between nature and society (Khagram et al., 2003). Although prioritization of ecological security at the global level began with the 1972 Stockholm Conference, the sustainable development paradigm gained acceptance with the publication of the Brundtland Report in 1987. The 1992 United Nations Conference on Environment and Development strongly endorsed the idea of sustainable development. It has become a global inspirational objective for simultaneously promoting environmental security and economic development. Sustainable development asserts that human well-being can be achieved through a well-focused interaction with the surrounding environment, utilizing appropriate institutions, organizations, and technologies (Zurlini & Müller, 2008).

Sustainable development affirms that humans should be able to ensure that it meets the needs of the present without compromising the ability of future generations to meet their own needs (WCED, 1987, p. 8). It demands that the developmental and environmental needs of the present generation should not be at the expense of future generations' developmental and environmental needs. It underlines the importance of intragenerational and intergenerational equity, fairness, and justice. Accordingly, "the genius of 'sustainability' lies in its ability to provide 'space' for serious attempts to grapple with the real, dynamic, and complex relationship among societies, economies, and natural environments, as well as between past, present, and the future" (Khagram et al., 2003, p. 296). Here, the question arises: what is to be sustained? Three basic elements must be sustained: society (comprising individuals and communities), the economy, and the environment. Based on these, the three pillars of sustainability—social, economic, and environmental—emerged (Elkington, 1999). Although sustainable development did not prioritize any of these three pillars, environmental sustainability is more equated with sustainable development, reducing its scope (Kingsbury et al., 2004).

5.1.2 MDGs

The world entered the third millennium with the declaration of the MDGs to create a shared future for all. The MDGs, *inter alia*, consider respect for nature as one of the central values that should guide international relations in the 21st century. It advances eight goals with 21 targets to be achieved by 2015. The seventh goal focuses on environmental sustainability, which has four targets. The first target addresses deforestation and global CO2 emissions. The second target addresses biodiversity loss and aims to reduce the percentage of loss by 2010. The third target seeks to halve the percentage of the population living without access to safe drinking water and basic sanitation. The fourth target aims to improve the lives of 100 million slum dwellers globally by 2020.

Since the number of goals was limited, the vision was restrictive regarding environmental security and climate change. The most significant limitation of the MDGs report on achieving goals is that it begins measuring success in 1990. The UN (2015) states that ozone-depleting substances have been virtually eliminated since 1990, that terrestrial and marine protected areas have increased significantly since 1990, that 1.9 billion people have gained access to safe drinking water since 1990, and that the proportion of people practicing open defecation has been substantially reduced since 1990. How can a policy perspective, which started to function in 2000, be evaluated based on the success rate of its previous decade? The UNGA (2015) claims that, through the SDGs, the world aims to achieve what the MDGs could not. It also affirms that the MDGs' desired objectives were not fully realized.

5.1.3 SDGs

The United Nations General Assembly (UNGA) (2015) adopted the 17 SDGs, with 169 targets, as part of its 2030 Agenda for sustainable development to transform the world. The global leaders affirmed that they will ensure "no one will be left behind" in this remarkable journey toward a sustainable future, striking a balance among the social, economic, and environmental dimensions of sustainable development. These goals aim to ensure a healthy environment by protecting the planet from degradation, sustainably managing resources, and taking prompt action against climate change. The SDGs are integrated, indivisible, and universal in nature. Compared to the MDGs, the SDGs give more attention to environmental sustainability and place a special focus on climate change. Omisore (2018) states that, from just one goal assigned to environmental security in the MDGs, almost all goals in the SDGs are somehow connected to environmental sustainability.

SDG 6, with eight targets, focuses on ensuring clean water and sanitation facilities for all. SDG 7, with five targets, advocates for increased use of renewable energy sources to reduce the burden on ecosystems. SDG 11, with ten targets, focuses on developing sustainable cities, transport systems, and human settlement planning to mitigate adverse impacts on cities. SDG 12, with 11 targets, affirms the importance of responsible consumption and production patterns for a sustainable society. SDG 13, with five targets, focuses explicitly on climate change. It acknowledges the role of UNFCCC in ensuring this objective. SDG 14, with 10 targets, affirms the importance of safeguarding life below water. SDG 15, with 12 targets, focuses on life on land. It particularly addresses biodiversity loss.

The SDGs are criticized for their vague language and escape clauses (Menton et al., 2020) and do not obligate members to specific actions (Easterly, 2015). Evaluating the environmental sustainability of consumption and production patterns, Hickel (2020) notes significant contradictions between the SDGs and the right to a healthy environment in practice. Although there are numerous rulings from the human rights tribunal regarding the states' obligation to protect the right to a healthy environment, the language of the SDGs does not adequately ensure these responsibilities. It is thus criticized as lacking a specific justice perspective (Menton et al., 2020). Lele (2017) states that SDG 6 focuses on availability and fails to guarantee intragenerational considerations. SDG 7 is criticized for focusing solely on the material domain without considering social aspects (Menton et al., 2020). SDG 11 often overlooks the inequitable distribution of the benefits and risks of urbanization (Sengupta & Sengupta, 2022). Although the circular economy is an efficient tool for achieving SDG 12, Tsai et al. (2025) note that most countries have not adequately implemented it, and there is no effective international mechanism to ensure its implementation. SDG 13 does not explicitly warn against fossil-fuel extraction and production (Menton et al., 2020). Achieving SDG 14 has been identified as one of the most challenging goals globally (Salvia et al., 2019). Krauss (2022) affirms that SDG 15 cannot meet the expectations of protecting the planet and its people, as no effective mechanism exists to identify and ensure biodiversity conservation and promotion.

5.2 Human Security Perspective: Core Components

Environmental issues pose significant threats to people's lives in various ways, affecting their livelihoods, health, resources, economy, and community life. However, they have a more adverse impact on vulnerable populations. Low-income individuals and communities are disproportionately affected by environmental issues. It does not mean that people in high-income societies are immune to environmental threats. Although all are affected by ecological problems, the poor and the marginalized are more vulnerable to these threats because they depend on nature for their livelihood. They live in ecologically sensitive areas due to affordability issues, fail to insure their homes and properties because of limited financial capacity, and often fail to receive proper compensation for their losses due to complex procedures. All these affect their security in environmental issues (Barnett, 2009). Environmental insecurity should be addressed through a people-centered perspective when it becomes a significant issue for people's peaceful living.

There is a close interconnection between environmental security and human security (UNDP, 1994). From the environmental security dimension, human security focuses on people's access to natural resources and their vulnerability to environmental degradation and climate change. It examines the root causes—the irresponsible human activities—that cause environmental issues and climate change. On the contrary, environmental threats adversely challenge human survival, individual well-being, and productivity. The human security perspective argues that human beings, their social relationships, and the ecosystem should be the primary objectives of environmental security policies and programs (Khagram et al., 2003). From this perspective, it attaches a human dimension to sustainable development, coining the term 'sustainable human development' (UNDP, 1994).

Khagram et al. (2003) argue that, like human development, human security focuses on the social dimension of sustainable development. However, human security has a broader framework of action. It also addresses the environmental and economic sustainability of the sustainable development objective, drawing on these dimensions. The human security approach emphasizes the importance of advancing a people-centered development model without undermining the potential dangers, such as environmental degradation, that people typically face from development projects.

5.2.1 People-Centered Approach

Sustainable development is less anthropocentric, prioritizing environmental security over economic growth. The most crucial concern in the sustainable development perspective was its focus on satisfying basic primary needs. The human security approach interlinks sustainable development with the human rights framework, affirming a people-centered perspective on development and environmental protection. It does not undermine the scope of sustainable development, but confirms that development and environmental security should become more humane (Khagram et al., 2003). It concerns "how people live and breathe in a society, how freely they exercise their many choices" (UNDP, 1994, p.23). People's living and breathing in a society depend on the appropriate circumstances and a secure environment. Without prioritizing the state's interests, it stands for individual well-being and sustainability (Adger, 2010).

5.2.2 Universalism of Life Claims

The universalism of life claims is rooted in the foundation of human rights (UNSG, 2010). It considers both human development and sustainability as its essential components. It addresses the insecurities and threats, such as poverty, inequality, and pollution, that people everywhere face (UNDP, 2022). The UNDP (1994, p. 13) states that the "universalism of life claims is the common thread that binds the demands of human development with the exigencies of development tomorrow, especially with the need for environmental preservation and regeneration for the future." It underscores that the present world has an ethical responsibility to maintain the ecosystem for future generations in the same way previous generations enjoyed it. Against the differentiating parameters, it underlines and promotes the life claims of all. Accordingly, it calls for intragenerational and intergenerational equity, peace, and justice.

5.2.3 Early Prevention Method

The human security perspective of environmental security affirms that the world does not need to address the threats that endanger the ecosystem's survival; instead, it should focus on adopting opportunities and possibilities that perfectly suit ensuring environmental sustainability (Khagram et al., 2003). The world was prioritizing a post-facto redressal mechanism to confront various ecological threats. In the event of sudden and unexpected natural disasters, a post-facto functioning system may be the only solution. However, concerning other crucial environmental insecurities, which are mostly anthropogenic, the world must focus on early prevention strategies (UNDP, 1994). It cannot simply wait for the insecurities to happen. Once the environment is harmed through anthropogenic activities, such as pollution, its effects persist for a long time, causing further insecurity to humans. The best method is to take necessary precautions to prevent humans from harming the environment.

Protection and enhancement are the most suitable early prevention methods concerning environmental security. The world is moving towards a specific culture where a secure environment will be viewed as a luxury. A safe environment is a crucial prerequisite for the well-being of all individuals. However, it has become unaffordable for many vulnerable people, as environmental insecurity mainly affects them. Numerous examples demonstrate that environmental protection and enhancement measures can enhance people's well-being, protect their fundamental rights, and foster socioeconomic growth (Khagram et al., 2003).

5.2.4 An Inclusive Approach

Human security is an inclusive policy approach that covers nearly all vital areas of human interaction, organized into seven categories and three freedoms (UNDP, 1994). Since threats to humans and the ecosystem are interconnected, encompassing global complexities and contemporary challenges, the world requires an interconnected and inclusive approach (UNSG, 2010). Human security affirms that promoting human dignity and environmental sustainability is achievable through an inclusive framework of protection and empowerment (UNDP, 1994) that utilizes both a bottom-up and a top-down approach.

5.3 Environmental Insecurity and Climate Change: A Human Security Response

Environmental degradation and climate change are dominating contemporary discussions on sustainability. As stated above, such a concentrated approach began only in the last quarter of the 20th century. While environmental security efforts began with the Stockholm Conference in 1972, international attempts to address climate change started in 1992 with the adoption of the UNFCCC. A significant effort to reduce carbon emissions and address climate change occurred in 1997 with the Kyoto Protocol, which came into effect in 2005. However, the exclusion of developing countries and the complexities concerning the understanding of carbon leakage undermined the scope and relevance of the Kyoto Protocol. Furthermore, countries such as the US and Canada have not incorporated the Kyoto Protocol into their domestic policies (King & van den Bergh, 2021). Since developing countries were not part of the Kyoto Protocol, it cannot be considered a global agreement on climate change. Accordingly, the first international agreement on climate change is the Paris Agreement of 2015 (Aldy et al., 2016). However, the lack of policy harmonization across countries and binding enforcement mechanisms obscures the value of climate justice (King & van den Bergh, 2021).

International institutions and organizations, such as the UN, UNDP, and World Bank, can only provide a basic global policy framework. It becomes effective only when states frame their domestic policies in accordance with the international guidelines. It affirms that a shift in domestic policy perspective, aligned with the international policy framework, can have far-reaching effects on environmental security and climate justice (King & van den Bergh, 2021). Both national and international policies should complement each other with an enhanced objective of sustainability.

5.3.1 A Pro-People Climate Justice

Anthropogenic risks to the natural environment are crucial concerns of human security. Most discussions on environmental security prioritize securing human beings from the side effects of environmental degradation. However, the climate justice perspective is about more than protecting people from environmental threats. It consists of all practical measures and policies that are beneficial to the whole world, including humans, other beings, and the ecosystem, by respecting the inherent value of the environment, prudently using the resources of nature, systematically utilizing the opportunities provided by the surrounding environment, and ensuring a deep connection between environment, society, and economy (Khagram et al., 2003). A pro-people climate justice perspective emphasizes the role of people in ensuring environmental security. If a state government does not prioritize environmental security in its policies and programs due to biased motives and intentions, individuals, NGOs, and communities should advance environmental protection programs (Jancar, 1993). History testifies to numerous instances where individuals and small groups have fought systematically for environmental and climate justice. Greta Thunberg, Elizabeth Wathuti, Sumak Gualinga, and Nita Ganguly are among the contemporary climate justice activists whose endeavors are welcomed by many policymakers. The Climate Action Network (CAN), Fridays for Future, Seed Mob, Rural Revive, and Narmada Bachao Andolan are notable movements that advocate for environmental protection and climate justice. Such increased involvement of individuals, movements, international organizations, and NGOs underscores the importance of global cooperation and collaboration (Khagram et al., 2003).

5.3.2 Justice for Present and Future Generations

Based on the progressive vision of sustainable development, the world should balance the developmental and environmental concerns of present and future generations (WCED, 1987). Justice implies fairness in words and actions. If the world disregards the concerns of a particular generation in pursuing the interests of other generations, it constitutes a specific violation of the principles of justice. When the present generation's activities lead to increased environmental degradation, it adversely affects future generations' fundamental right to a sustainable environment (UNDP, 1994). Similarly, the profit-oriented and unsustainable activities of specific countries and peoples increase the global carbon footprint, ozone depletion, and ecosystem contamination, denying the right to a healthy environment to millions (UNDP, 2022). The UNDP (1994) thus calls for both intragenerational and intergenerational justice and fairness in the pursuit of environmental security.

5.3.3 Justice for Marginalized Sections of Society

Climate justice should prioritize the intersection between climate change and social inequalities. Climate change is most unfairly experienced by the marginalized sections of society, who are the prime victims of structural violence (Porter et al., 2020). Mary Wollstonecraft (1792) stated that the world primarily requires justice, not charity (UNDP, 1994). Since the marginalized are more vulnerable to environmental insecurity, the rest of society should serve as ambassadors of justice and fairness towards them. UNGA (2015) clearly stated that the inspirational agenda for the 2030 world revolves around the principle of leaving no one behind. It becomes substantial only within the context of fair treatment towards the most vulnerable sections of society. "Global sustainability without global justice will always remain an elusive goal" (UNDP, 1994, p. 21). Given the drawbacks of pollution, the exploitation of natural resources, deforestation, urbanization, and ocean acidification, which disproportionately affect the lives of historically marginalized sections of society (UNDP, 2022), the world cannot ensure a sustainable future without a genuine march for justice.

5.3.4 Responsibility of the Industrialized Countries

Since it is explicitly affirmed that industrialized and developed countries are the largest producers of emissions and pollutants that contribute to climate change (Porter et al., 2020), they also bear the most significant responsibility to support the environmental security mission of developing countries, which are more vulnerable to the impacts of climate change (UNDP, 1994). The Environmental Kuznets Curve (EKC) indicates that pollution levels rise as countries undergo industrialization and economic growth, and reach their peak when they have achieved the highest levels of industrialization. However, pollution levels began to decline at the time of higher development levels. Based on the EKC, developed countries have begun to enter the decline stage of pollution. In contrast, developing countries are experiencing an increase in pollution as they move towards higher levels of industrialization (Chen X.M. et al., 2023). Industrialized countries have already reached a good level of economic development. However, developing countries are working to ensure human security for their citizens through development projects and policies. Although many prefer to follow the sustainable development paradigm, the associated financial burden often deters them. For environmental protection, they cannot sacrifice developmental programs. Similarly, if they do not adequately consider environmental protection measures alongside their development policies, it will cause numerous harms worldwide. This necessitates that developed countries significantly contribute to and support the sustainable development projects of developing countries (Aldy et al., 2016).

6 CONCLUSIONS

Environmental security and climate justice are essential for human well-being and the survival of ecosystems. Humans were eco-friendly beings. They protected and respected the surrounding nature, which is explicitly affirmed by all traditional religions and the cultural capital of almost all countries. However, the eco-friendly culture is sidelined by individuals' profit-oriented motives. The population explosion, industrialization, urbanization, scientific and technological advancements, and increased demands for natural resources have forced humans to exploit the ecosystem beyond its regenerative capacity. Ecological movements and human rights activists emphasized the importance of environmental security and the right to a healthy environment in the second half of the 20th century. This led to the formulation of the sustainable development paradigm in the Brundtland Report of 1987, which sought to balance developmental and environmental concerns for present and future generations. Various progressive international and national mechanisms emerged, such as the IPCC, UNFCCC, MDGs, and SDGs, promoting environmental security.

The UNDP's human security paradigm encompasses a broader framework that incorporates all vital concerns related to human-environment interaction across seven categories. It also prioritizes a people-centered developmental model, rooted in the universalism of life claims. Its early prevention strategies and empowerment concerns can potentially address environmental issues and climate change more significantly than other policies. The biased activities and unfair distribution of benefits and burdens have done enough damage to the ecosystem and people. From a people-centered perspective, the world should prioritize justice, fairness, and equity to advance people's security and the ecosystem. A pro-nature visionary approach should be developed to ensure environmental security and climate justice from a human security perspective.

7 LIMITATIONS AND RECOMMENDATIONS

The chapter has examined environmental insecurity and climate change from a global perspective, without specifically concentrating on a particular country or region. Country-specific studies or comparative studies across countries can be employed in future research to enhance global understanding of environmental security and climate justice. The study also did not focus on collecting primary data, as it would be more feasible for country-specific studies.

Environmental security and climate justice are global demands. All countries should equitably contribute to the promotion of environmental justice. However, developed countries are responsible for taking the lead, as they possess the power, wealth, technology, and resources to significantly influence other countries (Aldy et al., 2016). When the US government decided to withdraw from the Paris Agreement, it harmed global efforts for climate justice. This is because when the US becomes a "rogue state" (Stiglitz, 2017), i.e., a state that "actively blocks, stalls, or otherwise subverts an international process designed in the interests of all" (Broadhead, 2001, p. 462), it influences other similar countries to withdraw from international efforts for the realization of global well-being. When the world's primary polluter does not follow the guidelines of an International Agreement to protect its biased interests, it challenges the value and efficiency of international mechanisms. The world's countries, especially developed ones, should promote the objectives behind a common purpose (Broadhead, 2019).

Global leaders should consider revitalizing Gorbachev's proposal for a UN environmental security council, which could coordinate and implement various environmental security programs and policies (Trombetta, 2008). Just as the UN Security Council addresses many interstate conflicts, violations of peace and global order, and human rights issues, the UN Environmental Security Council should also have the power to take effective action in light of the Paris Agreement and other IPCC and UNFCCC conventions and agreements. However, its membership should be broader, encompassing representatives from both developed and developing countries and from all continents.

8 ETHICAL DECLARATION STATEMENT

The author has not used any AI tools, except the Grammarly Software for grammatical and typographical corrections, to write this chapter.

REFERENCES

Adger, W. N. (2010). Climate Change, Human Well-Being, and Insecurity. *New Political Economy*, *15*(2), 275–292. DOI: 10.1080/13563460903290912

Aldy, J., Pizer, W., Tavoni, M., Reis, L. A., Akimoto, J., Blanford, G., Carraro, C., Clarke, L. E., Edmonds, J., Iyer, G. C., & McJeon, H. C. (2016). Economic Tools to Promote Transparency and Comparability in the Paris Agreement. *Nature Climate Change*, *6*(11), 1000–1004. https://www.nature.com/articles/nclimate3106

Allen, R., & Bergin, A. (2009). Ocean Acidification: An Emerging Australian Environmental Security Challenge. *Australian Journal of Maritime and Ocean Affairs*, *1*(2), 49–56. DOI: 10.1080/18366503.2009.10815637

Barnett, J. (2009). Environmental Security. In Kitchin, R., & Thrift, N. (Eds.), *International Encyclopedia of Human Geography* (pp. 553–557). Elsevier., DOI: 10.1016/B978-008044910-4.00774-4

Blokhin, A. (2025, February 3). *The Five Countries that Produce the Most Carbon Dioxide (CO_2)*. Investopedia. https://www.investopedia.com/articles/investing/092915/5-countries-produce-most-carbon-dioxide-co2.asp#

Broadhead, L.-A. (2001). Canada as a Rogue State: Its Shameful Performance on Climate Change. *International Journal (Toronto, Ont.)*, *56*(3), 261–280. DOI: 10.1177/002070200105600305

Broadhead, L.-A. (2019). Environmental Insecurity: Another Case for Concept Change. *Peace and Conflict Studies*, *26*(1), 1. https://nsuworks.nova.edu/pcs/vol26/iss1/1

Burns, R. (2023, June 9). *Rising Sea Levels: Causes, Consequences, and Impact on India*. Give.do. https://give.do/blog/rising-sea-levels-causes-consequences-and-impact-on-india/#

Chen, X. H., Tee, K., Elnahass, M., & Ahmed, R. (2023). Assessing the Environmental Impacts of Renewable Energy Sources: A Case Study on Air Pollution and Carbon Emissions in China. *Journal of Environmental Management*, *345*, 118525. DOI: 10.1016/j.jenvman.2023.118525

Chen, X. M., Sharma, A., & Liu, H. (2023). The Impact of Climate Change on Environmental Sustainability and Human Mortality. *Environments, 10*(10), 165. https://www.mdpi.com/2076-3298/10/10/165

Church, J. A., Monselesan, D., Gregory, J. M., & Marzeion, B. (2013). Evaluating the Ability of Process-Based Models to Project Sea-Level Change. *Environmental Research Letters, 8*(1), 014051. https://iopscience.iop.org/article/10.1088/1748-9326/8/1/014051

Das, A., & Swain, P. K. (2024). Navigating the Sea Level Rise: Exploring the Interplay of Climate Change, Sea Level Rise, and Coastal Communities in India. *Environmental Monitoring and Assessment, 196*, 1010. https://link.springer.com/article/10.1007/s10661-024-13191-z

De Wilde, J. (2008). Environmental Security Deconstructed. In Brauch, H. G. (Ed.), *Globalization and Environmental Challenges: Reconceptualizing Security in the 21st Century* (pp. 595–602). Springer.

Dyer, H. (2001). Theoretical Aspects of Environmental Security. In Petzold-Bradley, E., Carius, A., & Vincze, A. (Eds.), *Responding to Environmental Conflicts: Implications for Theory and Practice* (pp. 67–81). Kluwer.

Easterly, W. (2015). The Trouble with the Sustainable Development Goals. *Current History (New York, N.Y.), 114*(775), 322–324. DOI: 10.1525/curh.2015.114.775.322

Elkington, J. (1999). *Cannibals with Forks: The Triple Bottom Line of 21st Century Business*. Oxford University Press.

Evans, S. (2021, October 5). *Analysis: Which Countries are Historically Responsible for Climate Change?* CarbonBrief. https://www.carbonbrief.org/analysis-which-countries-are-historically-responsible-for-climate-change/#

Gautam, B. P. S., Qureshi, A., Gwasikoti, A., Gondwal, M., Kumar, V., & Gondwal, M. (2024). Global Scenario of Plastic Production, Consumption, and Waste Generation and Their Impacts on Environment and Human Health. In Soni, R., Debbarma, P., Suyal, D. C., & Goel, R. (Eds.), *Advanced Strategies for Biodegradation of Plastic Polymers* (pp. 1–34). Springer., https://link.springer.com/chapter/10.1007/978-3-031-55661-6_1#DOI

Hickel, J. (2020). The Sustainable Development Index: Measuring the Ecological Efficiency of Human Development in the Anthropocene. *Ecological Economics, 167*, 106331. DOI: 10.1016/j.ecolecon.2019.05.011

Huo, J., & Peng, C. (2023). Depletion of Natural Resources and Environmental Quality: Prospects of Energy Use, Energy Imports, and Economic Growth Hindrances. *Resources Policy, 86*(A), 104049. DOI: 10.1016/j.resourpol.2023.104049

Jancar, B. (1993). The Environmental Attractor in the Former USSR: Ecology and Regional Change. In Lipschutz, R. D., & Conca, K. (Eds.), *The State and Social Power in Global Environmental Politics*. Columbia University Press.

Kabir, M., Habiba, U., & Farooqi, Z.-U.-R. (2021). Current Scenario of Environmental Pollution & Its Solution by Plantation in Changing Climate. *Journal of Sustainability Perspective*, 288-295. https://ejournal2.undip.ac.id/index.php/jsp/article/view/12016/6168

Khagram, S., Clark, W., & Raad, D. F. (2003). From the Environment and Human Security to Sustainable Security and Development. *Journal of Human Development*, *4*(2), 289–313. DOI: 10.1080/1464988032000087604

King, L. C., & van den Bergh, J. C. J. M. (2021). Potential Carbon Leakage Under the Paris Agreement. *Climatic Change*, *165*, 52. https://link.springer.com/article/10.1007/s10584-021-03082-4

Kingsbury, D., Remenyi, J., McKay, J., & Hunt, J. (2004). *Key Issues in Development*. Palgrave Macmillan.

Kolawole, A. S., & Iyiola, A. O. (2023). Environmental Pollution: Threats, Impact on Biodiversity, and Protection Strategies. In Izah, S. C., & Ogwu, M. C. (Eds.), *Sustainable Utilization and Conservation of Africa's Biological Resources and Environment* (pp. 377–409). Springer., https://link.springer.com/chapter/10.1007/978-981-19-6974-4_14

Krauss, J. E. (2022). Unpacking SDG 15, Its Targets and Indicators: Tracing Ideas of Conservation. *Globalizations*, *19*(8), 1179–1194. DOI: 10.1080/14747731.2022.2035480

Lamb, W. F., Mattioli, G., Levi, S., Roberts, T. J., Capstick, S., Creutzig, F., Minx, J. C., Müller-Hansen, F., Culhane, T., & Steinberger, J. K. (2020). Discourses of Climate Delay. *Global Sustainability*, *3*, e17. DOI: 10.1017/sus.2020.13

Lele, S. (2017). Sustainable Development Goal 6: Watering Down the Justice Concerns. *Wiley Interdisciplinary Reviews Water*, *4*(4), e1224. https://wires.onlinelibrary.wiley.com/doi/10.1002/wat2.1224

Lindsey, R. (2023, August 22). *Climate Change: Global Sea Level*. Climate.Gov. https://www.climate.gov/news-features/understanding-climate/climate-change-global-sea-level

Liu, H., Saleem, M. M., Al-Faryan, M. A. S., Khan, I., & Zafar, M. W. (2022). Impact of Governance and Globalization on Natural Resources Volatility: The Role of Financial Development in the Middle East North Africa Countries. *Resources Policy*, *78*, 102881. DOI: 10.1016/j.resourpol.2022.102881

Maza-Maza, J., Rodríguez-Caballero, E., Rodríguez-Lozano, B., & Cantón, Y. (2025). Impact of Deforestation on Ecosystem Functions of Ecuador's Tropical Dry Forest Under Different Protection Status. *Trees, Forests and People*, *22*, 101032. DOI: 10.1016/j.tfp.2025.101032

Menton, M., Larrea, C., Latorre, S., Martinez-Alier, J., Peck, M., Temper, L., & Walter, M. (2020). Environmental Justice and the SDGs: From Synergies to Gaps and Contradictions. *Sustainability Science*, *15*, 1621–1636. https://link.springer.com/article/10.1007/s11625-020-00789-8

Morgan, J., & Patomäki, H. (2021). Planetary Good Governance after the Paris Agreement: The Case for a Global Greenhouse Gas Tax. *Journal of Environmental Management*, *292*, 112753. DOI: 10.1016/j.jenvman.2021.112753

Nguyen, T. T., Grote, U., Neubacher, F., Rahut, D. B., Do, M. H., & Paudel, G. P. (2023). Security Risks from Climate Change and Environmental Degradation: Implications for Sustainable Land Use Transformation in the Global South. *Current Opinion in Environmental Sustainability*, *63*, 101322. DOI: 10.1016/j.cosust.2023.101322

Nor, B. A., & Yusof, Y. (2025). Environmental Degradation and Food Security in Somalia. *Discover Sustainability*, *6*, 75. https://link.springer.com/article/10.1007/s43621-024-00771-9

Omisore, A. G. (2018). Attaining Sustainable Development Goals in Sub-Saharan Africa: The Need to Address Environmental Challenges. *Environmental Development*, *25*, 138–145. DOI: 10.1016/j.envdev.2017.09.002

Popovski, V., & Stanchev, K. (2021). Extreme Poverty Eradication: Conceptual Evolution and Policy Changes. In Kakar, N., Popovski, V., & Robinson, N. A. (Eds.), *Fulfilling the Sustainable Development Goals on a Quest for a Sustainable World* (pp. 37–54). Routledge.

Porter, L., Rickards, L., Verlie, B., Bosomworth, K., Moloney, S., Lay, B., Latham, B., Anguelovski, I., & Pellow, D. (2020). Climate Justice in a Climate Changed World. *Planning Theory & Practice*, *21*(2), 293–321. DOI: 10.1080/14649357.2020.1748959

Robinson, D., & Igini, M. (2025, January 9). *15 Biggest Environmental Problems of 2025*. Earth.Org. https://earth.org/the-biggest-environmental-problems-of-our-lifetime/

Salvia, A. L., Filho, W. L., Brandli, L. L., & Griebeler, J. S. (2019). Assessing Research Trends Related to Sustainable Development Goals: Local and Global Issues. *Journal of Cleaner Production, 208*, 841–849. DOI: 10.1016/j.jclepro.2018.09.242

Sengupta, U., & Sengupta, U. (2022). SDG-11 and Smart Cities: Contradictions and Overlaps between Social and Environmental Justice Research Agendas. *Frontiers in Sociology*, *7*, 995603. https://www.frontiersin.org/journals/sociology/articles/10.3389/fsoc.2022.995603/full

Shi, S., & Yin, J. (2021). Global Research on Carbon Footprint: A Scientometric Review. *Environmental Impact Assessment Review*, *89*, 106571. DOI: 10.1016/j.eiar.2021.106571

Shue, H. (2014). *Climate Justice: Vulnerability and Protection*. Oxford University Press.

Siripipatthanakul, S., Phuangsuwan, P., Siripipattanakul, S., & Jaipong, P. (2025, a). Sustainable Tourism Based on Challenges, Tourist Experiences, and Host Communities: A Case Study of Ayutthaya, Thailand. In *Best Practices for Human Resources Management in Tourism and Hospitality* (pp. 361-394). IGI Global Scientific Publishing. DOI: DOI: 10.4018/979-8-3693-8779-5.ch014

Siripipatthanakul, S., Phuangsuwan, P., Siripipattanakul, S., Limna, P., Jaipong, P., & Chaiprakarn, S. (2025, b, September). Privacy and Information Security for Sustainable Community Among TikTok's Users. In *International Conference on Artificial Intelligence and Networking* (pp. 221-233). Singapore: Springer Nature Singapore. DOI: 10.1007/978-981-96-2015-9_15

Stiglitz, J. (2017, June 2). *Trump's Reneging on Paris Climate Deal Turns the US into a Rogue State*. The Guardian. https://www.theguardian.com/business/2017/jun/02/paris-climate-deal-to-trumps-rogue-america

Stoett, P., Scrich, V. M., Elliff, C. I., Andrade, M. M., Grilli, N. M., & Turra, A. (2024). Global Plastic Pollution, Sustainable Development, and Plastic Justice. *World Development*, *184*, 106756. DOI: 10.1016/j.worlddev.2024.106756

Sultana, F. (2022). Critical Climate Justice. *The Geographical Journal*, *188*(1), 118–124. https://rgs-ibg.onlinelibrary.wiley.com/doi/epdf/10.1111/geoj.12417

Tawiah, V. K., Zakari, A., & Khan, I. (2021). The Environment Footprint of China-Africa Engagement: An Analysis of the Effect of China–Africa Partnership on Carbon Emissions. *The Science of the Total Environment*, *756*, 143603. DOI: 10.1016/j.scitotenv.2020.143603

The Nature Conservancy. (2025). *How to Help Calculate Your Carbon Footprint.* The Nature Conservancy. https://www.nature.org/en-us/get-involved/how-to-help/carbon-footprint-calculator/

Trombetta, M. J. (2008). Environmental Security and Climate Change: Analyzing the Discourse. *Cambridge Review of International Affairs, 21*(4), 585–602. DOI: 10.1080/09557570802452920

Tsai, F-M., Mohamed, A.F., Bui, T.-D., Sethanan, K., & Tseng, M-L. (2025). Sustainable Development Goal 12 – Responsible Consumption and Production Challenges and Opportunities in World Regions: A Data-Driven Analysis on Circular Economy Practices. *Journal of Industrial and Production Engineering*, 1-47. DOI: 10.1080/21681015.2025.2532419

UN. (2015). *The Millennium Development Goals Report 2015.* https://www.un.org/millenniumgoals/2015_MDG_Report/pdf/MDG%202015%20rev%20(July%201).pdf

UNCHS. (United Nations Commission on Human Security). (2003). *Human Security Now.* UNESCO. https://unesdoc.unesco.org/ark:/48223/pf0000130589

UNDP. (1993, January 1). *Human Development Report: People's Participation.* https://hdr.undp.org/content/human-development-report-1993

UNDP. (1994, January 1). *Human Development Report: New Dimensions of Human Security.* https://hdr.undp.org/content/human-development-report-1994

UNDP. (2022, February 8). *Special Report on Human Security.* https://hdr.undp.org/content/2022-special-report-human-security

UNFCCC. (2015). *Adoption of the Paris Agreement—FCCC/CP/2015/L.9/rev.1.* https://unfccc.int/resource/docs/2015/cop21/eng/l09r01.pdf

UNGA. (2015). *Transforming Our World: The 2030 Agenda for Sustainable Development.* https://docs.un.org/en/A/RES/70/1

UNSG. (2010). *Human Security: Report of the Secretary-General.* UNGA A/64/701. https://docs.un.org/en/A/64/701

Wada, Y., Bierkens, M. F. P., de Roo, A., Dirmeyers, P. A., Famiglietti, J. S., Hanasaki, N., Konar, M., Liu, J., Schmied, H. M., Oki, T., Pokhrel, Y., Sivapalan, M., Troy, T. J., van Dijk, A. I. J. M., van Emmerik, T., van Huijgevoort, M. H. J., van Lanen, H. A. J., Vörösmarty, C. J., Wanders, N., & Wheater, H. (2017). Human-Water Interface in Hydrological Modelling: Current Status and Future Directions. *Hydrology and Earth System Sciences, 21*(8), 4169–4193. https://hess.copernicus.org/articles/21/4169/2017/

Wang, F., Xie, Y., Wang, L., Shuangshuang, L., & Jin, X. (2025). Mitigating Ice Sheets and Mountain Glaciers Melt with Geoengineering. *The Science of the Total Environment, 963*, 178450. DOI: 10.1016/j.scitotenv.2025.178450

WCED. (1987). *Our Common Future*. https://sustainabledevelopment.un.org/content/documents/5987our-common-future.pdf

WHO. (2025, February 18). *Biodiversity Loss.* https://www.who.int/news-room/fact-sheets/detail/biodiversity#

World Bank. (2025, October 10). *Circular Economy and Pollution Management.* https://www.worldbank.org/en/topic/pollution

Zurlini, G., & Müller, F. (2008). Environmental Security. In Jørgensen, S. E., & Fath, B. D. (Eds.), *Encyclopedia of Ecology* (pp. 1350–1356). Academic Press., DOI: 10.1016/B978-008045405-4.00707-2

KEY TERMS AND DEFINITIONS

Carbon Footprint: The carbon footprint is the amount of carbon dioxide released into the atmosphere by individuals, communities, institutions, or countries. Such a measurement helps address issues related to GHG emissions and global warming.

Climate Change: Climate change refers to long-term changes in climate patterns (temperatures and weather patterns). Anthropogenic climate change is one of the primary threats to human flourishing and the survival of the ecosystem.

Environmental Justice: Environmental justice is an inspirational call for fair and just interaction and respectful treatment of the ecosystem for a sustainable global future.

Human Rights: Human rights are the fundamental claims that humans inherit from their very human nature. Human rights are universal, indivisible, and interdependent.

People-Centered Approach: The people-centered approach places people at the center of a country's social, economic, and environmental policies.

SDGs: SDGs refer to the 17 well-designed goals of the UN's 2030 Agenda, which aims to transform the world into a sustainable future, leaving no one behind.

Sustainable Development: Sustainable development is a global attempt to balance the developmental and environmental needs of the present and future generations.

Chapter 8

The Legal Architecture of Energy Diplomacy: India-China Pathways Toward Green Governance and Sustainable Transitions

Abhinash Mohanta
https://orcid.org/0000-0002-8732-3202
KIIT University, India

Shaikh Sahanwaz Islam
https://orcid.org/0009-0006-2068-8959
Madhusudan Law University, India

Nikita Biswal
https://orcid.org/0009-0004-5215-2943
Christ University, Bengaluru, India

ABSTRACT

This chapter explores the evolving legal architecture of energy diplomacy through a comparative lens on India and China, focusing on how both states navigate the intersection of foreign policy, sustainability obligations, and clean technology governance. Drawing on a systematic review and thematic content analysis, it examines how both states navigate this intersection. It shows that energy diplomacy has shifted from a geopolitical tool to a legal and technological instrument shaping national and transnational green transitions. The analysis synthesizes treaty law, national frameworks, and case law to reveal gaps and convergence in climate accountability, technology transfer, and environmental justice. This chapter proposes "Green

DOI: 10.4018/979-8-3373-6821-4.ch008

Law Diplomacy" as a normative pillar of international governance, advocating for binding legal mechanisms over voluntary climate pledges to ensure equitable and sustainable transitions.

1. INTRODUCTION

Energy diplomacy is being played out today at the intersection of geopolitics, law, and technology, which reflects the need to secure energy supplies, diversify markets, and meet climate commitments. At its most basic level, energy diplomacy is the use of diplomatic tools, from bilateral agreements to multilateral negotiations, to secure stable access to energy resources and markets. Energy diplomacy has been analysed as a diplomatic activity that seeks to improve access to energy resources and markets, highlighting the connection between foreign policy objectives and energy outcomes. National goals like economic growth, energy security, and climate commitments are expressed through foreign policy, and energy diplomacy is the mechanism for their implementation. As a result, modern energy diplomacy affects trade, investment flows, supply contracts, strategic reserves, and collaborative research work. It is a mixed or blended field of foreign policy, combining international economic and energy law, based on trade rules, investment treaties, and regional market rules, to protect state interests. In the 21st century, a particularly strong sustainability dimension has been added to this agenda, as energy diplomacy increasingly balances supply security with climate change-related decarbonisation imperatives (Wang, 2025).

India and China are examples of these increasingly evolving pressures. Both are among the largest energy consumers and carbon emitters in the world, but their geopolitical positions and energy profiles differ. Historically, both have relied heavily on imported fossil fuels. In 2023, India imports about 4.5 million barrels of crude oil per day, and China imports about 11.3 million barrels per day, which makes it the largest importer of crude oil in the world. India also imported 264 million tonnes of coal and over 1.1 Tcf of liquefied natural gas, and China relied on imported coal and gas as well (Country Analysis Brief: India, 2025). Such dependencies influenced India's energy diplomacy with the Middle East and Russia, as well as China's traditional "oil diplomacy" in Africa and Central Asia. Consequently, both states increasingly turned to a number of international legal tools, including investment treaties and trade agreements with energy-related provisions, to secure energy access.

In recent years, both countries have undertaken a strategic shift toward renewable energy and clean technologies. India has set a non-fossil electricity generation target of 50% by 2030, energy independence by 2047, and net zero by 2070, and this will involve a massive expansion of solar, wind, hydro, and nuclear capacity (Deb &

Kailash, 2023). China has even more ambitious goals, aiming to peak emissions before 2030 and become carbon neutral by 2060, with projections that non-fossil energy will account for 80% of total consumption by that date. China now leads the world in renewable manufacturing, accounting for more than 40% of global renewable capacity by 2024 and investing around USD 818 billion in renewable technologies. India is also catching up in pace with record solar auctions, expanded hydroelectric and nuclear projects, and incentives for electric mobility. For both states, renewable autonomy is considered a way to reduce import vulnerability and increase strategic resilience.

Transitions revolve around technology policy. Technology in clean energy is considered strategic and secure. China's green revolution is a "national security multiplier," and India's National Solar and Hydrogen Missions are boosting local industry to achieve climate goals. Experts expect energy storage, grid infrastructure, electric vehicles, and renewable industrial advancements to balance energy security and sustainability. According to researchers, energy innovation requires a bridging technology strategy that balances security, equality, and sustainability through intellectual property frameworks, industrial incentives, and regulatory assistance.

The chapter examines how India and China's traditional foreign policy methods, focused on resources and economic linkages, connect with emerging sustainability and technology law regimes. How do their foreign policy actions affect international and domestic climate, clean energy, and technology governance laws? This includes examining their bilateral energy deals, multilateral climate forums, trade discussions, clean tech, and doctrinal obligations under international law. This study asks, "How do India and China's energy diplomacy strategies intersect with the legal framework of sustainability and technology governance?"

2. LITERATURE REVIEW

2.1 Legal Foundations of Energy Diplomacy

International law gives normative support for energy diplomacy today by incorporating climate obligations into cross-border energy relations. The Paris Agreement (2015) is the most significant case of this legal embedding, faithfully changing energy diplomacy by binding climate law. Article 6 makes voluntary cooperation, for example through international carbon markets, possible, allowing states to trade reductions against their nationally determined contributions (Article 6 of the Paris Agreement, 2016). Article 10 establishes a long-term vision for technology development and transfer to support low-emissions growth, and Article 13 increases transparency requirements by requiring states to report on mitigation efforts and support provided to increase accountability and trust (Horowitz, 2016). These provisions expand the 1992 UN Framework Convention on Climate Change, which requires developed countries to facilitate the transfer of clean energy technology to developing states (Lee et al., 2021). Together, the UNFCCC and the Paris Agreement make energy diplomacy a structured, rules-bound, multilateral process that is driven by environmental commitments rather than narrow geopolitical calculations.

The UN Charter (1945) provides the broader legal framework for all forms of diplomacy, including energy relations. Its basic principles of cooperation, peaceful resolution of disputes, and non-intervention determine the permitted scope of energy diplomacy. Article 1 promotes international cooperation in addressing economic and social problems; Article 2 affirms the sovereign equality of states and forbids the use or threat of force against another state. These principles demand that transboundary energy projects, including pipelines, LNG trade, grid interconnections, sanctions, and energy transition partnerships, be done through consent, dialogue, and respect for sovereignty. Energy diplomacy thus reflects the fundamental commitment of the Charter to peace and cooperative problem-solving, not unilateral coercion (Trahan, 2022).

A major legal turning point occurred in July 2025, when the International Court of Justice delivered an Advisory Opinion stating that states have binding obligations under customary international law, human rights treaties, and the UNFCCC to avoid hazardous climate change. This opinion was a head-turning moment, confirming that climate obligations are legal duties, not voluntary aspirations. The Court rejected claims from the United States, Japan, and Russia that the Paris Agreement functions as a self-contained regime separate from general international law, affirming that customary environmental law and human rights are fully applicable. Consequently, the Court clearly embedded the global climate regime within the broader framework of international law (Panek, 2025).

Huge effects on energy diplomacy. The responsibility for climate-related harm has crystallized, so failing to manage greenhouse gases that pollute transboundary areas could result in international liability. Second, the Opinion lets vulnerable governments and domestic litigants invoke climate commitments in international, arbitral, and constitutional courts. Third, investment disputes now center on climate issues, making energy-related arbitration more scrutinized due to environmental duties. This move requires significant polluters such as India and China to demonstrate compliance with climate laws rather than relying primarily on their climate goals.

For India, the ICJ Opinion strengthens its advocacy of common but differentiated responsibilities and historical accountability in the context of UNFCCC negotiations a long time ago. For China, the ruling has put limits on fossil fuel investments in the context of bilateral commitments like the Belt and Road, which could now come under more climate law scrutiny. Both states must therefore recalibrate their energy diplomacy around binding legal instruments that demonstrate environmental accountability, thereby supporting the Green Law Diplomacy framework advanced in this chapter (Nurgozhayeva, 2024).

An earlier attempt to create a rule-based framework for international energy cooperation through non-discriminatory trade and investment protections, including investor-state arbitration, is the Energy Charter Treaty (ECT, 1994). However, it still has a limited influence in Asia. Major regional powers such as India and China have not ratified the treaty, and Japan has withdrawn, leaving much of Asia's energy diplomacy to be carried out through bilateral and regional arrangements outside the ECT framework (Axelrod, 1996).

The World Trade Organization is another important legal cornerstone. Energy commodities and equipment are subject to the General Agreement on Tariffs and Trade, which means that WTO rules on non-discrimination, tariffs, quotas, and subsidies are fully applied (Olawuyi & Athwal, 2023). Although there is no dedicated WTO energy agreement, the GATT Article XX provides for environmental exceptions if they pass the requirements of necessity and non-discrimination. The WTO Secretariat has affirmed that members have policy space to implement measures in the climate sphere that are compatible with trade law.

The European Union's Carbon Border Adjustment Mechanism, which came into effect in October 2025, has intensified tensions. CBAM attaches a carbon price to imports like steel, cement, aluminium, fertilisers, and electricity to avoid carbon leakage. While justifiable on the part of the EU based on GATT Article XX, its compatibility with the WTO remains debated (WTO's Trade and Environment Week Signals New Climate-Trade Policy Momentum, 2025). India and China believe that CBAM contravenes the principle of common but differentiated responsibilities by placing carbon burdens on developing economies with lower historical emissions (Sambhav, 2024).

This dispute shows the relevance of Green Law Diplomacy. Rather than unilaterally imposed carbon actions, India and China are calling for a binding multilateral framework for carbon accountability under the aegis of the UNFCCC or the WTO, with uniform accounting standards, different timeframes, and safeguards against disguised protectionism. Recent simplifications to CBAM imply a lack of flexibility and help emphasise the need for multilateral solutions (Correa do Lago, 2025).

Energy diplomacy also intersects with intellectual property law. Climate treaties call for technology transfer; TRIPS flexibilities allow developed countries to access green technologies through compulsory licensing in developing countries. WTO leadership has called on states to make these flexibilities operational for climate-related inventions. Modern energy diplomacy, therefore, increasingly incorporates access to technology, IP regulation, investment, and finance in a single negotiating package.

These instruments design the architecture of green governance in energy policy as a whole. Green governance aligns with climate goals in energy transitions by integrating sustainability into trade rules, investment regimes, and diplomatic practices. OECD assessments show that climate priorities have become part of high-level regulatory and financial frameworks, making energy diplomacy part of a dense network of multilateral law that binds energy policy to sustainable development (OECD, 2025).

2.2 Evolution of National Energy Governance Frameworks

Over the years, India has evolved a complex legal architecture of its energy sector as a result of market reforms and environmental imperatives. Central to this framework is the Energy Conservation Act of 2001, which established the Bureau of Energy Efficiency and set standards for appliances and industrial processes to lower energy consumption. These measures are part of India's overall climate strategy and have shown tangible performance gains (e.g., saving about 28 million tonnes of oil equivalent in 2019-20). Complementing this statute, the renewable energy targets institutionalised in the National Electricity Policy, first issued in 2005, mandated the optimal utilisation of all energy resources and directed state regulators to enforce renewable purchase obligations. A flagship programme under this framework is the Jawaharlal Nehru National Solar Mission, launched in 2010 under the National Action Plan on Climate Change, which was intended to scale up solar capacity while reducing costs through preferential tariffs and mandatory procurement (Akoijam & Krishna, 2017). India originally aimed for 20 GW by 2022, but after revising the target to 100 GW, it has aggressively undershot this goal and is now on track to reach half a terawatt (GW) of non-fossil electricity by 2030. Collectively, these measures have resulted in a plurality of regimes that include overlapping statutes, regulatory guidelines, and institutional actors that are driving energy planning and environmental protection (The Energy Conservation (Amendment) Bill, 2022).

Another important milestone was the Electricity Act, 2003, which consolidated earlier power-sector laws and provided a clear impetus for renewable energy. In this regard, it requires state regulatory commissions to not only guarantee minimum procurement of renewable energy through Renewable Purchase Obligations but also ensure grid connectivity for renewable generators (Mukherjee et al., 2016). These provisions have been progressively strengthened, and penalties are now being imposed for non-compliance with centrally determined RPO targets. The Electricity (Amendment) Bill, 2022, goes further and aims to strengthen renewable investments by reforming power markets, promoting competition among distribution companies, securing payments, rationalising tariffs, and strengthening RPO enforcement. Together, these instruments stimulate private and foreign investment while maintaining social objectives such as rural electrification and affordability. Sector-specific incentives, such as tax benefits for solar manufacturing and rooftop subsidies, are in addition to general legislation. This layered legal pluralism includes central ministries and regulators, state governments, and independent agencies within national plans and international commitments, such as India's advanced plan, called its nationally determined contribution, under the Paris Agreement.

India has also used diplomacy and treaty-based means to advance the cause of clean energy at the international level. The International Solar Alliance, which was proposed by India and France at the 2015 Paris Climate Conference and committed to through a treaty in 2017, is one such approach. Headquartered in India, the ISA brings together over a hundred 'sunshine countries' to mobilise finance, technology, and best practices for solar deployment. This initiative strengthens India's domestic expansion of solar while promoting South-South cooperation and access to affordable clean energy. As of late 2023, the ISA had 116 signatories and 94 ratifications, underscoring the importance of linking energy security, access to technologies, and international legal cooperation to India's climate change goals (Jha, 2023).

China's energy governance has followed another path. Traditionally characterised by centralised planning, China for a long time had no overarching national energy statute but relied on sector-specific legislation, ministerial regulations, and Five-Year Plans. This changed in late 2024, when the National People's Congress passed China's first comprehensive Energy Law that will take effect in January 2025. The law emphasizes expanding the area of renewable energy, increasing the share of non-fossil energy consumption, and promoting the rational and efficient use of fossil energy, thereby largely codifying policy priorities already incorporated into the 14th Five-Year Plan (Policies & Action, 2025).

The Renewable Energy Law of 2005, which has undergone multiple amendments since 2009, established the basis for China's expansion of renewable energy and included feed-in tariffs, mandatory purchase of renewable energy by the electricity grid, and a special renewable energy fund (Schuman & Lin, 2012). Along with challenges such as grid curtailment, the framework enabled the rapid growth of wind and solar capacity. In recent years, China has shifted from supply-side subsidies to demand-side requirements; for example, the new draft rules for 2024 on the minimum consumption of renewable energy by industrial users indicate a convergence of objectives for energy security and low-carbon demand (Policies & Action, 2025).

Internationally, China pursues its energy interests through the Belt and Road Initiative, a network of bilateral and multilateral agreements that support infrastructure and energy investment. While not a single treaty, the BRI is a quasi-legal structure of memoranda, joint ventures, and financial arrangements driven by Chinese state entities. Since 2016-17, China has been trying to "green" the BRI by guiding its policies to encourage low-carbon and environmentally responsible projects, including green cooperation agreements and the Belt and Road Green Development Coalition (Zhai, 2021). Critics, however, say that continued support for coal and big hydropower projects puts these promises at risk, making the "Green Belt and Road" story controversial (Petersmann, 2020).

Energy governance in India is pluralistic and federal, with many laws, regulators, and market mechanisms that support constitutional development aspirations and international climate commitments. Planning organizations and state-owned companies enforce top-down targets in China's model. Despite their institutional differences, both countries believe that energy security and the implementation of large-scale sustainable technologies are essential. India emphasizes the role of renewables in reducing imports and meeting climate goals, whereas China incorporates clean energy into its industrial policy and carbon neutrality goal. Thus, while institutional pathways differ, both systems emphasize energy diversity, technical innovation, and environmental sustainability as fundamental features of modern energy governance (Policies & Action, 2025).

2.3 Bilateral and Regional Energy Diplomacy in Practice

India's and China's interactions with third-country energy markets are characterised by a mix of competition and cooperation, organised through investment treaties, memoranda of understanding, and political agreements. In Iran, China sought long-term access to oil resources through state-owned national oil companies, backed by the China-Iran BIT (2000) (International Investment Agreements Navigator, Iran, 2025). India, by contrast, did not have a formal BIT but relied instead on strategic political arrangements such as the 2003 New Delhi Declaration. In Myanmar, competition came in the form of China's oil and gas pipelines to bypass the Strait of Malacca, and in India's pipeline connectivity in its Northeast. China and Myanmar concluded a BIT in 2001, and India followed suit with its own BIT in 2008 (in force 2009). India also signed MoUs on power and renewable energy cooperation with Myanmar in 2016, which highlights India's preference for flexible diplomatic instruments over binding treaties (Paing, 2016).

In Africa, China has often adopted "oil-for-infrastructure" arrangements at the state level, which have been labelled neo-mercantilist (Kong, 2009). Indian firms like ONGC Videsh and Reliance have also invested across Africa, operating under India's BITs or MoUs with states such as Ethiopia and Sudan. These agreements will often cover dispute settlement by ICSID or UNCITRAL arbitration. However, India's experience with investor-state dispute settlement led it to terminate 58 older BITs and adopt a more restrictive Model BIT, which narrowed the scope of investor claims and strengthened regulatory autonomy (Vidyarthi, 2020).

More than bilateral competition, India and China have joint commissions and energy dialogues. Both are part of working groups on energy and climate technologies and of trilateral arrangements, including cooperation frameworks among India, Iran, and Afghanistan. Renewable energy cooperation has also spread. In 2015, India and Iran signed an agreement to develop solar projects in Chabahar; in 2016, India and Myanmar signed a renewable energy MoU (Paing, 2016). BITs like the China-Iran BIT (2000) and the China-Myanmar BIT (2001) provide for arbitration mechanisms for such ventures (International Investment Agreements Navigator, Iran, 2025). India's BITs, including the 2008 Myanmar treaty, likewise provide for fair treatment and ISDS, although recent Indian practice has favoured domestic courts and government-to-government consultations. In practice, political negotiation is the dominant method of dispute management, even when arbitration clauses exist, in contrast to India's conservative position and China's more robust ISDS orientation.

At the regional level, India and China are leaders of different multilateral energy platforms. India has a BIMSTEC Energy Centre under a Host Country Agreement signed in November 2024, which will make Bengaluru a regional energy hub (India Inks Agreement to Establish BIMSTEC Energy Centre in Bengaluru, 2024). India also plays a central role in the Indian Ocean Rim Association, which regularly convenes ministerial-level discussions on energy security and renewable energy. The IORA Renewable Energy Ministers' Forum of 2014 was the first to illustrate India's support for Indian Ocean cooperation ("Fish: Indian Ocean - IORA Ministerial Blue Economy Conference", 2015).

The Shanghai Cooperation Organisation and the so-called 'ASEAN+3' primarily serve as channels for China's regional energy engagement. The organisation of the SCO Energy Club was established under the 2013 MoU to facilitate energy security and technology exchange between Eurasian producers and consumers (SCO Secretary-General Vladimir Norov's Report at the 2020 Global Energy Interconnection Conference (Asia Energy and Power Conference), 2020). ASEAN+3's APAEC 2021-25 and 2026-30 energy roadmaps emphasize cross-border grid interconnection and clean energy investment. These tools are political promises, not treaties. SCO Energy Club is based on consultative memoranda, IORA on non-binding cooperation, and BIMSTEC on institutional structures. The tiny dispute resolution procedures mainly defer to national authorities.

Dispute settlement in these frameworks is essentially political rather than judicial. ASEAN+3 uses consultative methods akin to APEC, whereas the SCO Energy Club offers exclusively intergovernmental consultation. By contrast, traditional BITs have express ISDS clauses. India's 2016 Model BIT dramatically narrows the scope of arbitration, excluding taxation and government procurement and strengthening consent requirements (Vidyarthi, 2020). In practice, energy disputes very rarely go to arbitration. For instance, the issue of Nigeria's termination of oil-for-infrastructure contracts with Chinese companies in 2007 was resolved through political rather than ISDS procedures (Kong, 2009).

Iran sanctions demonstrate how national energy interests and international law interact. India and China relied on Iranian crude until the 2010s. UN Security Council resolutions 1929 and 2231, along with US-led financial restrictions, hindered trade. India outwardly opposed unilateral sanctions but decreased Iranian imports after 2012, diversified its supply sources, and delayed significant projects. Sanctions legislation halted the Iran-Pakistan-India gas pipeline and GAIL's 2005 $22 billion LNG contract. India secured a sanctions reprieve for the Chabahar Port project in 2016, allowing limited collaboration despite zero crude imports. Indian contracts were renegotiated to maintain long-term connectivity against UN wishes (Cheema, 2012).

Overall, India and China's adoption of external energy diplomacy is carried out against a dense structure of treaties, MoUs, regional platforms, and sanctions regimes. Combining elements of competition and coordination, their engagements abroad are constrained by international law and influenced by investment frameworks, as both countries strive to secure access to energy and technology.

2.4 Energy Diplomacy and Technology Policy: Legal Mechanism for Green Innovation

Intellectual property rights (IPR) and licensing arrangements are key to the cross-border spread of renewable energy technologies. Many of the basic clean energy technologies (e.g., basic solar panel and wind turbine designs) are already in the public domain or comprise incremental innovations patented by various companies. As John Barton has observed, major players in photovoltaics account for around 60 per cent of the market, and this situation often makes the availability of key technologies dependent on competition rather than exclusion. Manufacturers in developing countries are thus often able to obtain reasonable licensing terms, especially where patents are spread among multiple owners. Tata BP Solar in India and Suntech Power in China, for example, have entered global markets as licensees of advanced photovoltaic technology from foreign partners, showing how licensing and joint ventures can facilitate the transfer of renewable know-how (Barton, 2009). At the multilateral level, the existing WTO TRIPS Agreement contains provisions aimed at encouraging technology transfer, notably Article 66.2, which requires developed members to incentivise technology transfer to the least developed countries to strengthen their technological capacity (Watal & Caminero, 2019).

Meanwhile, energy diplomacy has expanded to include artificial intelligence and digital control in smart grids, presenting legal issues that current frameworks cannot address. AI systems increasingly manage electricity distribution in real time, including load balancing, renewable integration, and grid stability, and they regulate cross-border energy flows. The EU's 2024 AI Act classifies AI control of essential energy infrastructure as "high risk". Impact evaluations, openness, human oversight, and extensive tracking of AI choices are required by the Act, and operators must retain human control and explain algorithm results to stakeholders. This paradigm defines the legal responsibility for AI-based energy systems (Bouquet et al., 2023).

India and China lack adequate energy infrastructure and AI regulations. India's Digital Personal Data Protection Act, 2023, applies only to personal data, not operational grid data or critical sector AI governance. The 2017 Chinese Cybersecurity Law prioritises data localisation and security but not openness or accountability in AI-controlled grid systems. These regulatory loopholes are crucial for OSOWOG, the Asian Power Grid, and bilateral interconnections. One state using opaque AI technologies and another following EU-style transparency requirements will disagree on grid management, data access, liability, and investment losses. Current bilateral energy contracts do not normally cover grid failures allegedly induced by artificial intelligence-based choices in another jurisdiction. There may be contract disputes and difficulties with state liability (Volkova et al., 2024).

India and China should build a bilateral Information Technology (AI) governance framework that sets high standards for shared energy infrastructure to enhance Green Law Diplomacy. This might involve pre-deployment risk evaluations of AI systems operating across cross-border grids, uniform reporting on algorithmic choices that impact energy flows, and collaborative audit methods to ensure compliance. Inspired by the EU's high-risk model, the framework may require AI systems that control more than 10% of cross-border electricity to comply with third-party audits, maintain real-time monitoring logs, be able to override in the event of a malfunction, and document the decision-making process. Embedding AI accountability in energy diplomacy would ensure efficiency gains are not obfuscated and allow India and China to lead responsible AI-based energy governance in accordance with G20 digital resilience conversations (Mihăilescu, 2025).

In practice, however, clean technology transfer has relied more on specific licensing, subsidies, and direct investment than on treaty-based mechanisms. Patent law at once limits and facilitates green innovation. While patents provide exclusive rights, competition in the marketplace, plus the active support of states, can still lead to affordable pricing and local adoption. Legal frameworks matter, but so too do industrial policy and finance.

Trade agreements with stricter IP protections have raised concerns. Many recent bilateral and plurilateral trade deals, especially North-South agreements, feature "TRIPS-plus" clauses that exceed WTO regulations (Cheng, 2022). Such clauses can lengthen patent durations or improve enforcement for renewable energy technologies, increasing costs and limiting access. Global South actors responded with various models. UNCTAD and South-South platforms advocate "green technology banks" and patent-free invention pools for climate adaption. Recent UNCTAD proposals encourage poor nations to engage on patent-free technology transfer and build South-based green technology banks that consider key climate technologies as quasi-public commodities (South-South Cooperation for Climate Adaptation and Sustainable Development, 2022). India and South Africa have actively opposed TRIPS-plus provisions in climate-related negotiations, favouring Doha-style declarations that provide flexibility in licensing and finance.

India has been especially vocal in framing these debates in terms of the principle of common but differentiated responsibilities and the concept of technology equity. At COP30, India declared that access to technology is a right and not a bargaining tool and reiterated that CBDR remains the bedrock of the climate regime, while urging the developed countries to fulfil historical commitments relating to finance, technology transfer, and capacity building (India Reaffirms Commitment to Equity, Climate Justice and Multilateralism at COP30, in Belem, Brazil, 2025). Domestically, India's 2016 National IPR Policy captures this approach in asserting compliance with the TRIPs Agreements while stressing the need to utilise the flexibilities available to protect development and, particularly, the public interest, including compulsory licensing where needed (Srinivas, 2018).

China has taken a different, but no less strategic, path by prioritising the development of domestic green technology. Heavy state investment in renewable R&D and manufacturing, combined with joint-venture requirements, has helped Chinese firms lock in key patents. The 14th Five-Year Plan for Science and Technology lists advanced renewable energy as a national priority supported by coordinated pilot projects and subsidies, led by state agencies (14th FYP Energy Technology Innovation, 2025). China uses this industrial capacity to advance its diplomatic objectives through South-South cooperation programmes such as the Belt and Road Initiative and Africa's Solar Belt, thereby exerting diplomatic leverage in international standard-setting through technological leadership.

Taken together, modern technology policies are quasi-diplomatic tools in multilateral energy governance. States increasingly enter into innovation roadmaps, licensing strategies, and patent portfolios to shape negotiations at forums such as the IEA and COP. Therefore, energy diplomacy today goes beyond fossil fuels and includes patents, digital governance, and legal frameworks for innovation, which have become central leverage points in the struggle for national interests and global decarbonisation.

2.5 Transitional Energy Infrastructure and Digital Governance

Cross-border power infrastructure has become one of the defining characteristics of modern energy diplomacy. India's "One Sun, One World, One Grid" (OSOWOG), jointly announced with the United Kingdom at COP26, prescribes a global renewable electricity network that aims to balance supply across time zones. Prime Minister Modi presented OSOWOG as a means of mitigating storage demands, cutting carbon emissions, and lowering energy costs while creating new opportunities for international collaboration (Jain, 2023). The initiative has since received the support of over eighty countries. China's parallel proposal, the Global Energy Interconnection (GEI), introduced by President Xi in 2015, seeks to connect remote renewable generation zones to major demand centres via ultra-high-voltage transmission lines (Alvarez et al., 2022). While OSOWOG is promoted through the International Solar Alliance and a growing coalition of states, GEI is promoted through the Global Energy Interconnection Development and Cooperation Organisation, a Beijing-based entity with more than 500 members (Huang et al., 2019). Both projects are considered strategically important by many, although some analysts warn that a GEI-dominated grid might reproduce geopolitical dependencies similar to those in other Chinese infrastructure projects (Zissler, 2023).

The creation of such transnational grids requires effective legal frameworks to regulate cross-border connectivity. The ASEAN Power Grid, which was first proposed in 2007, is headed towards an updated regional agreement, with the respective member states looking forward to signing by late 2025 (Energising Asean: Power Grid Deal Set for Oct 2025 Signing, Sarawak to Play Key Role, 2025). Negotiations focus on harmonising technical standards, regulatory practices, and environmental safeguards, and on establishing a new financing facility to mobilise investment in transmission infrastructure. Bilateral and trilateral arrangements facilitate these regional efforts. Existing interconnections, such as those between India and Bhutan and Laos and Thailand, as well as the CASA-1000 project, are dictated by memoranda of understanding for tariffs, scheduling, data exchange, and dispute resolution. While OSOWOG and GEI remain in the conceptual stage, their eventual implementation will necessitate binding treaties, financing instruments, and intergovernmental agreements. In South Asia, in contrast, OSOWOG's first phase builds on already existing India-Bhutan-Bangladesh-Nepal-Myanmar links, while China has advanced GEI corridors within Asia and Eurasia (Sharma, 2022). The ASEAN Power Grid presents the legal avenues likely to influence the development of future OSOWOG and GEI arrangements.

Cybersecurity of digital data governance has been integrated into cross-border energy systems. Smart grids and advanced metering create vast amounts of sensitive data, and regulators are placed in the position of adding privacy and security protections to the energy sector. In Europe, electricity consumption data is considered personal information under the General Data Protection Regulation as a result of the 2019 Electricity Directive, which subjects smart meter data to that regulation. India has taken a similar route with the Forum of Regulators 2015 Model Smart Grid Regulations, which require utilities to prioritise the protection of consumers' data (H. et al., 2025). This framework has been reinforced by the Digital Personal Data Protection Act, 2023, which brings in consent and data minimisation requirements. Consequently, stringent conditions apply to the cross-border transfer of personal data related to energy issues. China has designated energy as a "critical sector" under its 2017 Cybersecurity Law, subject to extensive security and localisation obligations, including the requirement to store critical data domestically under Article 37 (Ai-min et al., 2018). While China's regime focuses more on sovereignty than on individual consent, both India and China seem to be moving towards GDPR-inspired governance models.

Diplomatic engagement is also adopting cybersecurity norms related to transnational energy grids. India hosted the 8th India-EU Cyber Dialogue in March 2025, which discussed threats to critical information infrastructure and cooperation on resilience (Eighth India-EU Cyber Dialogue (March 20, 2025), 2025). Similar discussions take place through G20 digital forums and Asia-Pacific cybersecurity efforts, contributing to the progressive convergence of standards. India and the European Union have discussed cooperation in areas of secure telecom and 6G networks that are essential for grid control, while the discussion on the future of the power grid in the South-East Asian countries (ASEAN), which is expected to be signed off soon, is said to have inclusion for cyber resilience. China claims that its legal regime aligns with international best practices, and its 2021 Data Security Law and the Personal Information Protection Law contain provisions similar to those of the GDPR, including data classifications and lawful data processing (Creemers, 2022). India's DPDP Act also shows potential parallels with GDPR principles, suggesting a shared normative path.

Artificial intelligence and digital sustainability are emerging as key topics in energy governance. Both India and China are implementing AI-powered smart grid systems to improve the integration and operational efficiencies of renewable energy. Machine learning tools, fed by smart meters, predict demand and detect equipment failures, raising questions about data ownership, algorithm accountability, and the environmental impact of digital infrastructure. India's draft AI ethics guidelines from 2021 and China's rules on AI-generated content from 2022 are indicative of emerging concerns about transparency and supervision. Energy regulations could soon require audits of AI-based dispatching and pricing systems. Both countries also recognise the energy intensity of data centres, as evidenced by India's Green Data Centre policy and China's "dual carbon" strategy to foster low-emission computing infrastructure. As cross-border grids grow, they will transmit not only electricity but also data subject to evolving digital legislation.

These technological systems are increasingly being intertwined with global climate commitments. As participants in the Paris Agreement, India and China incorporate climate goals into grid development and digital planning. OSOWOG and the International Solar Alliance are making a very clear point about Paris alignment with the One Sun Declaration. National electricity laws in both countries now include environmental and digital considerations, such as the use of digital tools for emissions monitoring and renewable integration. India also often replicates foreign regulatory models, and China relies on ISO and UN standards. ASEAN's power grid framework is based on its 2007 agreement and could learn from the EU's Clean Energy Package. As OSOWOG and GEI evolve, their legal architectures are likely to become more aligned with international norms on energy trade, data protection, cybersecurity, and environmental governance (Glende & Wolter, 2022).

Scholarship on energy diplomacy falls into three categories. The first recognizes energy diplomacy as a foreign policy tool for supply security, market access, and geopolitical influence, emphasizing fossil-fuel-driven statecraft and strategic control of strategic resources. A second body of research on energy diplomacy examines international economic law, trade laws, investment treaties, and dispute settlement processes for cross-border energy flows. Since the Paris Agreement made energy transitions a legal and diplomatic obligation rather than a domestic policy decision, subsequent research has linked energy diplomacy to sustainability, climate governance, and technology policy.

Comparative studies of India and China are few in this burgeoning literature. India's multilateral climate diplomacy and China's Belt and Road Initiative energy investments are generally studied separately without integrating the climate's legal, technological, and governance aspects. Green governance and sustainable transitions scholarship emphasizes technical standards, intellectual property regimes, and digital infrastructure, but rarely links them to energy diplomacy as a legal architecture. This chapter addresses this gap by comparing India's and China's green governance and sustainable energy transitions, using international law, domestic energy governance, and technology policy.

Figure 1. Conceptual Framework

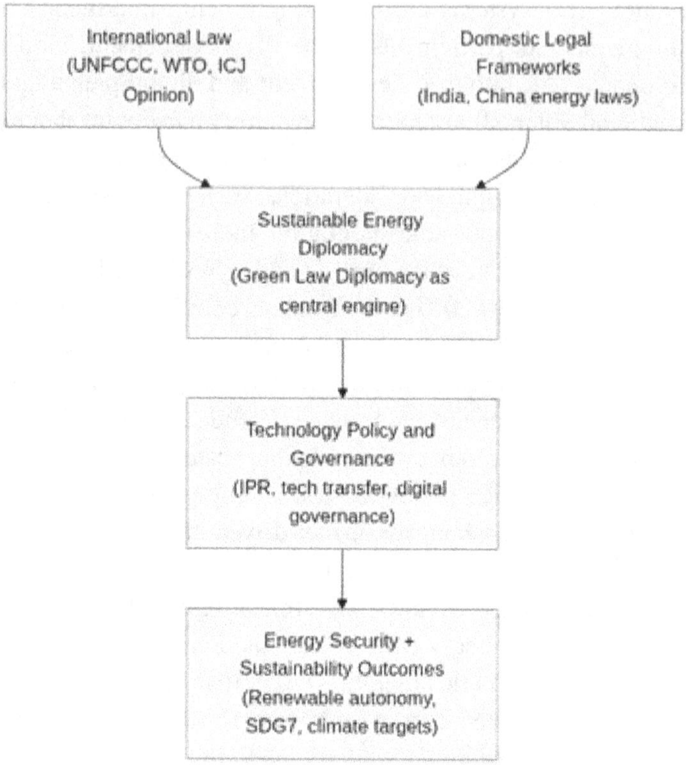

Figure 1 links international and domestic law, technology governance, and diplomatic practice to sustainable energy outcomes. Green Law Diplomacy sits at the center, connecting legal norms and technology policy to energy security and climate goals.

3. METHODOLOGY

According to Siripipatthanakul et al. (2025; a & b), a qualitative review should employ a systematic review and thematic content analysis.

Thus, this chapter adopts a qualitative, doctrinal, and comparative methodology and follows a systematic doctrinal review of international legal instruments, national laws, policy documents, and judicial materials relevant to energy diplomacy and sustainable transitions. Primary sources include international treaties, such as the

UNFCCC and the Paris Agreement, WTO treaties, advisory opinions of the International Court of Justice, regional cooperation frameworks, and local energy and environmental laws in India and China.

Secondary sources include peer-reviewed academic literature, policy reports from international organisations, and governmental strategy documents. A comparative content analysis is conducted to identify similarities and differences in the diplomatic practices of India and China regarding energy security, sustainable development, and technology governance. Case illustrations, such as India's multilateral solar diplomacy and China's Belt and Road energy projects, are used to put doctrinal findings in perspective. This methodology enables the chapter to produce analytically based insights rather than empirical insights, as is the practice in qualitative legal research within international governance studies.

4. RESULTS

This section summarises the major legal and policy findings that have emerged from the doctrinal and comparative analyses undertaken in the present chapter. These findings include implications of the 2025 ICJ Advisory Opinion on legally binding climate obligations; legal dynamics of trade-climate interaction under WTO law and the EU's Carbon Border Adjustment Mechanism; contrasting national legal frameworks for energy transitions in India and China; and governance gaps in digital and technological dimensions of energy infrastructure. Collectively, these findings highlight the role that legal instruments, judicial norms, and policy frameworks play in structuring contemporary energy diplomacy and sustainable transition strategies.

Figure 2. Comparative pathways of India and China

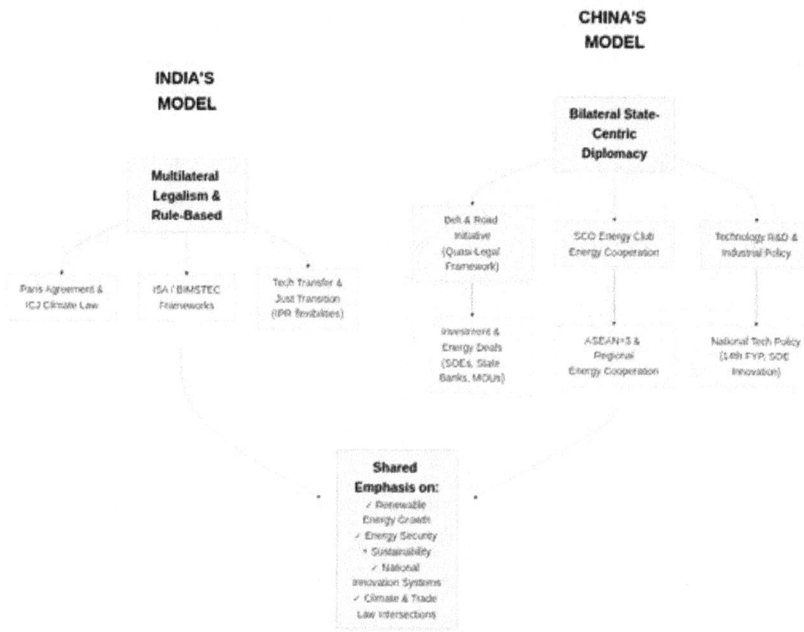

Figure 2 shows sustainable energy diplomacy regarding differing emphases: India's multilateral, rule-based legal model versus China's bilateral, state-led approach. Shared areas reflect common goals of renewable autonomy and sustainable transitions.

5. DISCUSSION

5.1 Doctrinal and Jurisprudential Gaps in Sustainable Energy Diplomacy

This chapter synthesizes the legal and policy chapters on green governance and sustainable energy diplomacy. According to the comparative study, India's model emphasizes multilateral legalism and norm-based collaboration, whereas China's model emphasizes bilateral and state-based cooperation under an administrative and industrial policy framework. Despite these institutional disparities, both states increasingly use legislative norms, technology governance, and institutional processes to reconcile energy security and sustainability. This section examines how doctrinal gaps, weak enforcement, and jurisdictional limits fail to reform accountability in transnational energy governance and the potential of judicial and human rights-based pathways for sustainable energy diplomacy.

The necessity of climate and energy changes is acknowledged, but no international convention requires "clean energy diplomacy" compliance. The UNFCCC and Paris Agreement function on voluntary, nationally decided contributions, participation and punishments have lagged. Thus, energy diplomacy relies on bilateral agreements and ad hoc collaboration, with no legal requirement to promote renewable energy or limit fossil fuel investment. Although greenhouse gas emissions and air and water pollution cause transboundary damage, this legal void gives states freedom in energy policy. Conflict between sovereign autonomy and transnational regulation creates a governance deficit in which major emitters have few external restraints and vulnerable governments have no effective remedies.

National courts and human rights bodies have started to address this gap by applying existing legal frameworks. A landmark example is *Urgenda Foundation v. State of the Netherlands* (2019), in which the Dutch Supreme Court held that climate mitigation is a human right (Meguro, 2020). The Court ruled that Articles 2 and 8 of the European Convention on Human Rights include a duty for the state to prevent dangerous climate change, including the significant risks to life and welfare from unchecked emissions. It forced the Netherlands to cut emissions by at least 25–40 percent by 2020 compared with 1990 levels, establishing that states cannot invoke sovereignty to evade responsibility when human rights are threatened.

A similar process took place in *Ashgar Leghari v. Federation of Pakistan* (2015), in which the Lahore High Court dealt with a question of delay in implementing the national climate policy by the government (Barritt & Sediti, 2019). The Court found that climate change threatened water, food, and energy security, especially for disadvantaged populations, and that state inactivity infringed fundamental rights to life and dignity. By creating a Climate Change Commission to supervise government activities, the Court imposed a constitutional requirement on state agencies, not treaty law. Both Urgenda and Leghari show how courts have made climate change a human-centered legal problem. They have little territorial influence and do not force other nations to lower emissions or compensate for cross-border harm.

An Advisory Opinion by the International Court of Justice may realign this balance. The ICJ confirmed states' legal obligation to adopt preventive measures against climate change and recognised the legal relevance of a 1.5-degree Celsius target (Schaugg et al., 2025). It held that the failure to take appropriate measures, including continued fossil fuel production, the issuance of new exploration licences, subsidies, or ineffective regulation, could constitute an internationally wrongful act. The Court emphasised that the obligation extends to both states and private actors to exercise due diligence to control emissions. Although advisory, the opinion provides authoritative guidance on what major emitters should do: that is, do the utmost to avoid foreseeable transboundary harm.

Major doctrinal gaps remain despite this advancement. The ICJ Opinion relies on general principles and human rights legislation, but has not created multinational enforcement frameworks. Noncompliance or due diligence penalties are not included in the Paris Agreement's temperature goals. There is no global energy transition treaty. Some experts, such as Philippe Sands, have advocated a Paris-aligned interpretation of trade and investment treaties, but states have generally opposed external adjudication of energy policy. Thus, there is no enforceable framework for clean energy diplomacy or an organization that may censure or compel states to invest in energy.

At the domestic level, legal accountability usually finishes at the national borders. India's Supreme Court has developed one of the most progressive environmental jurisprudences in the world through its interpretation of Article 21 of the Constitution, which protects the right to a clean environment. Through the *M.C. Mehta* line of cases, the Court laid down the polluter-pays principle and absolute liability for hazardous industries, particularly in the 1987 Oleum Gas Leak case (Mishra, 2023). Subsequent judgements enforced reform or closure of industries polluting the environment and focused on the joint responsibility between the state and its citizens (Chowdhury, 2019). In 2024, the Court recognised the right to live with dignity in the face of adverse climate impacts, thereby strengthening domestic environmental accountability (Ekhator, 2025).

India's law doesn't cover transboundary harm. Neither local legislation nor constitutional recognition of extraterritorial environmental rights imposes accountability for emissions impacting neighboring countries. China has comparable issues. It has rapidly increased sustainable energy and set carbon neutrality targets, but environmental protection is primarily an administrative policy rather than a rights-based commitment. International energy investments under China's Belt and Road Initiative create accountability concerns because environmental and social screening for transboundary implications is voluntary (Heggelund, 2021).

Overall, neither India nor China has legislative structures that guarantee accountability for cross-border energy decisions. Domestic methods address harm at the source, while political rather than legal instruments dominate transboundary pollution and global emissions. Human rights jurisprudence and the ICJ's Advisory Opinion give normative guidance, but gaps remain until they are codified. Major emitters have sovereign protection, whereas vulnerable populations have structural disadvantages. Urgenda and Leghari offer hope, but a comprehensive legal regime for clean energy diplomacy that balances sovereignty and global duty is still far off.

5.2 Energy Diplomacy, Human Rights, and Environmental Justice

The human rights framework has gained prominence in global energy policy as international institutions recognize energy as part of the rights to health, food, and a decent standard of living. UN organizations have interpreted Article 11 of the International Covenant on Economic, Social, and Cultural Rights to include essential energy services. In General Recommendation No. 34, the CEDAW Committee links rural women's welfare to electricity and encourages states to focus on sustainable and renewable energy generation, including expanding the grid to remote areas. Goal 7 of the 2030 Agenda requires "affordable, reliable, sustainable, and modern energy for all." The human rights framework for energy governance requires states to transition to renewable energy while ensuring equity, participation, and development. Energy diplomacy is anticipated to support climate mitigation, socioeconomic rights, and the UN right to development philosophy.

However, global clean energy goals and Global South development claims conflict. Article 1 of the ICESCR and the 1986 UN Declaration on the Right to Development give governments development sovereignty. India and China often frame energy changes this way. In climate discussions, India has said it has successfully avoided a commitment to a coal phase-out schedule, suggesting that worker rights, energy security, and SDG 7 universal access targets may prevent a speedy phase-out (Narula, 2023). Under the Belt and Road Initiative, China has made energy investments abroad to build partner countries. Critics say such projects haven't always considered community rights or environmental protections. Social mobilization and financial pressure forced Chinese developers to strengthen protections when Myanmar's Myitsone Dam was suspended due to local opposition and concerns that most of the electricity would be exported to China (Kirchherr et al., 2017). Energy diplomacy struggles to balance development ambitions with international responsibilities for environmental and labor justice.

Chinese overseas energy projects are another source of these frictions. Studies have shown that before the Myitsone controversy, Chinese hydropower developers followed only the host state standards, but then began to consider global environmental and social safeguards (Kirchherr et al., 2017). Nonetheless, controversies remain. A Human Rights Watch report on Guinea's Souapiti Dam found that the frequency of inadequate compensation and insufficient resettlement for roughly 16,000 displaced people led to severe disruption of their livelihoods ("We're Leaving Everything Behind": The Impact of Guinea's Souapiti Dam on Displaced Communities, 2020). Across Africa, Chinese financing has also gone toward new coal-fired power plants. In countries such as Ghana, Kenya, Malawi, and Zimbabwe, Chinese investors reportedly account for a large share of proposed coal capacity. Civil society groups say this is a double standard, as China makes a big push at home for renewables but exports fossil-fuel infrastructure abroad. In Zimbabwe, large-scale coal projects have been criticised as inconsistent with Paris Agreement commitments (Mambondiyani, 2020). Even projects associated with the clean energy supply chains have raised fears. A survey carried out in 2023 identified 102 allegations of human rights and environmental harm in the context of Chinese overseas investments in transition minerals, with the majority of the cases involving local communities (Unpacking Clean Energy: Human Rights Impacts of Chinese Overseas Investment in (Transition Minerals, 2023). These findings have put China's energy diplomacy under greater scrutiny, incentivised stronger safeguards, and reminded host states of their own protective duties.

India's discourse on just transition reflects similar concerns. Coal remains at the core of job creation, industrial development, and rural electrification, making sudden phase-outs politically and socially sensitive. India has thus evaded legislation to enact a binding coal exit in favour of growth and universal access under SDG 7. At subsequent COP meetings, Indian representatives have stressed the importance of development goals and the need to come before rapid coal retirement (Narula, 2023). Domestically, labour and social security legislation provides for notice, compensation, or re-employment in the event of plant closures, which is consistent with the principles of the International Labour Organization's Just Transitions. National electricity plans involve an orderly transition, with the old coal power plants phased out over time, with renewables taking over as new power plants are built. Courts and policymakers have come to frame energy policy in rights-based terms, linking electrification, rooftop solar programmes, and sugarcane clean cooking initiatives with poverty eradication and women's welfare. Government announcements on reskilling funds for displaced mine workers provide further support for the view that decarbonisation must be socially inclusive, in line with the Paris Agreement's emphasis on decent work and leaving no one behind (Just Transition of the Workforce and the Creation of Decent Work and Quality Jobs, 2020).

Energy diplomacy affects environmental justice lawsuits. Domestic courts now apply constitutional rights to energy access and environmental harm in line with international standards. India enforces pollution and environmental resource control through public interest litigation under Article 21 of the Constitution. These verdicts validate SDG and ICESCR pledges and show that government actions can be reviewed for their compliance with rights. Rarely, international complaints and human rights cases send diplomatic messages outside domestic tribunals. Judicial references to the Paris Agreement fair transition principles show a state's international compliance. Rights-based litigation establishes cross-border energy governance human rights rules alongside formal diplomacy.

5.3 Reform Pathways and Policy Recommendations

Energy diplomacy reform must incorporate sustainability into international and local law. The language on "sustainable energy" in bilateral investment treaties and other international investment agreements needs to change. Treaties recognize the state's power to regulate for legitimate objectives. Environmental regulations do not contravene CETA Article 8.9 (1). Investment treaties of the future can require investors to directly address environmental challenges. The Morocco-Nigeria BIT mandates stricter environmental impact assessment standards for the host state. Reform proposals also favour renewable energy investors and exclude fossil fuel investments from investor-state dispute settlement (Lieser, 2024). The IIA reform "toolbox" of UNCTAD allows treaty formulations that promote sustainable energy investment, cooperative R&D, and technical support. Climate objectives may be the focus of investment promotion if treaty language requires collaboration on low-carbon infrastructure and faster procedures for renewable projects (Trends in the Investment Treaty Regime and a Reform Toolbox for the Energy Transition, 2022).

Beyond treaty drafting, structured South-South legal cooperation can enhance green governance. A dedicated South-South Energy Law Forum, hosted under UNDP or another multilateral platform, could assist developing states in harmonising legal standards for sustainable energy governance. UN officials have stated that numerous solutions are available within the Global South itself (Li & Nascimento, 2021). A forum may bring together Indian, Chinese, African, and Latin American jurists and policymakers to produce model legal terms on renewable investment, technology transfer, subsidies, and dispute resolution. This strategy would complement regional cooperation mechanisms, such as energy integration within the Shanghai Cooperation Organization, and help build uniform legal norms for emerging countries.

Domestic legal reform is equally important. National legislation should reflect accountability mechanisms for energy diplomacy. In China, the proposed 2025 Energy Law emphasises the use of renewable energy and places obligations on "clean and efficient" use of coal through advanced technologies (You, 2025). India could make similar reforms by modifying existing laws, such as the Electricity Act or the Environment Protection Act, to require that international energy agreements be subject to parliamentary oversight and be consistent with climate commitments. Model provisions could require bilateral and regional energy accords to align with obligations under the Paris Agreement. Domestic laws could also operationalise Article 4.1(c) of the UNFCCC by providing support for technology transfers and harmonising national policy with international climate goals (Trends in the Investment Treaty Regime and Reform Toolbar for the Energy Transition, 2022).

For technology governance, states must report climate action and support, including technology transfer, under Article 13 of the Paris Agreement. Domestic laws may require green technology agreement registries and emissions reporting. International academics propose a Green Technology Transfer Agreement to legalize renewable research, manufacture, and deployment without IP concerns. A UN-sponsored pact may address South-South technology transfer, including China's Global South cooperation programs, and expand on UNFCCC and IRENA. Regional monitoring institutions strengthen accountability; SCO and BIMSTEC promote energy cooperation and may incorporate sustainability. Energy working groups should monitor data transparency, environmental performance, and cross-border project emissions. BIMSTEC's energy centre and grid interconnection initiatives offer joint audits and peer reviews. These regional frameworks would support sustainable energy diplomacy. Sustainability in investment treaties, South-South law coordination, country-level governance, and regional monitoring are needed for reform. This can align energy diplomacy, climate transparency, and technology transfer for accountability and long-term decarbonization.

6. CONCLUSION

The comparison indicates that India and China approach sustainable energy diplomacy differently. India prioritizes multilateralism, legality, and cooperation. India's G20 presidency encouraged "green growth" and the International Solar Alliance and Coalition for Disaster Resilient Infrastructure. In global forums, including the UNFCCC, G20, BRICS, and BIMSTEC, India prioritizes equity through common but differentiated duties and universal, rules-based commitments. However, China's strategy is bilateral and state-driven. Beijing employs funding from the Belt and Road Initiative and government-to-government agreements rather than multilateral institutions. China funded most climate change programs bilaterally between 2013 and 2022, according to the World Resources Institute. These approaches affect regional energy transitions legally and diplomatically.

Technology policy impacts bilateral energy diplomacy. Renewables and low-carbon infrastructure require technology transfer, standards-setting, and supply chain security. Scholars argue energy diplomacy increasingly involves technology, infrastructure, and transport. India supports African solar and green hydrogen research. State-led industrial policy makes China the world leader in solar panels, batteries, and electric vehicles. According to scholarly research, including a green development seminar at the University of California, Los Angeles, Chinese companies lead many clean tech fields, supported by local industrial rules that have led to tariffs and export prohibitions. These incidents highlight the importance of trade, IP, and investment agreements in clean tech legislation. The digital governance sector has grown significantly. G20 energy initiatives include smart grids, cybersecurity, and digital infrastructure. Sustainable energy diplomacy requires technology and digital governance since climate diplomacy today incorporates cross-border data flows and cyber standards.

This progression demonstrates an international systemic problem. Sustainable energy diplomacy goes beyond voluntary agreements. The Paris Agreement's politically and legally flexible "pledge and review" model relies on non-binding nationally specified contributions. Scholars think such agreements are unenforceable under international law, making long-term investment risky. History shows that agreements without compliance measures fail. Climate club proposals with standard criteria and punishments demonstrate the necessity for enforceable norms. Effective energy diplomacy involves legal integration, including technological norms, verifiable methodologies, and law-based accountability.

In this context, "Green Law Diplomacy" might be the fourth pillar of global environmental governance, alongside mitigation, adaptation, and finance. Green Law Diplomacy would require governments to incorporate climate goals into trade, investment, and technology agreements, making legal institutions essential to achieving these goals. Foreign ministries should develop green diplomacy teams to align domestic climate laws with global duties. International organizations should expand their mandates to support legal response initiatives, such as integrating commerce with climate goals and protecting key energy infrastructure through cyber rules. Recent G20 comments highlighting technology transfer and capacity building suggest that developing governments are already doing so.

Green law diplomacy needs institutional support. Future UNFCCC conferences should create a working committee to design and implement enforceable RE technology, technology transfer, and digital governance frameworks. This would fill gaps in the Paris system, which prioritizes mitigation and financing but lacks requirements for technology cooperation. Coordination between the WTO and the UNFCCC is essential. WTO environmental and trade debates. These efforts offer promise for trade-environmental harmony. A unified mechanism might harmonize definitions of environmental goods, address carbon border measures, and incorporate environmental factors into dispute settlement, reducing conflict and enabling CBDR-aligned solutions.

Regional platforms enable action. India's BIMSTEC Energy Centre could become a regional agency for monitoring sustainable energy investment, while the Shanghai Cooperation Organization's Energy Club could be formalized through contractual commitments. Diplomatic training must evolve. Diplomats could construct legally sound sustainability frameworks by including climate, energy, and technology legislation in their courses. Climate ambition must be translated into permanent change through diplomacy and legal mechanisms.

7. LIMITATIONS AND FUTURE RESEARCH DIRECTIONS

While this chapter offers a detailed doctrinal and comparative study of energy diplomacy and green governance in India and China, it is limited by its reliance on legal texts, policy documents, and secondary sources rather than empirical field data. The lack of stakeholder interviews or project-level assessments limits our understanding of the dynamics of implementation and institutional practice. Future research might benefit from adopting a mixed-method approach that includes case studies, regulatory impact assessments, and interviews with policymakers and industry stakeholders. This would enhance our understanding of technology governance and its relationship to transnational energy infrastructure. Further comparative work involving other emerging economies would also make the analysis of sustainable energy diplomacy in the Global South more thorough.

REFERENCES

14th FYP Energy Technology Innovation. (2025, June 24). https://www.iea.org/policies/15718-14th-fyp-energy-technology-innovation

Advancing Climate Justice at the International Court of Justice. (2025, November). https://www.ciel.org/project-update/advancing-climate-justice-at-the-icj/

Ai-min, Q., Shao, G., & Zheng, W. (2018). Assessing China's Cybersecurity Law. *Computer Law & Security Review*, *34*(6), 1342. DOI: 10.1016/j.clsr.2018.08.007

Akoijam, A. S., & Krishna, V. V. (2017). Exploring the Jawaharlal Nehru National Solar Mission (JNNSM): Impact on innovation ecosystem in India. *African Journal of Science, Technology, Innovation and Development*, *9*(5), 573. DOI: 10.1080/20421338.2017.1359466

Alvarez, D. del B., Sasakawa, A., Yamaguchi, K., & Wang, J. (2022). Strategic priorities for regional power connectivity in Asia in the overlapping of subregional initiatives and China's global energy interconnection. *Asia Pacific Business Review*, *30*(2), 376. DOI: 10.1080/13602381.2022.2093528

Article 6 of the Paris Agreement. (2016). https://unfccc.int/process-and-meetings/the-paris-agreement/article6

Axelrod, R. S. (1996). The European Energy Charter Treaty. *Energy Policy*, *24*(6), 497. DOI: 10.1016/0301-4215(96)00022-5

Barritt, E., & Sediti, B. (2019). The Symbolic Value of Leghari v Federation of Pakistan: Climate Change Adjudication in the Global South. *King's Law Journal : KLJ*, *30*(2), 203. DOI: 10.1080/09615768.2019.1648370

Barton, J. H. (2009, March 18). *Patenting and Access to Clean Energy Technologies in Developing Countries*.

Bouquet, P., Jackson, I., Nick, M., & Kaboli, A. (2023). AI-based forecasting for optimised solar energy management and smart grid efficiency. *International Journal of Production Research*, *62*(13), 4623. DOI: 10.1080/00207543.2023.2269565

Cheema, S. A. (2012, April 29). *India's Approach to Sanctions on Iran*. https://www.e-ir.info/2012/04/29/indias-approach-to-sanctions-on-iran/

Cheng, W. (2022, November 15). *International Clean Technology Diffusion: Pathways and Prospects*. https://www.southcentre.int/wp-content/uploads/2022/11/SV243_221115.pdf

Chowdhury, N. (2019). Constitutionally Shackled. *The Story of Environmental Jurisprudence in India, 159*. Advance online publication. DOI: 10.1007/978-981-13-9065-4_14

Corrêa do Lago, A. A. (2025, September 25). *COP30: An Integrated Vision for Climate Change and Trade*. https://tessforum.org/latest/cop30-an-integrated-vision-for-climate-change-and-trade *Country Analysis Brief: India*. (2025). [U.S. Department of Energy]. U.S. Department of Energy. https://www.eia.gov/international/content/analysis/countries_long/India/pdf/India.pdf

Creemers, R. (2022). China's emerging data protection framework. *Journal of Cybersecurity, 8*(1). Advance online publication. DOI: 10.1093/cybsec/tyac011

Deb, K., & Kailash, S. (2023, September 5). *Mainstreaming the Energy Transition in India's Policy Framework.*

Eighth India-EU Cyber Dialogue (March 20, 2025). (2025, March 21). https://www.mea.gov.in/pressreleases.htm?dtl/39250/Eighth+IndiaEU+Cyber+Dialogue+March+20+2025

Ekhator, E. (2025). Environmental Justice in the Global South. *Palgrave Studies in Sustainable Futures, 595*. Advance online publication. DOI: 10.1007/978-3-031-55387-5_30

Energising Asean: power grid deal set for Oct 2025 signing, Sarawak to play key role. (2025, May 25). https://www.scoop.my/news/258075/energising-asean-power-grid-deal-set-for-oct-2025-signing-sarawak-to-play-key-role/

Fish: Indian Ocean - IORA Ministerial Blue Economy Conference. (2015). *Africa Research Bulletin Economic, Financial, and Technical Series, 52*(8). DOI: 10.1111/j.1467-6346.2015.06637.x

Glende, E., & Wolter, M. (2022). Extra-long distance ultra-high voltage direct current. *2022 IEEE Sustainable Power and Energy Conference (iSPEC)*, 1. DOI: 10.1109/ispec54162.2022.10033043

H., B., Altaf Q., Nabab, A., M., & Asmae, B. (2025). Smart Grid Security and Protection. *Lecture Notes in Electrical Engineering*. DOI: 10.1007/978-981-96-0824-9

Heggelund, G. (2021). China's climate and energy policy: At a turning point? *International Environmental Agreement: Politics, Law and Economics, 21*(1), 9. DOI: 10.1007/s10784-021-09528-5

Horowitz, C. A. (2016). Paris Agreement. *International Legal Materials, 55*(4), 740. DOI: 10.1017/s0020782900004253

Huang, J., Lin, C., Zhou, H., Xu, Z., & Lin, C. (2019). Research on key technologies of deduction of multinational power trading in the context of Global Energy Interconnection. *Global Energy Interconnection, 2*(6), 560. DOI: 10.1016/j.gloei.2020.01.010

India inks agreement to establish Bimstec Energy Centre in Bengaluru. (2024, November 10).

India Reaffirms Commitment to Equity, Climate Justice and Multilateralism at CoP30, in Belém, Brazil. (2025, November 11). https://www.pib.gov.in/PressReleasePage.aspx?PRID=2189018

International Investment Agreements Navigator. (2025, May 27). https://investmentpolicy.unctad.org/international-investment-agreements/countries/98/iran-islamic-republic-of

Jain, A. K. (2023). Sustainable Energy: One Sun-One World-One Grid. *Green Energy and Technology, 91.* Advance online publication. DOI: 10.1007/978-981-99-0216-3_6

Jha, V. (2023). *International Solar Alliance: Bridging the Gap* [Center for Social and Economic Progress]. Center for Social and Economic Progress. https://csep.org/reports/international-solar-alliance-bridging-the-gap

Just Transition of the Workforce, and the Creation of Decent Work and Quality Jobs. (2020). [UNFCCC]. UNFCCC. https://unfccc.int/sites/default/files/resource/Just%20transition.pdf

Kirchherr, J., Matthews, N., Charles, K., & Walton, M. J. (2017). "Learning it the Hard Way": Social safeguards norms in Chinese-led dam projects in Myanmar, Laos and Cambodia. *Energy Policy, 102*, 529. DOI: 10.1016/j.enpol.2016.12.058

Kong, B. (2009). *China's International Petroleum Policy.* DOI: 10.5040/9798400625749

Lee, W.-J., Juskenaite, I., & Mwebaza, R. (2021). Public–Private Partnerships for Climate Technology Transfer and Innovation: Lessons from the Climate Technology Centre and Network. *Sustainability, 13*(6), 3185. DOI: 10.3390/su13063185

Lieser, A. (2024, September 11). *Bilateral Investment Treaties as a Tool for Global Climate Governance?* https://voelkerrechtsblog.org/bilateral-investment-treaties-as-a-tool-for-global-climate-governance/

Mambondiyani, A. (2020, December 9). *China's billion-dollar dirty coal projects.* https://gga.org/chinas-billion-dollar-dirty-coal-projects/

Meguro, M. (2020). State of the Netherlands v. Urgenda Foundation. *The American Journal of International Law, 114*(4), 729. DOI: 10.1017/ajil.2020.52

Mihăilescu, M. (2025). Digital Governance Assisted by Artificial Intelligence in the Energy Sector. *Advances in Computational Intelligence and Robotics, 1*. Advance online publication. DOI: 10.4018/979-8-3373-3186-7.ch001

Mishra, S. (2023). *M.C. Mehta And Anr vs. Union of India & Ors*. THE JOURNAL OF UNIQUE LAWS AND STUDENTS., DOI: 10.59126/v2i4a4

Mukherjee, S., Dhingra, T., & Sengupta, A. (2016). Status of Electricity Act, 2003: A systematic review of literature. *Energy Policy*, *102*, 237. DOI: 10.1016/j.enpol.2016.12.001

Narula, K. (2023, June 23). *India Must Begin Paving the Way to Ending Coal*. https://www.geopoliticalmonitor.com/india-must-begin-paving-the-way-to-ending-coal

Nurgozhayeva, R. (2024). The Belt and Road Initiative and Sustainability: A Driving Force for Change or a Missed Opportunity? *Chinese Journal of Environmental Law*, *8*(2), 119. DOI: 10.1163/24686042-12340120

OECD. (2025). Government at a Glance 2025. *Government at a Glance*. DOI: 10.1787/0efd0bcd-en

Olawuyi, D. S., & Athwal, E. I. (2023). Energy in WTO law and policy. *Edward Elgar Publishing eBooks*, 477. DOI: 10.4337/9781800882867.00034

Paing, T. H. (2016, August 30). *Burma, India Agree to Cooperate on Defense, Energy, Infrastructure*. https://www.irrawaddy.com/news/burma/burma-india-agree-to-cooperate-on-defense-energy-infrastructure.html

Panek, S. (2025, September 22). *The ICJ's Advisory Opinion on Climate Change: Insights and Impact*. https://www.standrewslawreview.com/post/the-icj-s-advisory-opinion-on-climate-change-insights-and-impact

Petersmann, E. (2020). *International Settlement of Trade and Investment Disputes Over Chinese 'Silk Road Projects' Inside the European Union*. 45. DOI: 10.1007/978-3-030-46000-6_4

Policies & action. (2025, November 6). /climateactiontracker.org/countries/china/policies-action

Sambhav, A. (2024). LEGAL ANALYSIS OF CARBON BORDER ADJUSTMENT MECHANISM WITHIN THE WORLD'S TRADE ORGANIZATION FRAMEWORK. SSRN *Electronic Journal*. DOI: 10.2139/ssrn.5034538

Schaugg, L., Jones, N., & Qi, J. (2025, July 28). *Historic International Court of Justice Opinion Confirms States' Climate Obligations*. https://www.iisd.org/articles/deep-dive/icj-advisory-opinion-climate-change

Schuman, S., & Lin, A. (2012). China's Renewable Energy Law and its impact on renewable power in China: Progress, challenges and recommendations for improving implementation. *Energy Policy, 51*, 89. DOI: 10.1016/j.enpol.2012.06.066

SCO Secretary-General Vladimir Norov's report at the 2020 Global Energy Interconnection Conference (Asia Energy and Power Conference). (2020, September 2). https://eng.sectsco.org/20201102/688470.html

Sharma, K. R. (2022). One Sun, One World, One Grid. *IPEM Law Journal, 6*(6), 97. DOI: 10.61691/ipem_law.6.2022.97-102

Siripipatthanakul, S., Chaiprakarn, S., Rungsiyanont, S., & Phuangsuwan, P. (2025, a). A Case Study of Sustainable Development in Hospitals in Thailand Using ChatGPT. In *Sustainable Development for Hospitals and Clinics* (pp. 1-32). IGI Global Scientific Publishing. DOI: DOI: 10.4018/979-8-3693-9800-5.ch001

Siripipatthanakul, S., Phuangsuwan, P., Siripipattanakul, S., Limna, P., Jaipong, P., & Chaiprakarn, S. (2025, b, September). Sustainable Development Adopting Digital Intelligence Among University Students. In *International Conference on Artificial Intelligence and Networking* (pp. 211-220). Singapore: Springer Nature Singapore. DOI: 10.1007/978-981-96-2015-9_14

SOUTH–SOUTH COOPERATION for Climate Adaptation and Sustainable Development. (2022). [UNCTAD Report]. UNITED NATIONS CONFERENCE ON TRADE AND DEVELOPMENT. https://unctad.org/system/files/official-document/tcsgdsinf2022d1_en.pdf

Srinivas, K. (2018). Innovations and business incubation for startups for effective technology transfer. *ORYZA- An International Journal on Rice, 55*, 90. DOI: 10.5958/2249-5266.2018.00011.5

The Energy Conservation (Amendment) Bill, 2022. (2022). https://prsindia.org/billtrack/the-energy-conservation-amendment-bill-2022

Trahan, J. (2022). Vetoes and the UN Charter: The obligation to act in accordance with the 'Purposes and Principles' of the United Nations. *Journal on the Use of Force and International Law, 9*(2), 243. DOI: 10.1080/20531702.2022.2059882

Trends in the Investment Treaty Regime and Reform Toolbar for the Energy Transition. (2022). [UNCTAD Report]. UNCATD. https://unctad.org/system/files/official-document/diaepcbinf2023d4_en.pdf

Unpacking clean energy: Human rights impacts of Chinese overseas investment in transition minerals. (2023). [Unpacking clean energy: Human rights impacts of Chinese overseas investment in transition minerals]. Business & Human Rights Resource Centre. https://media.business-humanrights.org/media/documents/2023_China_TM_briefing.pdf

Vidyarthi, A. (2020, August 2). *Revisiting India's Position to Not Join the ICSID Convention*. https://legalblogs.wolterskluwer.com/arbitration-blog/revisiting-indias-position-to-not-join-the-icsid-convention

Volkova, A., Hatamian, M., Anapyanova, A., & de Meer, H. (2024). Being Accountable is Smart: Navigating the Technical and Regulatory Landscape of AI-based Services for Power Grid. *arXiv*. DOI: 10.48550/ARXIV.2408.01121

Wang, M. (2025, July 28). *"Build first, break later": How China's green transformation was built to last*. https://www.weforum.org/stories/2025/07/chinas-green-transformation

Watal, J., & Caminero, L. (2019). Least-Developed Countries. *Transfer of Technology and the TRIPS Agreement, 199*. Advance online publication. DOI: 10.1007/978-981-13-2856-5_10

"We're Leaving Everything Behind" The Impact of Guinea's Souapiti Dam on Displaced Communities. (2020, April 16). https://www.hrw.org/report/2020/04/16/were-leaving-everything-behind/impact-guineas-souapiti-dam-displaced-communities

WTO's Trade and Environment Week signals new climate-trade policy momentum. (2025, July 18). https://www.c2es.org/2025/07/wtos-trade-and-environment-week-signals-new-climate-trade-policy-momentum/

You, X. (2025, January 8). *China has a new sweeping energy law. Here's what you need to know*. https://www.ciphernews.com/articles/china-has-a-new-sweeping-energy-law-heres-what-you-need-to-know/

Zhai, T. (2021). Environmental Challenges, Opportunities, and Policy Implications to Materialize China's Green Belt and Road Initiative. *Sustainability*, *13*(18), 10428. DOI: 10.3390/su131810428

Zissler, R. (2023). *International Power Interconnections Progress and Help Countries in the Energy Crisis*. https://www.renewable-ei.org/en/activities/column/REupdate/20230119.php

KEY TERMS AND DEFINITIONS

Common but Differentiated Responsibilities (CBDR): A UNFCCC principle assigning greater climate obligations to historically high-emitting developed states.

Customary International Law: Binding norms recognized by widespread state practice, applicable to climate responsibility post-ICJ 2025.

Digital Energy Infrastructure: Smart grids and AI-driven energy systems that require cybersecurity and data governance law.

Energy Diplomacy: The use of foreign policy tools to secure energy access and shape energy markets.

Environmental Justice: The principle that energy transitions must respect the rights and participation of affected communities.

Green governance: It refers to the integration of environmental sustainability principles into legal, institutional, and policy frameworks to guide decision-making, regulate economic activity, and ensure accountability in the pursuit of sustainable development and climate objectives.

SDG7: SDG 7 aims to ensure access to affordable, reliable, sustainable, and modern energy for all by 2030.

South-South Cooperation: Legal and policy collaboration among developing countries on energy, trade, and sustainability.

Technology Transfer: Legal and policy mechanisms for sharing clean energy innovations, often involving IPR and multilateral mandates.

Transnational Governance: Regulatory coordination across states on shared infrastructure and cross-border environmental impacts.

KEY TERMS AND DEFINITIONS

Chapter 9
Shaping Sustainable Growth With AI in Technology-Driven Organizations

Pongsakorn Limna
https://orcid.org/0000-0002-7448-5261
Pathumthani University, Thailand

ABSTRACT

This chapter explores the transformative role of artificial intelligence (AI) in shaping green and sustainable growth within technology-driven organizations. The study highlights how AI enhances operational efficiency, optimizes resource utilization, and facilitates transitions toward circular economy practices. The findings demonstrate that AI applications significantly reduce environmental impacts while supporting economic competitiveness. Governance and policy frameworks emerged as critical enablers of responsible AI adoption, ensuring transparency, accountability, and alignment with environmental, social, and governance (ESG) standards. Moreover, the integration of AI with emerging technologies was found to generate new opportunities for predictive governance and sustainability innovation. While barriers such as regulatory fragmentation, high implementation costs, and ethical concerns persist, the study underscores AI's potential as a cornerstone of organizational transformation toward long-term ecological resilience and sustainable development.

DOI: 10.4018/979-8-3373-6821-4.ch009

1. INTRODUCTION

The global shift toward sustainability has placed increasing pressure on organizations to adopt innovative strategies that balance economic performance with environmental and social responsibility. As businesses grapple with climate change, resource scarcity, and rising stakeholder expectations, technology has emerged as both a driver of growth and a tool for achieving sustainability objectives (Le, 2022; Martínez-Peláez et al., 2023; Odeyemi et al., 2024). Artificial intelligence (AI) has emerged as one of the most transformative forces shaping the trajectory of modern organizations, driving innovation, efficiency, and competitiveness across industries. As global economies confront the urgent challenges of climate change, environmental degradation, and resource scarcity, organizations are increasingly compelled to align their technological strategies with sustainability goals. The convergence of AI and sustainability has given rise to a paradigm where technology-driven organizations not only pursue growth but also contribute to environmental stewardship and social responsibility. This dual objective has positioned AI as a critical enabler of green transformation, offering solutions that balance economic performance with ecological resilience (Abdulmuhsin et al., 2025; Alamandi, 2025; Erturk et al., 2025; Khokhar, 2025; Siripipatthanakul et al., 2025).

AI's capacity to analyze complex datasets, optimize resource use, and enhance decision-making allows organizations to integrate sustainability into their core strategies rather than treating it as an external or supplementary initiative. From reducing carbon footprints through energy-efficient operations to advancing circular economy models and enabling predictive maintenance, AI fosters systemic efficiencies that contribute to long-term sustainability. Additionally, AI-driven innovations—such as smart grids and green supply chain management—exemplify how technology can serve as a catalyst for reducing environmental impact while maintaining profitability. For organizations striving to remain competitive in the digital age, embedding AI into sustainability initiatives is no longer a choice but a necessity (Ojadi et al., 2023; Papagiannidis et al., 2023; Rajaperumal & Columbus, 2025; Zhang et al., 2025).

In today's digital economy, technology-driven organizations function as critical testing grounds and proof-of-concept environments for harnessing AI to advance sustainable growth. The adoption of AI tools reflects a broader shift in corporate governance, where digital transformation intersects with corporate social responsibility (CSR) and environmental, social, and governance (ESG) imperatives. This evolution is reinforced by regulatory pressures, stakeholder expectations, and the growing recognition that sustainable practices enhance brand reputation, attract investment, and secure long-term market relevance. The integration of green AI—approaches that not only optimize processes but also minimize the environmental costs of computing—further underscores the need for balance between technological advancement and ecological responsibility (Bolón-Canedo et al., 2024; Camilleri, 2024; Kannan et al., 2024; Popescu & Yu, 2024; Sklavos et al., 2024; Yu et al., 2025).

Given its significance, green and sustainable growth with AI in technology-driven organizations is a critical topic to study. This review article explores the multidimensional role of AI in shaping green and sustainable growth within technology-driven organizations. By analyzing existing literature and case evidence, the article highlights pathways through which AI can drive greener operations, support corporate sustainability agendas, and foster innovation aligned with global environmental priorities. Ultimately, this study underscores AI's potential to reshape organizational landscapes, transforming them into engines of sustainable development in an increasingly interconnected and resource-constrained world.

1.1 Research Objective

The objective of this study is to examine how AI contributes to shaping green and sustainable growth in technology-driven organizations.

1.2 Research Question

How does AI enable green and sustainable growth in technology-driven organizations through governance mechanisms, policy frameworks, and organizational strategies?

2. LITERATURE REVIEW

2.1 Green and Sustainable Growth in Technology-Driven Organizations

Sustainable growth is grounded in the principle of balancing environmental protection, social equity, and economic prosperity, often referred to as the triple bottom line. In the contemporary digital economy, technology-driven organizations are uniquely positioned to play a key role in advancing sustainability agendas. These organizations are not only responsible for developing cutting-edge technologies but also serve as testbeds for integrating digital innovation with environmentally conscious practices. Digital transformation, encompassing cloud computing, big data analytics, and platform-based ecosystems, has been instrumental in accelerating efficiency and reducing resource consumption across industries. By embedding sustainability considerations into their digital transformation strategies, organizations can reduce operational inefficiencies, optimize energy use, and minimize carbon footprints (Feroz et al., 2021; Kannan & Gambetta, 2025; Priya, 2024; Wei & Shen, 2025; Wu et al., 2024).

The role of technology-driven organizations extends beyond internal efficiency gains; they influence global supply chains, consumer behavior, and industry norms. Many firms are leveraging digital ecosystems to implement sustainable procurement, trace supply chain emissions, and adopt green logistics systems. Moreover, technology-driven sectors are central to realizing international sustainability goals, such as the United Nations Sustainable Development Goals (SDGs), which emphasize innovation and responsible consumption. The growing emphasis on CSR and ESG metrics has further encouraged organizations to adopt sustainability as a strategic imperative rather than a peripheral activity. In this sense, technology-driven organizations act as critical engines for shaping greener pathways to economic growth (Hariyani et al., 2024; Kandpal et al., 2024; Khan et al., 2025; Osei et al., 2025; Ullah et al., 2025).

2.2 Artificial Intelligence as a Catalyst for Sustainability

AI has emerged as a transformative force in the pursuit of sustainable development. Its capabilities in data processing, predictive analytics, optimization, and decision-making provide opportunities for organizations to minimize waste, reduce energy use, and optimize resource allocation. AI applications in sustainability are evident across multiple domains. For example, AI-enabled smart grids enhance the integration of renewable energy sources by predicting energy demand and adjusting supply in real time, thereby reducing reliance on fossil fuels. In transportation, AI algorithms optimize traffic flows and logistics routes, significantly lowering fuel consumption and carbon emissions (Agrawal et al., 2023; AlZubi & Galyna, 2023; Ojadi et al., 2024; Olawade et al., 2024; Miller et al., 2024; Ukoba et al., 2024).

In organizational contexts, AI is reshaping product design, production processes, and waste management. Machine learning models are used to forecast demand, helping firms minimize overproduction and reduce material waste. Similarly, AI tools in agriculture enable precision farming by monitoring soil health, predicting crop yields, and reducing water and fertilizer use, which directly support green growth objectives. The integration of AI into circular economy practices also highlights its catalytic role, as organizations adopt AI-driven platforms for waste tracking, product recycling, and resource recovery (Adewusi et al., 2024; Dhanya et al., 2023; Lanzalonga et al., 2024; Lin & Wei, 2023; Oladapo et al., 2024; Sen et al., 2025).

AI's potential extends beyond operational improvements to strategic decision-making. By analyzing vast datasets, AI systems can identify sustainability risks, model long-term environmental impacts, and support compliance with ESG standards. Some organizations are already deploying AI-enabled sustainability dashboards to track emissions and assess carbon footprints, providing real-time insights that guide green investments. In this regard, AI does not only serve as a technological tool but also as a strategic driver of sustainability-oriented innovation (Greif et al., 2025; Naim, 2024; Popescu & Yu, 2024; Younas & Kassim, 2025).

2.3 Governance and Policy Frameworks for AI-Enabled Sustainability

The integration of AI into sustainability agendas is closely linked to governance structures and policy frameworks. Global governance mechanisms such as the Paris Agreement, the United Nations' SDGs, and ESG standards set the normative foundation for sustainability, while regional and national policies provide regulatory frameworks that encourage or constrain AI adoption for green purposes. Governments worldwide are increasingly promoting digital governance policies that align with environmental objectives, such as supporting AI applications in climate monitoring, emission tracking, and renewable energy integration (Andrade & Vasquez, 2024; Ghosh et al., 2025; Hossain et al., 2025; Işık et al., 2025; Mutambara, 2025; Shafik, 2025; Siripipatthanakul et al., 2025; Wang et al., 2025; Xiao & Xiao, 2025).

In technology-driven organizations, governance mechanisms often take the form of internal policies, codes of conduct, and ESG reporting frameworks that embed AI tools into sustainability practices. For instance, AI is used to automate compliance monitoring, ensuring organizations meet regulatory requirements on carbon emissions, waste management, and resource efficiency. Blockchain integration further enhances transparency by providing immutable records of environmental data. However, governance challenges remain. Concerns about algorithmic bias, data privacy, and the opacity of AI decision-making raise questions about accountability and trust in AI-driven sustainability systems. These ethical and regulatory issues underscore the importance of robust governance frameworks that balance innovation with responsibility (Chauhan & Sahoo, 2024; Cheong, 2024; Kumar & Suthar, 2025; Lee et al., 2025; Mukherjee, 2025; Rauf & Baolei, 2025; Suljic, 2025; Tettey et al., 2025).

Public–private partnerships play a vital role in shaping governance for AI-enabled sustainability. Collaborative initiatives bring together governments, technology firms, research institutions, and civil society to create regulatory sandboxes, pilot programs, and industry standards. Such collaborations enable experimentation with AI solutions in controlled environments, fostering innovation while mitigating risks. Furthermore, international organizations and industry consortia are working toward harmonizing AI and sustainability regulations, but disparities across regions highlight the fragmented nature of global governance. This gap underscores the need for coordinated policies that ensure AI is effectively aligned with sustainability goals across borders and industries (Babatunde et al., 2025; Liu et al., 2025; Mallik et al., 2025; Mohieldin et al., 2025; Moraes, 2025; Mutambara, 2025; Kulothungan et al., 2025).

2.4 Challenges and Emerging Trends

Despite its promise, the integration of AI into green and sustainable growth strategies faces significant challenges. From an organizational perspective, barriers include the high costs of AI adoption, the lack of specialized expertise, and resistance to cultural change within firms. Many organizations struggle to justify the initial investment in AI infrastructure when short-term profitability pressures dominate corporate agendas. Moreover, smaller firms often lack the resources to implement sophisticated AI systems, leading to disparities in adoption rates across industries (Hangl et al., 2023; Ikpe, 2024; Kar et al., 2021; Madanchian & Taherdoost, 2025; Papagiannidis et al., 2023; Sánchez et al., 2025; Yigitcanlar et al., 2021).

Technological limitations also persist, particularly in terms of interoperability and scalability. AI systems require large volumes of high-quality data, yet environmental datasets are often fragmented, inconsistent, or inaccessible. Furthermore, AI algorithms may not be easily adaptable to diverse regional contexts, limiting their effectiveness in global sustainability efforts. On the policy front, regulation frequently lags behind technological innovation. The lack of standardized guidelines for AI in sustainability creates uncertainty, discourages investment, and leads to inconsistent adoption across jurisdictions (Alenezi & Akour, 2025; Buczynski et al., 2022; Ficili et al., 2025; Jørgensen et al., 2025; Srivastava & Maity, 2023).

Emerging trends, however, demonstrate the growing potential of AI in sustainability transitions. The integration of AI with other digital technologies, such as blockchain, the Internet of Things (IoT), and digital twins, is creating new opportunities for real-time environmental monitoring, predictive governance, and smarter green infrastructure. AI-driven supply chains are increasingly capable of automating sustainability audits and ensuring compliance with ESG requirements. Moreover, generative AI is being explored for its role in sustainability innovation, including designing green materials, simulating policy impacts, and supporting knowledge-sharing platforms that accelerate global climate solutions (Abdessadak et al., 2025; Bahaw et al., 2025; Bibri et al., 2024; Chauhan & Sahoo, 2024; Hernandez et al., 2025; Jaiswal & Pandey, 2024; Orenuga et al., 2024; Sah et al., 2024; Tripathi et al., 2024).

2.5 Synthesis of Related Research

Salamzadeh et al. (2022) highlight that rapid technological advances have led to the emergence of New Technology-Based Firms (NTBFs) focused on technology-driven business models. A key concern is the accelerated pace of technology adoption, which shortens the life cycle of technology-based goods and services. Technological progress fosters greater integration among communities, businesses, goods, and services, reflected in the growing number of NTBFs striving to enhance international relations. Addressing sustainability dimensions amid technological change is increasingly recognized as a global priority. While some studies emphasize the environmental impacts of technology, others identify technology as a primary driver of innovation, which can contribute to sustainability. The study examines international NTBFs to identify factors, key drivers, and uncertainties affecting sustainability-oriented innovations, and to develop future scenarios for these firms. The scenario analysis underscores the importance of advancing digital technologies and expanding international networks through digital platforms, which accelerate knowledge and innovation development, facilitate growth in target markets, reduce costs, and promote sustainability. By identifying key drivers and uncertainties, the research provides NTBF owners with insights to enhance sustainability-oriented innovation, increasing awareness of opportunities and informing strategic interventions at both the conceptualization and project management levels.

Suljic (2025) highlights that AI-driven digitalization demands a paradigm shift in leadership, moving beyond conventional frameworks to address disruptive innovation, ethical governance, and sustainability. Digital transformation is framed not only as a technological transition but as an organizational realignment that requires leaders with vision, adaptability, and the capacity to foster innovation while embedding sustainability as a strategic priority. Drawing on insights from leading consultancies such as McKinsey & Company, Boston Consulting Group, and Deloitte, as well as case studies of industry pioneers including Microsoft, Siemens, and JPMorgan Chase, the study underscores the advanced leadership skills needed to integrate frontier technologies such as AI, machine learning, blockchain, and financial technology (FinTech) within both multinational corporations and SMEs. Central to this leadership model is establishing ethical AI governance frameworks that ensure transparency, accountability, and equity, particularly in sectors such as healthcare, finance, and law enforcement, where algorithmic decisions have significant societal consequences. The research further identifies cross-functional collaboration as critical for organizational agility, enabling innovation through interdisciplinary synergies. Siemens' AI-enabled energy optimization systems illustrate how sustainability can be aligned with technological progress, reinforcing the broader potential of responsible digital strategies. This study concludes that strategic leadership remains the decisive factor in securing resilience, competitiveness, and transformative impact in the AI era.

Kanellopoulou et al. (2025) emphasize that the rapid rise of AI presents significant challenges that may impede adoption and limit the realization of innovation opportunities. Given the historically low success rates of digitalization projects, a structured approach to AI-driven business transformation becomes essential. To address this need, the study introduces the "APSS" methodology, a four-phase framework encompassing Awareness, Piloting, Scaling, and Sustainability, designed to guide organizations through successful and enduring AI implementations. The methodology provides a systematic progression: cultivating organizational awareness, piloting AI applications to meet specific business objectives, scaling successful initiatives, and embedding AI within long-term strategies to ensure sustained impact. Grounded in both theoretical foundations and practical insights, the APSS approach offers a comprehensive roadmap for mitigating risks, fostering cross-functional collaboration, and driving innovation. A case study of a leading technology-driven organization in Greece demonstrates the methodology's practical applicability and effectiveness in overcoming adoption barriers while generating measurable business value. By integrating change management, open innovation, and human-centric principles, this study offers a replicable and academically grounded framework for organizations seeking to navigate AI adoption successfully.

3. METHODOLOGY

According to the studies of Siripipatthanakul et al. (2025; a & b), the book's chapter is suggested to use a systematic review and thematic content analysis.

Thus, this chapter employed a systematic review to explore governance challenges and policy-driven strategies shaping the integration of AI and other emerging technologies into green governance and sustainability initiatives within technology-driven organizations. To ensure methodological rigor, the review followed the PRISMA (Preferred Reporting Items for Systematic Reviews and Meta-Analyses) guidelines, which provided a transparent, replicable, and comprehensive framework for identifying, screening, and selecting relevant literature. This structured process minimized potential bias and strengthened the robustness of the findings.

A purposive sampling strategy was applied to capture literature addressing the intersection of technology policy, green governance frameworks, and AI-driven sustainability practices. This targeted approach enabled the review to highlight research on regulatory instruments, governance mechanisms, policy innovations, and technological interventions that facilitate sustainability transitions. Sources included peer-reviewed journal articles, scholarly papers, conference proceedings, books, book chapters, and editorial notes indexed in Scopus, Web of Science (WoS), and Google Scholar. To capture contemporary debates and advancements, the review was limited to studies published between 2020 and 2025.

The search strategy combined multiple keyword clusters such as "AI in sustainability policy," "technology governance," "green governance," "digital transformation for sustainability," "circular economy and AI," "environmental regulation and technology," "ESG policy instruments," and "AI-enabled governance frameworks." Inclusion criteria required that selected studies explicitly address governance-related sustainability issues with a clear connection to technology and AI policy, including regulatory compliance in digital systems, policy support for AI-enabled green innovation, and governance mechanisms integrating technologies such as blockchain and IoT into sustainable frameworks. Conceptual analyses, empirical studies, case studies, and systematic or narrative reviews were all considered. Exclusion criteria filtered out technical engineering articles without governance or policy relevance and non-scholarly sources such as blogs, news reports, or unreviewed white papers.

The finalized corpus was examined using content analysis to systematically identify, code, and categorize recurring concepts and patterns. The analysis emphasized the relationships between AI-driven technology policy instruments and their impact on green governance outcomes. To enhance interpretability, findings were supplemented with visual representations of data, including a word cloud generated with simplewordcloud.com to illustrate the frequency and prominence of key concepts across the literature. This methodological approach provides a comprehensive foundation for understanding how AI and technology policy frameworks shape green governance structures, influence stakeholder behavior, and drive sustainable outcomes. By combining systematic literature selection, rigorous content analysis, and visual mapping, the study highlights how AI-enabled governance and innovation policies can act as powerful catalysts for advancing green and sustainable growth in technology-driven organizations.

4. RESULTS

The findings demonstrate that AI is a critical enabler of sustainable growth in technology-driven organizations. AI enhances operational efficiency, optimizes resource use, and supports the transition toward circular economy practices. The results also reveal the pivotal role of governance and policy frameworks in ensuring responsible adoption while addressing barriers such as ethical risks and regulatory fragmentation. Figure 1 presents a word cloud that illustrates the most frequent and significant concepts identified in the analysis.

Figure 1. Word Cloud of Key Themes from on AI-Enabled Sustainable Growth in Technology-Driven Organizations

Figure 1 visually represents the most prominent concepts and themes emerging from the study's results on AI and sustainable growth in technology-driven organizations. The size of each word indicates its relative frequency and importance in the dataset, highlighting the central role of certain terms. At the core, "artificial," "intelligence," and "AI" dominate the cloud, reflecting the study's emphasis on AI as the primary driver of green and sustainable transformation. Closely associated terms include "sustainability," "environmental," and "organizations," underscoring the focus on how AI is embedded in corporate sustainability strategies and environmental stewardship practices. Other significant clusters include "resource," "energy," and "efficiency," pointing to AI's role in optimizing resource use, reducing waste, and enabling energy-efficient solutions such as smart grids and predictive analytics. The prominence of "governance," "policy," "accountability," and "transparency" underscores the importance of institutional frameworks and ethical considerations in the responsible deployment of AI in sustainability initiatives. Additionally, terms like "circular," "product," "opportunities," and "integration" suggest AI's role in advancing circular economy practices, sustainable product design, and cross-sector integration. Words such as "challenges," "risks," and "concerns" also appear, signaling that while AI presents transformative opportunities, barriers such as policy fragmentation, ethical issues, and adoption disparities remain critical considerations.

The findings of this systematic review reveal that AI has become a central enabler of green and sustainable growth within technology-driven organizations in the digital economy. Across the reviewed literature, AI consistently demonstrated its capacity to optimize resource utilization, enhance operational efficiency, and minimize environmental impacts. In manufacturing, AI-enabled predictive analytics facilitated more efficient energy consumption and waste reduction, while in logistics, AI-supported route optimization significantly decreased greenhouse gas emissions. These applications highlight AI's potential to move beyond efficiency gains toward transformative contributions to sustainability by embedding ecological considerations into core business functions. Technology-driven organizations that integrated AI into their sustainability agendas were found to achieve a dual benefit of maintaining competitiveness while contributing meaningfully to environmental stewardship.

A second important result concerns the role of AI in advancing circular economy practices. The evidence showed that AI-powered platforms support waste tracking, recycling, and resource recovery, making circular models more feasible and scalable. For example, machine learning algorithms were applied to monitor material flows and predict opportunities for reuse, while AI-driven decision systems facilitated sustainable product design. These capabilities enable organizations to close resource loops and extend product lifecycles, directly addressing sustainability challenges such as overproduction and resource depletion. Moreover, in the agriculture and energy sectors, AI applications enabled precision farming and the integration of renewable energy into smart grids, reinforcing AI's role as a catalyst for sustainability transitions in industries traditionally associated with high environmental impact.

The results also emphasized the critical influence of governance and policy frameworks in shaping AI-enabled sustainability practices. Studies highlighted that organizations operating under strong ESG mandates, regulatory incentives, and transparent governance structures were more likely to adopt AI-driven green strategies effectively. Public–private collaborations emerged as a particularly important mechanism, providing platforms for experimentation through regulatory sandboxes, pilot projects, and shared knowledge. These governance models not only reduced risks associated with AI adoption but also aligned corporate actions with global sustainability standards, such as the United Nations SDGs. However, challenges were identified, including fragmented policies across jurisdictions, insufficient harmonization of standards, and ethical concerns related to transparency and accountability in AI decision-making, underscoring the need for more coherent global frameworks.

Finally, emerging trends across the literature point to the increasing integration of AI with complementary digital technologies, creating new opportunities for sustainable innovation. The combination of AI with blockchain, IoT, and digital twins enabled advanced applications such as real-time environmental monitoring, automated sustainability audits, and predictive governance models. These integrations not only improved organizational responsiveness to environmental risks but also enhanced transparency and accountability in sustainability reporting. Additionally, the rise of generative AI was noted for its potential to accelerate sustainability innovation, particularly in the development of eco-friendly materials and in simulating policy interventions. These findings suggest that AI is not merely a supportive tool but a transformative force in reshaping technology-driven organizations into active engines of sustainable growth.

5. DISCUSSIONS

The findings of this study underscore the transformative role of AI in enabling green and sustainable growth within technology-driven organizations. The evidence suggests that AI not only enhances operational efficiency and resource optimization but also supports broader sustainability transitions through governance, policy alignment, and cross-sectoral integration. These insights contribute to an expanding body of literature that positions AI as both a technological enabler and a strategic driver of sustainability-oriented innovation.

The results affirm that AI is a critical mechanism for embedding sustainability into organizations' core operations. By optimizing resource allocation and enabling predictive analytics, AI facilitates reductions in energy consumption, emissions, and waste (Agrawal et al., 2023; Miller et al., 2024). These findings are consistent with prior research that highlights the ability of AI-enabled systems such as smart grids and precision agriculture to minimize ecological footprints while maintaining profitability (Adewusi et al., 2024; Ukoba et al., 2024). Importantly, this convergence of ecological responsibility and economic viability reflects the triple bottom line principle, in which environmental stewardship, social equity, and profitability are pursued in tandem (Feroz et al., 2021; Khan et al., 2025).

The findings emphasize AI's capacity to advance circular economy practices. Evidence from the review illustrates that AI-powered platforms enable waste tracking, recycling, and resource recovery, thereby extending product life cycles and reducing overproduction (Lanzalonga et al., 2024; Oladapo et al., 2024; Purushothaman et al., 2025; Singh, 2025). This aligns with earlier studies suggesting that machine learning and predictive models can support industrial symbiosis and sustainable product design (Lin & Wei, 2023; Sen et al., 2025; Zhao et al., 2025). By embedding intelligence into material flows, organizations can transition from linear consumption models toward regenerative systems that mitigate resource depletion.

The study highlights governance and policy frameworks as pivotal factors shaping the adoption and effectiveness of AI for sustainability. Organizations operating within strong ESG structures were better positioned to deploy AI responsibly and transparently. This finding aligns with existing research identifying governance mechanisms—such as ESG reporting frameworks, regulatory incentives, and public–private partnerships—as critical enablers of AI-enabled sustainability (Cheong, 2024; Ghosh et al., 2025; Siripipatthanakul et al., 2025 c). However, the results also reveal persistent governance challenges, including fragmented policies, a lack of harmonized standards, and ethical concerns regarding transparency and accountability (Camilleri, 2024; Mukherjee, 2025). These limitations point to the need for coordinated global frameworks that balance innovation with responsibility (Kulothungan et al., 2025).

Finally, the integration of AI with emerging technologies such as blockchain, IoT, and digital twins is shown to open new frontiers for sustainable innovation. Prior studies confirm that these combinations enhance environmental monitoring, automate compliance processes, and enable predictive governance models (Hemdan et al., 2023; Chauhan & Sahoo, 2024; Bibri et al., 2024; Abdessadak et al., 2025). Moreover, the potential of generative AI to design green materials and simulate sustainability policies underscores its capacity to accelerate transitions to low-carbon futures (Amirkhizi et al., 2025; Bahaw et al., 2025; Jaiswal & Pandey, 2024). These synergies suggest that AI should be viewed not as a standalone tool but as a cornerstone of digital ecosystems that foster resilient and sustainable growth pathways.

In sum, the discussion affirms that AI offers significant opportunities for technology-driven organizations to achieve sustainable growth. Yet, realizing this potential requires overcoming barriers related to cost, expertise, governance, and regulatory fragmentation (Hangl et al., 2023; Madanchian & Taherdoost, 2025). Addressing these challenges through strategic investments, capacity building, and policy harmonization will be essential for ensuring that AI-driven sustainability transitions are inclusive, scalable, and ethically grounded.

6. CONCLUSION

This chapter has demonstrated that AI plays a pivotal role in shaping sustainable growth in technology-driven organizations by integrating ESG imperatives into digital transformation strategies. The findings highlight how AI enhances operational efficiency, reduces resource consumption, and enables predictive decision-making that supports sustainability transitions across multiple sectors. Applications such as smart grids, precision agriculture, and sustainable logistics showcase the potential of AI not only to optimize performance but also to mitigate ecological impacts in ways that were previously unattainable. A significant conclusion drawn from this review is that AI's transformative power extends beyond technological innovation to strategic governance and organizational adaptation. Firms that align their AI adoption with ESG mandates, international sustainability goals such as the United Nations SDGs, and regulatory frameworks are better positioned to achieve a dual outcome: competitive advantage and meaningful environmental stewardship.

Fostering sustainable growth in technology-driven enterprises necessitates the integration of AI at both strategic and operational levels to enhance efficiency, innovation, and long-term resilience. Artificial intelligence technologies, including predictive analytics, machine learning, and automation, enable enterprises to enhance resource utilization, reduce waste, and refine operations, thereby directly supporting environmental and economic sustainability objectives. Utilizing AI for data-driven decision-making enables firms to comprehend and reduce their environmental impact, from streamlining supply chains to improving product lifecycles. Furthermore, AI propels the creation of innovative sustainable products and services, such as intelligent energy grids and precision agriculture instruments, thereby unlocking new markets and remaining relevant in a resource-aware environment. Integrating AI into the fundamental business plan enables technology companies to create a robust synergy between profitability and purpose, cultivating an agile culture that can adapt to rapid technological and environmental changes while ensuring responsible growth.

Furthermore, the convergence of AI with complementary technologies such as blockchain, IoT, and digital twins creates synergies that magnify sustainability benefits, allowing for real-time monitoring, transparent reporting, and enhanced resilience. However, the study also reveals that AI's integration into sustainability initiatives is not without challenges. Issues such as high implementation costs, fragmented governance structures, lack of harmonized global policies, and ethical concerns surrounding transparency and accountability remain pressing. Thus, while AI offers transformative potential, its impact on sustainable growth will depend on the extent to which organizations, policymakers, and stakeholders can collaboratively address these barriers. In conclusion, AI must be viewed not simply as a technological tool but as a cornerstone of organizational transformation toward green and sustainable growth. Its adoption provides organizations with opportunities to embed sustainability into their core strategies, but these opportunities must be supported by coherent governance, cross-sectoral collaboration, and forward-looking policy frameworks to achieve global environmental and social objectives.

7. IMPLICATIONS

The findings of this study carry important implications for practice, policy, and theory. Practical implications suggest that technology-driven organizations should embed AI into sustainability strategies as a central, rather than peripheral, element of corporate governance. Companies can leverage AI to achieve operational efficiencies—such as optimized energy management and reduced emissions—while simultaneously enhancing transparency and accountability through data-driven reporting mechanisms. The study also underscores the need for businesses to invest in cross-functional teams capable of integrating AI solutions into sustainability operations, thereby reducing silos and enhancing organizational agility. Moreover, policy implications highlight the vital role of governments and international bodies in creating harmonized regulatory frameworks for AI-enabled sustainability. Current policy fragmentation hampers cross-border collaboration and discourages investment in green technologies. Therefore, coordinated global governance is required to set standards, ensure ethical AI deployment, and incentivize innovation through regulatory sandboxes, tax benefits, or public–private partnerships. Such measures would not only accelerate AI adoption but also reduce disparities between large corporations and SMEs, which often face greater barriers in adopting advanced technologies. Furthermore, this study's contribution to the broader discourse on the intersection of digital transformation and sustainability has theoretical implications. By framing AI as both a technological enabler and a governance instrument, the research enriches theoretical perspectives on sustainable innovation and responsible digitalization. The integration of AI into circular economy models, ESG frameworks, and sustainability reporting systems expands the conceptual understanding of how emerging technologies can drive systemic environmental and organizational change in this digital economy. Overall, the study calls for a holistic approach where AI is integrated not only into technical processes but also into governance and policy structures that define long-term sustainability trajectories.

8. LIMITATIONS

Despite its contributions, this study is subject to several limitations. First, as a systematic literature review, the findings are constrained by the scope of the included studies, which were limited to publications from 2020 to 2025. While this ensures contemporary relevance, it may exclude earlier foundational research that could provide additional context to the evolution of AI and sustainability integration. Second, the reliance on published literature introduces potential publication bias, as studies with positive outcomes are more likely to be disseminated than those reporting failures or challenges. This may skew the evidence toward an overly optimistic view of AI's role in sustainability. Third, the study's focus on governance and organizational strategies means that the technical dimensions of AI implementation, such as algorithmic optimization and software scalability, were addressed only tangentially. A more technical examination might provide deeper insights into the environmental costs of AI deployment, such as carbon emissions associated with data centers and model training. Fourth, while this study acknowledges regional disparities in governance and adoption, the analysis is largely global and may overlook localized cultural, political, and economic contexts that shape AI-enabled sustainability practices in specific countries or industries. Finally, the review methodology, while systematic and rigorous, inherently limits causal inferences. The relationships identified between AI adoption, governance, and sustainability outcomes remain largely correlational, requiring empirical testing in real-world organizational contexts to validate and expand the conclusions.

9. FUTURE RESEARCH

Future research should build on these findings by addressing both theoretical gaps and practical challenges. One promising direction involves conducting empirical case studies of organizations that have successfully embedded AI into sustainability practices. Such studies could provide rich insights into best practices, success factors, and context-specific barriers. Comparative research across industries and regions would also shed light on the heterogeneity of AI adoption and the varying impacts of regulatory, cultural, and economic conditions. Another area for future exploration lies in the intersection of AI and the circular economy. While this study highlights AI's role in resource recovery and waste reduction, further research could evaluate its capacity to support large-scale circular ecosystems across global supply chains. Similarly, the integration of AI with blockchain, IoT, and digital twins merits deeper investigation, particularly regarding how these technologies can collectively enhance transparency, accountability, and predictive governance for sustainability transitions. Scholars should also examine the ethical and environmental costs of AI adoption, including energy-intensive model training, algorithmic bias, and privacy risks. Research into "green AI" approaches—such as energy-efficient computing and low-carbon model design—will be crucial for ensuring that AI adoption itself does not undermine sustainability goals. Furthermore, longitudinal studies could assess the long-term impact of AI-enabled governance frameworks on corporate behavior, stakeholder trust, and overall progress toward international sustainability targets such as the SDGs. Finally, interdisciplinary research bridging management, computer science, and environmental studies will be essential for developing holistic frameworks that integrate technological innovation with ethical, regulatory, and ecological considerations. Such research would not only enrich academic knowledge but also provide actionable guidance for policymakers and organizational leaders navigating the complexities of AI-enabled sustainable growth.

10. ACKNOWLEDGEMENT

The author expresses his sincere appreciation to the International College of Pathumthani University and the Research and Innovation Center of Pathumthani University for their invaluable support, guidance, and encouragement throughout this work.

11. DECLARATION

This study is not subject to competing interests, as stated by the author.

REFERENCES

Abdessadak, A., Ghennioui, H., Thirion-Moreau, N., Elbhiri, B., Abraim, M., & Merzouk, S. (2025). Digital twin technology and artificial intelligence in energy transition: A comprehensive systematic review of applications. *Energy Reports*, *13*, 5196–5218. DOI: 10.1016/j.egyr.2025.04.060

Abdulmuhsin, A. A., & Hussein, H. D., AL-Abrrow, H., Masa'deh, R. E., & Alkhwaldi, A. F. (2025). Impact of artificial intelligence and knowledge management on proactive green innovation: The moderating role of trust and sustainability. *Asia-Pacific Journal of Business Administration*, *17*(3), 765–795. DOI: 10.1108/APJBA-05-2024-0301

Adewusi, A. O., Asuzu, O. F., Olorunsogo, T., Iwuanyanwu, C., Adaga, E., & Daraojimba, D. O. (2024). AI in precision agriculture: A review of technologies for sustainable farming practices. *World Journal of Advanced Research and Reviews*, *21*(1), 2276–2285. DOI: 10.30574/wjarr.2024.21.1.0314

Agrawal, R., Majumdar, A., Kumar, A., & Luthra, S. (2023). Integration of artificial intelligence in sustainable manufacturing: Current status and future opportunities. *Operations Management Research*, *16*(4), 1720–1741. DOI: 10.1007/s12063-023-00383-y

Alamandi, M. (2025). Sustainable innovation management: Balancing economic growth and environmental responsibility. *Sustainability*, *17*(10), 4362. DOI: 10.3390/su17104362

Alenezi, M., & Akour, M. (2025). AI-driven innovations in software engineering: A review of current practices and future directions. *Applied Sciences (Basel, Switzerland), 15*(3), 1344. DOI: 10.3390/app15031344

AlZubi, A. A., & Galyna, K. (2023). Artificial intelligence and internet of things for sustainable farming and smart agriculture. *IEEE Access : Practical Innovations, Open Solutions, 11*, 78686–78692. DOI: 10.1109/ACCESS.2023.3298215

Amirkhizi, P. J., Pedrammehr, S., Pakzad, S., & Shahhoseini, A. (2025). Generative artificial intelligence in adaptive social manufacturing: A pathway to achieving industry 5.0 sustainability goals. *Processes (Basel, Switzerland), 13*(4), 1174. DOI: 10.3390/pr13041174

Andrade, A. C. C., & Vasquez, M. C. S. (2024). The compatibility between SDGs and the EU regulatory framework of AI. *Journal of Ethics and Legal Technologies, 6*(1), 111–144. DOI: 10.14658/pupj-JELT-2024-1-6

Babatunde, O. B., Okonji, C. T., Olanihun, Z. S., & Daniel, H. A. (2025). Public-private partnerships in advancing AI-based logistics. *NIU Journal of Humanities, 10*(2), 87–100. DOI: 10.58709/niujhu.v10i2.2199

Bahaw, P., Forgenie, D., Sadiq, G., & Sookhai, S. (2025). Generative AI for business sustainability: Examining usability, usefulness, and triple bottom line impacts in small and medium enterprises. *Sustainable Futures : An Applied Journal of Technology, Environment and Society, 100815*. Advance online publication. DOI: 10.1016/j.sftr.2025.100815

Bibri, S. E., Huang, J., Jagatheesaperumal, S. K., & Krogstie, J. (2024). The synergistic interplay of artificial intelligence and digital twin in environmentally planning sustainable smart cities: A comprehensive systematic review. *Environmental Science and Ecotechnology, 20*, 100433. DOI: 10.1016/j.ese.2024.100433

Bolón-Canedo, V., Morán-Fernández, L., Cancela, B., & Alonso-Betanzos, A. (2024). A review of green artificial intelligence: Towards a more sustainable future. *Neurocomputing, 599*, 128096. DOI: 10.1016/j.neucom.2024.128096

Buczynski, W., Steffek, F., Cuzzolin, F., & Jamnik, M., & SAHAKIAN, B. (2022). Hard law and soft law regulations of artificial intelligence in investment management. *Cambridge Yearbook of European Legal Studies, 24*, 262–293. DOI: 10.1017/cel.2022.10

Camilleri, M. A. (2024). Artificial intelligence governance: Ethical considerations and implications for social responsibility. *Expert Systems: International Journal of Knowledge Engineering and Neural Networks*, *41*(7), e13406. DOI: 10.1111/exsy.13406

Chauhan, M., & Sahoo, D. R. (2024). Towards a greener tomorrow: Exploring the potential of AI, blockchain, and IoT in sustainable development. *Nature Environment and Pollution Technology*, *23*(2), 1105–1113. DOI: 10.46488/NEPT.2024.v23i02.044

Cheong, B. C. (2024). Transparency and accountability in AI systems: Safeguarding wellbeing in the age of algorithmic decision-making. *Frontiers in Human Dynamics*, *6*, 1421273. DOI: 10.3389/fhumd.2024.1421273

Dhanya, D., Kumar, S. S., Thilagavathy, A., Prasad, D. V., & Boopathi, S. (2023). Data analytics and artificial intelligence in the circular economy: Case studies. In Mishra, B. (Ed.), *Intelligent Engineering Applications and Applied Sciences for Sustainability* (pp. 40–58). IGI Global Scientific Publishing., DOI: 10.4018/979-8-3693-0044-2.ch003

Erturk, A., Özşahin, M., & Simovic, V. (Eds.). (2025). *Human-centric, sustainable, and resilient organizations in the digital age*. IGI Global Scientific Publishing., DOI: 10.4018/979-8-3693-8181-6

Feroz, A. K., Zo, H., & Chiravuri, A. (2021). Digital transformation and environmental sustainability: A review and research agenda. *Sustainability*, *13*(3), 1530. DOI: 10.3390/su13031530

Ficili, I., Giacobbe, M., Tricomi, G., & Puliafito, A. (2025). From sensors to data intelligence: Leveraging IoT, cloud, and edge computing with AI. *Sensors (Basel)*, *25*(6), 1763. DOI: 10.3390/s25061763

Ghosh, A., Saini, A., & Barad, H. (2025). Artificial intelligence in governance: Recent trends, risks, challenges, innovative frameworks and future directions. *AI & Society*. Advance online publication. DOI: 10.1007/s00146-025-02312-y

Greif, L., Röckel, F., Kimmig, A., & Ovtcharova, J. (2025). A systematic review of current AI techniques used in the context of the SDGs. *International Journal of Environmental Research*, *19*(1), 1–36. DOI: 10.1007/s41742-024-00668-5

Hangl, J., Krause, S., & Behrens, V. J. (2023). Drivers, barriers and social considerations for AI adoption in SCM. *Technology in Society*, *74*, 102299. DOI: 10.1016/j.techsoc.2023.102299

Hariyani, D., Hariyani, P., Mishra, S., & Sharma, M. K. (2024). A literature review on green supply chain management for sustainable sourcing and distribution. *Waste Management Bulletin, 2*(4), 231–248. DOI: 10.1016/j.wmb.2024.11.009

Hemdan, E. E. D., El-Shafai, W., & Sayed, A. (2023). Integrating digital twins with IoT-based blockchain: Concept, architecture, challenges, and future scope. *Wireless Personal Communications, 131*, 2193–2216. DOI: 10.1007/s11277-023-10538-6

Hernandez, J. C. R., Villa-Enciso, E., Cardona-Acevedo, S., Valencia, J., & Velasquez Salas, S. (2025). Smart innovation for a circular economy: A systematic review of emerging trends and the future of AI in the sustainable economy. *Sustainability, 17*(13), 5793. DOI: 10.3390/su17135793

Hossain, I., Haque, A. M., & Ullah, S. A. (Eds.). (2025). *Intersecting environmental governance with technological advancements*. IGI Global Scientific Publishing., DOI: 10.4018/979-8-3693-7001-8

Ikpe, E. O. (2024). Adoption and implementation of artificial intelligence in small businesses in selected developing countries. *Journal of Health. Applied Sciences and Management, 8*(1), 26–34. DOI: 10.4314/johasam.v8i1.3

Işık, C., Ongan, S., & Islam, H. (2025). Driving energy transition through artificial intelligence: Integrating economic, environmental, social, and governance (ECON-ESG) factors in OECD countries. *Journal of the Knowledge Economy*. Advance online publication. DOI: 10.1007/s13132-025-02687-9

Jaiswal, S., & Pandey, S. (2024). Exploring generative AI as a catalyst for sustainability: Strategies for waste and energy reduction. *Management Insight, 20*(1), 14–26. DOI: 10.21844/mijia.20.1.2

Jørgensen, B. N., Gunasekaran, S. S., & Ma, Z. G. (2025). Impact of EU laws on AI adoption in smart grids: A review of regulatory barriers, technological challenges, and stakeholder benefits. *Energies, 18*(12), 3002. DOI: 10.3390/en18123002

Kandpal, V., Jaswal, A., Gonzalez, E. D. R. S., & Agarwal, N. (Eds.). (2024). *Sustainable energy transition: Circular economy and sustainable financing for environmental, social and governance (ESG) practices*. Springer., DOI: 10.1007/978-3-031-52943-6

Kanellopoulou, D., Giannakopoulos, G., & Karkaletsis, V. (2025). Accelerating AI-powered digital innovation through "APSS": A novel methodology for sustainable business AI transformation. *Journal of Open Innovation, 11*(1), 100495. DOI: 10.1016/j.joitmc.2025.100495

Kannan, H., Rodriguez, R. V., Paprika, Z. Z., & Ade-Ibijola, A. (Eds.). (2024). *Exploring ethical dimensions of environmental sustainability and use of AI.* IGI Global Scientific Publishing., DOI: 10.4018/979-8-3693-0892-9

Kannan, S., & Gambetta, N. (2025). Technology-driven sustainability in small and medium-sized enterprises: A systematic literature review. *Journal of Small Business Strategy*, *35*(1), 129–157. DOI: 10.53703/001c.126636

Kar, S., Kar, A. K., & Gupta, M. P. (2021). Modeling drivers and barriers of artificial intelligence adoption: Insights from a strategic management perspective. *Intelligent Systems in Accounting, Finance & Management*, *28*(4), 217–238. DOI: 10.1002/isaf.1503

Khan, R. Z., Razak, L. A., & Premaratne, G. (2025). Green growth and sustainability: A systematic literature review on theories, measures and future directions. *Cleaner and Responsible Consumption*, *17*, 100274. DOI: 10.1016/j.clrc.2025.100274

Khokhar, M. (Ed.). (2025). *AI-driven business model innovation.* IGI Global Scientific Publishing., DOI: 10.4018/979-8-3693-9571-4

Kulothungan, V., Mohan, P. R., & Gupta, D. (2025, May). AI regulation and capitalist growth: Balancing innovation, ethics, and global governance. In *2025 IEEE 11th Conference on Big Data Security on Cloud (BigDataSecurity)* (pp. 39-45). IEEE. https://doi.org/DOI: 10.1109/BigDataSecurity66063.2025.00020

Kumar, D., & Suthar, N. (2025). Balancing innovation and regulation: Ethical frameworks for AI policy and governance. In Youssef, A., & Arslan, A. (Eds.), *Navigating the Intersection of AI Policy, Technology, and Governance* (pp. 171–200). IGI Global Scientific Publishing., DOI: 10.4018/979-8-3373-1210-1.ch007

Lanzalonga, F., Marseglia, R., Irace, A., & Biancone, P. P. (2024). The application of artificial intelligence in waste management: Understanding the potential of data-driven approaches for the circular economy paradigm. *Management Decision.* Advance online publication. DOI: 10.1108/MD-10-2023-1733

Le, T. T. (2022). How do corporate social responsibility and green innovation transform corporate green strategy into sustainable firm performance? *Journal of Cleaner Production*, *362*, 132228. DOI: 10.1016/j.jclepro.2022.132228

Lee, S. U., Perera, H., Liu, Y., Xia, B., Lu, Q., Zhu, L., Cairns, J., & Nottage, M. (2025). Integrating ESG and AI: A comprehensive responsible AI assessment framework. *AI and Ethics*, •••, 1–28. DOI: 10.1007/s43681-025-00741-5

Lin, K. Y., & Wei, S. H. (2023). Advancing the industrial circular economy: The integrative role of machine learning in resource optimization. *Journal of Green Economy and Low-Carbon Development*, *2*(3), 122–136. DOI: 10.56578/jgelcd020302

Liu, Z., Wu, C., & Xu, X. (2025). The role of artificial intelligence in sustainable development and industrial transformation. *Asia Pacific Economic and Management Review*, *2*(2), 1–6. DOI: 10.62177/apemr.v2i2.185

Madanchian, M., & Taherdoost, H. (2025). Barriers and enablers of AI adoption in human resource management: A critical analysis of organizational and technological factors. *Information (Basel)*, *16*(1), 51. DOI: 10.3390/info16010051

Mallik, S., Mathivanan, S. K., Sangeetha, S., & Soufiene, B. O. (Eds.). (2025). *Real-world applications of AI innovation*. IGI Global Scientific Publishing., DOI: 10.4018/979-8-3693-4252-7

Martínez-Peláez, R., Ochoa-Brust, A., Rivera, S., Félix, V. G., Ostos, R., Brito, H., & Mena, L. J. (2023). Role of digital transformation for achieving sustainability: Mediated role of stakeholders, key capabilities, and technology. *Sustainability*, *15*(14), 11221. DOI: 10.3390/su151411221

Miller, T., Durlik, I., Kostecka, E., Łobodzińska, A., & Matuszak, M. (2024). The emerging role of artificial intelligence in enhancing energy efficiency and reducing GHG emissions in transport systems. *Energies*, *17*(24), 6271. DOI: 10.3390/en17246271

Mohieldin, M., Gonzalez-Perez, M. A., & Zahran, M. (2025). AI governance, regulation, and ethical implementation. In *AI-Powered Sustainable Business. Palgrave Studies in Moral and Mindful Approaches to Leadership and Business* (pp. 81-102). Palgrave Macmillan. https://doi.org/DOI: 10.1007/978-3-031-93357-8_5

Moraes, T. (2025). Regulatory sandboxes for trustworthy artificial intelligence – Global and Latin American experiences. *International Review of Law Computers & Technology*, *39*(1), 55–74. DOI: 10.1080/13600869.2024.2351674

Mukherjee, B. N. (2025). Navigating AI governance: National and international legal and regulatory frameworks. In Youssef, A., & Arslan, A. (Eds.), *Navigating the Intersection of AI Policy, Technology, and Governance* (pp. 201–224). IGI Global Scientific Publishing., DOI: 10.4018/979-8-3373-1210-1.ch008

Mutambara, A. G. O. (Ed.). (2025). *Deploying artificial intelligence to achieve the UN sustainable development goals: Enablers, drivers and strategic framework*. Springer., DOI: 10.1007/978-3-031-88423-8

Naim, A. (Ed.). (2024). *Harnessing high-performance computing and AI for environmental sustainability*. IGI Global Scientific Publishing., DOI: 10.4018/979-8-3693-1794-5

Odeyemi, O., Usman, F. O., Mhlongo, N. Z., Elufioye, O. A., & Ike, C. U. (2024). Sustainable entrepreneurship: A review of green business practices and environmental impact. *World Journal of Advanced Research and Reviews, 21*(2), 346–358. DOI: 10.30574/wjarr.2024.21.2.0461

Ojadi, J. O., Odionu, C., Onukwulu, E., & Owulade, O. (2024). Big data analytics and AI for optimizing supply chain sustainability and reducing greenhouse gas emissions in logistics and transportation. *International Journal of Multidisciplinary Research and Growth Evaluation, 5*(1), 1536–1548. DOI: 10.54660/.IJMRGE.2024.5.1.1536-1548

Ojadi, J. O., Onukwulu, E., Odionu, C., & Owulade, O. (2023). AI-driven predictive analytics for carbon emission reduction in industrial manufacturing: A machine learning approach to sustainable production. *International Journal of Multidisciplinary Research and Growth Evaluation, 4*(1), 948–960. DOI: 10.54660/.IJMRGE.2023.4.1.948-960

Oladapo, B. I., Olawumi, M. A., & Omigbodun, F. T. (2024). AI-driven circular economy of enhancing sustainability and efficiency in industrial operations. *Sustainability, 16*(23), 10358. DOI: 10.3390/su162310358

Olawade, D. B., Wada, O. Z., David-Olawade, A. C., Fapohunda, O., Ige, A. O., & Ling, J. (2024). Artificial intelligence potential for net zero sustainability: Current evidence and prospects. *Next Sustainability, 4*, 100041. DOI: 10.1016/j.nxsust.2024.100041

Orenuga, A., Oyeyemi, B. B., & John, A. O. (2024). AI and sustainable supply chain practices: ESG goals in the US and Nigeria. *International Journal of Future Engineering Innovations, 1*(1), 127–137. DOI: 10.54660/IJFEI.2024.1.1.127-137

Osei, A., Zhu, N., Borgi, H., & Cobbinah, J. (2025). Empowering sustainable production: Firm governance, finance strategies, and tech-innovation in advancing circular economy for SDG 12 in the US, China and Japan. *Sustainable Development, 33*(4), 5282–5307. DOI: 10.1002/sd.3390

Papagiannidis, E., Enholm, I. M., Dremel, C., Mikalef, P., & Krogstie, J. (2023). Toward AI governance: Identifying best practices and potential barriers and outcomes. *Information Systems Frontiers, 25*, 123–141. DOI: 10.1007/s10796-022-10251-y

Popescu, C. R., & Yu, P. (Eds.). (2024). *Intersecting environmental social governance and AI for business sustainability*. IGI Global Scientific Publishing., DOI: 10.4018/979-8-3693-1151-6

Priya, M. R. (2024). The triple bottom line of green transitions: Assessing the economic, social, and environmental impacts of sustainable development goals by 2030. In Yıldırım, S., Yıldırım, D. C., Demirtaş, I., & Kandpal, V. (Eds.) *Green Transition Impacts on the Economy, Society, and Environment* (pp. 182-201). IGI Global Scientific Publishing. https://doi.org/DOI: 10.4018/979-8-3693-3985-5.ch011

Purushothaman, R., Alamelu, R., & Sudha, M. (2025). Leveraging solutions for advancing circular economy strategies: An overview. In Pujari, P., Khan, S. A., Kumar, A., & Naim, A. (Eds.), *Fostering Economic Diversification and Sustainable Business Through Digital Intelligence* (pp. 283–304). IGI Global Scientific Publishing., DOI: 10.4018/979-8-3693-8492-3.ch014

Rajaperumal, T. A., & Columbus, C. C. (2025). Transforming the electrical grid: The role of AI in advancing smart, sustainable, and secure energy systems. *Energy Informatics*, *8*, 51. DOI: 10.1186/s42162-024-00461-w

Rauf, F., & Baolei, Q. (2025). Enhancing ESG disclosure through AI adoption: The moderating role of institutional quality. *Applied Economics*, •••, 1–19. DOI: 10.1080/00036846.2025.2547117

Sah, B. P., Hasan, M. M., Shofiullah, S., & Faysal, S. A. (2024). AI-driven IoT and blockchain integration in industry 5.0: A systematic review of supply chain transformation. *Innovatech Engineering Journal*, *1*(1), 99–116. DOI: 10.70937/itej.v1i01.12

Salamzadeh, A., Hadizadeh, M., Rastgoo, N., Rahman, M. M., & Radfard, S. (2022). Sustainability-oriented innovation foresight in international new technology based firms. *Sustainability*, *14*(20), 13501. DOI: 10.3390/su142013501

Sánchez, E., Calderón, R., & Herrera, F. (2025). Artificial intelligence adoption in SMEs: Survey based on TOE–DOI framework, primary methodology and challenges. *Applied Sciences (Basel, Switzerland)*, *15*(12), 6465. DOI: 10.3390/app15126465

Sen, M., Garai, S., Dutta, S., Rakshit, D., Roy, A., Das, R., & Das, S. (2025). Harnessing artificial intelligence for circular economy in agriculture: Reducing waste and maximising efficiency. In Garwi, J., Dzingirai, M., & Masengu, R. (Eds.), *Integrating Agriculture, Green Marketing Strategies, and Artificial Intelligence* (pp. 317–362). IGI Global Scientific Publishing., DOI: 10.4018/979-8-3693-6468-0.ch011

Shafik, W. (Ed.). (2025). *Factoring technology in global sustainability: A focus on the sustainable development goals*. Springer., DOI: 10.1007/978-981-96-7299-8

Singh, A. (2025). AI-driven innovations for enabling a circular economy: Optimizing resource efficiency and sustainability. In Ozen, E., Singh, A., Taneja, S., Rajaram, R., & Davim, J. P. (Eds.) *Innovating Sustainability Through Digital Circular Economy* (pp. 47-64). IGI Global Scientific Publishing. https://doi.org/DOI: 10.4018/979-8-3373-0578-3.ch003

Siripipatthanakul, S., Phuangsuwan, P., Siripipattanakul, S., & Jaipong, P. (2025, a). Sustainable Tourism Based on Challenges, Tourist Experiences, and Host Communities: A Case Study of Ayutthaya, Thailand. In *Best Practices for Human Resources Management in Tourism and Hospitality* (pp. 361-394). IGI Global Scientific Publishing. DOI: DOI: 10.4018/979-8-3693-8779-5.ch014

Siripipatthanakul, S., Rakpathum, W., Siripipattanakul, S., Kaewnaknaew, C., Phuangsuwan, P., & Jaipong, P. (2025, b). The Future Artificial Intelligence (AI) for the Hotel and Hospitality Industry: The Case of Hotel's Human Resource Management (HRM). In *Best Practices for Human Resources Management in Tourism and Hospitality* (pp. 395-422). IGI Global Scientific Publishing. DOI: DOI: 10.4018/979-8-3693-8779-5.ch015

Siripipatthanakul, S., Phuangsuwan, P., & Limna, P. Muthmainnha, Siripipattanakul, S., Vui, C. N., & Jaipong, P. (2025, c). Artificial intelligence (AI) influencing sustainable governance: Governance adopting AI. In Swaroop, A., Virdee, B., Correia, S. D., & Polkowski, Z. (Eds.) *Proceedings of Data Analytics and Management. ICDAM 2024. Lecture Notes in Networks and Systems*. Springer. https://doi.org/DOI: 10.1007/978-981-96-3372-2_38

Sklavos, G., Theodossiou, G., Papanikolaou, Z., Karelakis, C., & Ragazou, K. (2024). Environmental, social, and governance-based artificial intelligence governance: Digitalizing firms' leadership and human resources management. *Sustainability*, *16*(16), 7154. DOI: 10.3390/su16167154

Srivastava, A., & Maity, R. (2023). Assessing the potential of AI–ML in urban climate change adaptation and sustainable development. *Sustainability*, *15*(23), 16461. DOI: 10.3390/su152316461

Suljic, V. (2025). Strategic leadership in AI-driven digital transformation: Ethical governance, innovation management, and sustainable practices for global enterprises and SMEs. *SBS Journal of Applied Business Research*, *13*(1), 17–35. DOI: 10.70301/JOUR/SBS-JABR/2025/13/1/2

Tettey, D. J., Chrisben, D., Victoria, M. C., Chukwubuikem, E. P., & Abdullahi, A. (2025). Responsible AI deployment in sustainable project execution: Ensuring transparency, carbon efficiency and regulatory alignment. *International Journal of Science and Research Archive, 14*(3), 1686–1705. DOI: 10.30574/ijsra.2025.14.3.0920

Tripathi, S., Bachmann, N., Brunner, M., Rizk, Z., & Jodlbauer, H. (2024). Assessing the current landscape of AI and sustainability literature: Identifying key trends, addressing gaps and challenges. *Journal of Big Data, 11*, 65. DOI: 10.1186/s40537-024-00912-x

Ukoba, K., Olatunji, K. O., Adeoye, E., Jen, T. C., & Madyira, D. M. (2024). Optimizing renewable energy systems through artificial intelligence: Review and future prospects. *Energy & Environment, 35*(7), 3833–3879. DOI: 10.1177/0958305X241256293

Ullah, A., Pandey, J., & Masengu, R. (Eds.). (2025). *Impacts of technology on operations management: Adoption, adaptation, and optimization*. IGI Global Scientific Publishing., DOI: 10.4018/979-8-3693-6205-1

Wang, Q., Li, Y., & Li, R. (2025). Integrating artificial intelligence in energy transition: A comprehensive review. *Energy Strategy Reviews, 57*, 101600. DOI: 10.1016/j.esr.2024.101600

Wei, J., & Shen, Y. (2025). Impact and mechanism of digital transformation on performance in manufacturing firms. *Innovation and Green Development, 4*(1), 100205. DOI: 10.1016/j.igd.2025.100205

Wu, S., Cheng, P., & Yang, F. (2024). Study on the impact of digital transformation on green competitive advantage: The role of green innovation and government regulation. *PLoS One, 19*(8), e0306603. DOI: 10.1371/journal.pone.0306603

Xiao, Y., & Xiao, L. (2025). The impact of artificial intelligence-driven ESG performance on sustainable development of central state-owned enterprises listed companies. *Scientific Reports, 15*, 8548. DOI: 10.1038/s41598-025-93694-y

Yigitcanlar, T., Mehmood, R., & Corchado, J. M. (2021). Green artificial intelligence: Towards an efficient, sustainable and equitable technology for smart cities and futures. *Sustainability, 13*(16), 8952. DOI: 10.3390/su13168952

Younas, A., & Kassim, A. A. (Eds.). (2025). *Transforming business through digital sustainability models*. IGI Global Scientific Publishing., DOI: 10.4018/979-8-3373-0608-7

Yu, X., Fan, L., & Yu, Y. (2025). Artificial intelligence and corporate ESG performance: A mechanism analysis based on corporate efficiency and external environment. *Sustainability, 17*(9), 3819. DOI: 10.3390/su17093819

Zhang, L., Innab, N., Shuhidan, S. M., Pan, Y., Zhang, Y., Som, H. M., & Alasbali, N. (2025). Artificial intelligence-driven internet of things-based green supply chain for carbon reduction in sustainable manufacturing. *Journal of Environmental Management, 389*, 126170. DOI: 10.1016/j.jenvman.2025.126170

Zhao, L., Sun, Y., & Xiao, G. (2025). A novel machine learning-based method for identifying Industrial Symbiosis opportunities. *Journal of Cleaner Production, 513*, 145607. DOI: 10.1016/j.jclepro.2025.145607

KEY TERMS AND DEFINITIONS

AI in Technology-Driven Organizations: Artificial intelligence in technology-oriented organizations revolutionizes operations by automating tasks, facilitating superior data-driven decisions, improving customer experiences through personalization, and promoting innovation, thereby rendering companies more efficient, competitive, and agile by integrating intelligence into strategy, workflows, and value creation, although implementation necessitates the management of costs, complexity, and security.

Artificial Intelligence (AI): Artificial Intelligence (AI) refers to the emulation of human intelligence in machines, allowing them to learn, reason, solve problems, perceive, and comprehend language to execute tasks that generally necessitate human cognition, such as data analysis, pattern recognition, decision-making, and the automation of intricate processes, with prevalent instances including virtual assistants, recommendation systems, and autonomous vehicles. AI systems acquire knowledge from extensive datasets, enabling them to improve performance over time without explicit reprogramming for each new scenario, thereby increasing accuracy and efficiency across all sectors.

ESG (Environmental, Social, and Governance): ESG (Environmental, Social, and Governance) is a framework employed by investors and corporations to evaluate a company's non-financial performance, emphasizing its environmental impact (E), treatment of individuals (S), and internal governance practices (G), thereby facilitating the measurement of sustainability, risk management, trust-building, and the generation of long-term value beyond mere financial returns. It is an essential instrument for responsible investing, demonstrating how companies function ethically and sustainably in domains such as climate change, diversity, and corporate accountability.

Organizational transformation: Organizational transformation is a comprehensive, systemic alteration of a company's strategy, structure, culture, and operations to align with new objectives or environments, transcending mere modifications to redefine its functioning and value delivery, often driven by technological advancements, market dynamics, or internal demands for efficiency and innovation. It is a multifaceted, continuous process that entails synchronizing individuals, procedures, and technology to transition from the current state to a preferred future state, thereby augmenting resilience and competitiveness.

Sustainable Growth: Sustainable growth refers to the expansion of activities or the economy in a manner that is enduring, harmonizing financial prosperity with environmental stewardship and social equality, thereby enabling future generations to meet their needs. In a corporate context, it typically denotes the highest growth rate a corporation may achieve without incurring further debt or issuing new shares, determined by profitability and retained earnings. It pertains to being consistent, ethical, and accountable in the long run, rather than merely pursuing short-term profits.

Chapter 10
The Net Zero Transition for Green HRM and Governance Policies

Syamsul Hadi
https://orcid.org/0000-0001-7226-4978
Universitas Sarjanawiyata Tamansiswa, Indonesia

Penpim Phuangsuwan
https://orcid.org/0000-0003-2371-7998
University of Phayao, Thailand

Lin Dar Ong
https://orcid.org/0000-0003-1182-4323
Faculty of Business and Economics, Universiti Malaya, Malaysia

Saizal Pinjaman
https://orcid.org/0000-0002-6790-2688
Universiti malaysia Sabah, Malaysia

Auliya Rosiana
Universitas Sarjanawiyata Tamansiswa, Indonesia

Ambar Lukitaningsih
https://orcid.org/0009-0002-9847-9242
Universitas Sarjanawiyata Tamansiswa, Indonesia

ABSTRACT

This chapter aims to synthesise conceptual progress in Net Zero Human Resource Management (HRM) published between 2019 and 2025. Moreover, it examines how green employment, training, and performance appraisal are integrated with upper-level green governance policies to facilitate low-carbon organizational management approaches. Fifty articles were selected from an initial total of 578 records according to the PRISMA guidelines. Citation chaining and query transformation were used to include recent empirical studies, with a particular focus on those from emerging economies. This review shows a positive relationship between green human capital and green recruitment and training. By contrast, the green performance appraisal process is far more problematic for a variety of reasons, not least ongoing measurement difficulties.

DOI: 10.4018/979-8-3373-6821-4.ch010

1. INTRODUCTION

Global imperative for moving to a "Net Zero" trajectory has driven the reimagining of human capital strategies to address global challenges like climate change and the continued exhaustion of resources (Hien, 2025; Shaikh et al., 2025). In such a context, Net Zero HRM is shifting from an administrative adaptation to a high-profile set of skills dedicated to orienting organizational activity within the boundaries set by the natural environment. Discussion in academia has progressed from traditional "Green HRM" to include green recruitment, training, and performance appraisals as pillars of sustainable business models (DEV, 2025; Faisal, 2023; Roscoe, et al., 2019). This process mirrors a growing realization that HR functions are instrumental in promoting environmentally friendly behaviour and forms part of the "soft" infrastructure for organisational sustainability performance (Hakuduwal, & Kc, 2025; Yong et al., 2020). In particular, empirical studies show that successful green HRM practices can enhance environmental, economic, and social performance. Meanwhile, industries like manufacturing/service, healthcare, or banking are getting into practice on strategies to comply with stringent regulations and pressure from stakeholders (Akbar, et al., 2025). Similar orientation is noted in the Nepalese setting (Hakuduwal, & Kc, 2025). With increasing pressure on companies to demonstrate environmental integrity, the incorporation of net-zero principles into HR operations has shifted from a reactive to a strategic necessity.

Notwithstanding, the empirical evidence continues to accumulate, with some studies showing that successful green HRM practices can enhance environmental, economic, and social performance (Akbar et al., 2025; Hakuduwal & Kc, 2025). A consistent and general problem remains unsettled in the present literature. Studies still tend to adopt a largely "silo" approach, treating recruitment, training, and performance appraisal as separate, discrete practices rather than as part of an integrated strategy (Rahman, Abidah Saad, et al., 2025; Shahrulnizam). Consequently, such assimilation of research has been particularly thin in elucidating the interaction between these HR strategies and a variety of green governance policies institutionalized within the organizational milieu (Gaikwad et al., 2025; Pawar, 2025; Roscoe et al., 2019). This fragmentation leads to a strategic blind spot in which HR initiatives and practices cannot connect to systemic-level results that reflect the macro governance and institutional agenda (Rahman, Abidah Saad, et al., 2025; Shahrulnizam). In addition, barriers to implementation, such as costs and cultural sensitivities, and challenges in obtaining standardised measures, continue to prevent a full embrace of these policies, especially in emerging and public markets. However, in the absence of a coherent HR framework to support these actions, standalone HR methods frequently fall short of the scalability needed to make a net-zero difference.

Furthermore, the discussion among scholars on the relative impact of various green HRM dimensions on sustainability outcomes is ongoing and inconclusive. However, some research advocates that green recruitment is the key corporate entry point to becoming an eco-attractive employer, while others stress training and performance management to pave the way for substantial long-term social and environmental commitment through behavioral modification (Hien, 2025; Yong et al., 2020). On the other hand, without strong performance appraisal systems attached to incentives, staff training programs might not always be followed by consistent pro-environmental behavior (Veerasamy, et al., 2023). In addition, the variation in measures and lack of standardized, evidence-based metrics to show green performance continue to confound knowledge, making it difficult to determine the real impact of such interventions (Rahman, Abidah Saad, et al., 2025). Because of these inconsistencies, organizations do not have a clear understanding of how to formulate cohesive Net Zero HRM strategies and tend to use business resources ineffectively when it comes to increased organizational resilience (DEV, 2025; Rosyafah, et al., 2025).

It's during this hiatus that the present systematic review offers a unique perspective by developing a policy-sensitive conceptual framework that nests HR-centred sustainability interventions within broader green governance infrastructures (Faisal, 2023; Shaikh et al., 2025). Based on the theoretical background of RBV and AMO, this study explains how a bundle of HRM practices can simultaneously leverage human capital to create a sustainable competitive advantage (Hien, 2025; Shahzad et al., 2024; Veerasamy et al., 2023). More specifically, the AMO model explains that green recruitment secures the 'Ability' element by selecting qualified individuals. In contrast, training positively influences this ability, and performance appraisals offer employees the 'Motivation' and 'Opportunity' to perform tasks in an environmentally friendly manner (Akbar et al., 2025; Hadi, Purnasari, et al., 2025; Hadi, Sulistyawati, et al., 2025). Finally, this review is expected to provide a strategic roadmap for both academics and practitioners by integrating recent empirical trends and contrasting the varied governance systems through which HRM could most effectively support the Net Zero agenda (Kumari & Sharma, 2025; Muhammad et al., 2025). While Net Zero has propelled HRM to the strategic radar, existing literature lacks an integrated view in which recruitment, training, and performance management are treated as isolated interventions, severed from overarching mechanisms of governance. This fragmented perspective, combined with inconsistent findings regarding the role of different HR dimensions relative to the strength of one mediating mechanism or another, hinders organizations from achieving systemic sustainability effects. As a result, this study seeks to fill these important gaps by integrating fragmented empirical evidence into an organized set of findings. Based on this, we will present how core HR practices can be linked to green governance policies from AMO and

RBV theories, steering a firm towards the strategic use of human capital to make the organization sustainable.

2. LITERATURE REVIEW

2.1 Integrating AMO and RBV Frameworks

This chapter joint use of RBV and AMO dimensions to understand organisations' transition towards Net Zero. According to the RBV, HC is a resource that meets the VRIN criteria and represents a source of an organization's survival capabilities (DEV, 2025; Hien, 2025). In the context of the transition to a low-carbon economy, RBV provides insight into how "green competitive advantage" is contingent on the organization's ability to accumulate and embed environmental knowledge in routine human resource conduct. Another lens of explanation for this resource is offered by the AMO framework. Based on SDT and the relationship between green HRM practices and employee engagement, we propose that Net Zero HRM practices can be deployed along three related paths: increasing employees' ability through green recruitment and training mechanisms; strengthening motivation using green performance appraisal and reward systems; and creating opportunity by enhancing employees' involvement in environmental decision-making (Faisal, 2023; Veerasamy, et al., 2023). The combination of RBV and AMO reveals a holistic picture, illustrating that HRM practices are more than just the shape of an individual's behaviour; they contribute significantly to reconfiguring an organization's resource base in response to global climate challenges.

2.2 Green HRM: The Soft Infrastructure for Net Zero

Green human resource management, or GHRM, is increasingly recognized as "soft infrastructure" on the path to achieving net-zero emissions. GHRM is the conscious application of environmental dimensions across the stages of an employee's recruitment through exit. Recent studies show that green recruitment acts as a "gatekeeping" role where it only recruits candidates who possess the organisation's environmental values and attitude towards (Muhammad et al., 2025; Shaikh et al., 2025). Green training is a form of skill diversity-building that enhances employees' technical and behavioral competencies, including resource efficiency knowledge and eco-innovation capabilities (Shahzad, et al., 2024; Veerasamy, et al., 2023). The third dimension, green performance appraisal, plays a monitoring role by translating organizational environmental goals into requirements for individual action (Rahman, Saad, et al., 2025). Collectively, these GHRM practices go beyond the inculcation of environmental awareness and also contribute to organizational resilience by developing a strong and coherent green ID (Rosyafah, et al., 2025).

2.3 Green Governance Policies as Strategic Enablers

The formal structures and policy context for implementing HRM practices are set out in Green Governance. Without strong governance, the GHRM programs will always remain broken and shallow. Recent studies suggest that good green governance is the one with a strong visible top management commitment, transparency in environmental reporting and allocation of adequate resources (Shahrulnizam, et al., 2024; Shaikh et al., 2025). According to the literature, the "alignment gap" between strategic-level policy perspectives and HR practices at the organizational level has also been one of the most researched topics (Gaikwad, et al., 2025; Rahman, Abidah Saad, et al., 2025). Green governance closes this gap by setting forth rules that obligate each strategic HR decision to be made with ecological compatibility in mind. It has also been found that the governance system that encourages employee participation is superior to a purely top-down approach in developing long-run compliance (Kumari, & Sharma, 2025). The result is that green governance serves as a strategic enabler, providing a stable legal and managerial base upon which the continued work of greening HRM throughout the organization can be built.

3. METHODOLOGY

This research follows an SLR with the PRISMA protocol, which offers strong principles of transparency and attention to reproducibility (Gaikwad et al., 2025; Shaikh et al., 2025; Siripipatthanakul et al., a-c). The first stage in the research process is "Query Transformation, where the two questions of "Burst Pressure" and "Mechanical Strength" are transformed into precise search terms. A broad set of primary databases (over 270 million publications) was searched, focusing on 2019–2025, to reflect both historical Green HRM underpinnings and recent Net Zero strategic advances (Hien, 2025; Roscoe). We categorised the keywords in three thematic clusters: Net Zero HRM (the other two were HR practices such as Recruitment, Training, and Performance management) and green governance policies, according to Hadi et al., (2025), for quality-oriented content analysis and review, the search had been restricted by rigorous key criteria, so that all studies included in this study were based on high-quality sources (Hadi et al., 2020). Only peer-reviewed papers that covered HR practices and governance policies and were written in English were considered. Others excluded were conference abstracts, publication reviews with no environmental focus, and HRM studies that were not relevant. An initial search identified 578 studies, including 483 from the search and an additional 95 via "citation chaining. The summary of the selection process is shown in Table 1.

Table 1. Stage selection of articles

Stages Selection	**Number of Articles**	**Description**
Initial Identification	483	Records found via database search using transformed queries.
Citation Chaining	95	Additional works were found through backward and forward citation tracking.
Total Identification	578	The total pool of candidate papers before screening.
Screening & Eligibility	(528)	Records excluded based on relevance scores or mismatched topics.
Included Core Studies	50	Final set of articles selected for in-depth thematic synthesis.
Initial Identification	483	Records found via database search using transformed queries.

After screening, 50 studies with relevant scores were selected. The four main factors that trigger examination in the aforementioned papers are employee training, performance-related appraisals, greening the hiring process, and the role and impact of green governance. However, no generalisations could be made across different industrial sectors, such as banking and manufacturing, or across geography to identify patterns and gaps in research on this issue.

4. RESULTS

4.1 Publication Trend

This article charts the field of research on Net Zero Human Resource Management (HRM) by analysing 50 key studies. The analysis is based on the temporal development of publications, the geographic distribution, with a focus on emerging economies, and the breadth of industrial sectors investigated. An examination of the literature reveals an exponential increase in academic attention to Net Zero-based HRM. This development can be divided into three phases:

Figure 1. Net Zero literature: Publication Trend (2019-2025)

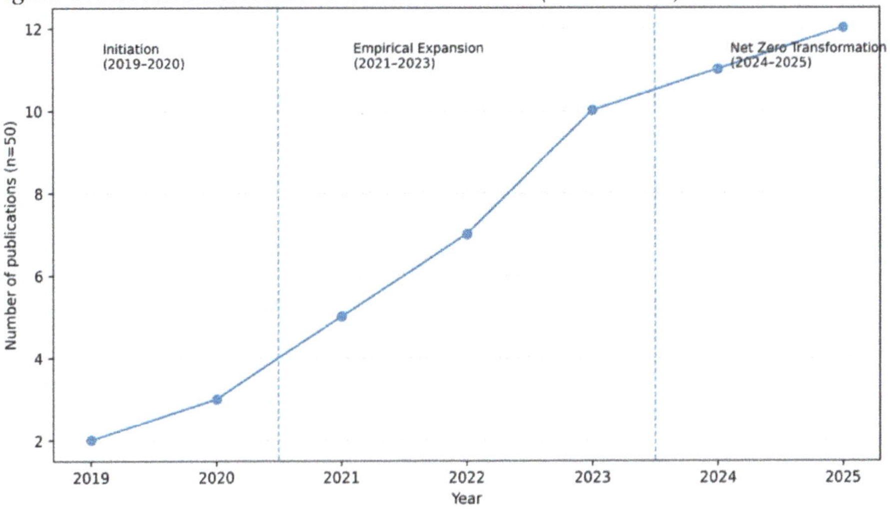

1. Conceptual Initiation Phase (2019–2020): Articles from this period, which tackle problems similar to those addressed by (Roscoe, et al., 2019) and (Yong et al., 2020), laid the theoretical foundation for Green HRM. Emphasis was placed on the discourse around green HR and on an analysis of its early impact on corporate environmental performance.
2. Phase of Empirical Expansion (2021– Service providers made more serious empirical progress during this time, going beyond the principles level. It was in this phase that other potentially mediating/moderating mechanisms have been studied, such as employee engagement (Faisal, 2023), green innovation (Veerasamy, et al., 2023) and pro-environmental organizational culture.
3. Net Zero Transformation Phase (2024–2025): The latest literature indicates the global concern for net -zero emission goals. Recent research, such as Akbar et al. (2025), Hien (2025), and Shaikh et al. (2025)It embeds green governance directly into its policies and compares the efficacy of environmental policies across organizations in achieving decarbonization targets.

4.2 Geographic Distribution: Focus on Emerging Economies

Geographically, the Net Zero HRM literature has been heavily concentrated, with an unusually high percentage of studies from less-developed nations. This empirical distribution profile differs from the pattern in the mainstream HRM literature, which is heavily Western-bound.

Figure 2. Geographic Distribution

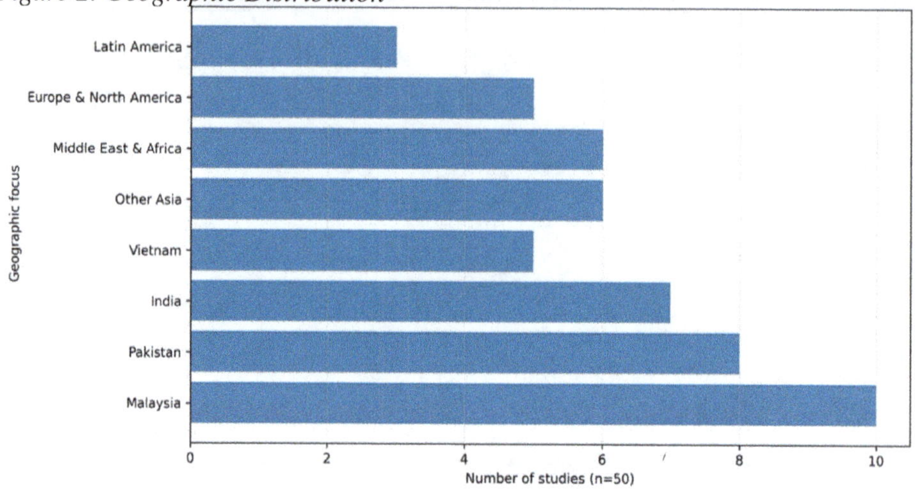

The contrast with conventional management writing is dramatic, as the chart shows. The top bars, or the largest areas, of the graph belong to Emerging Asian Economies (namely, Malaysia, Pakistan, India, and Vietnam). This is a sign that, for this particular area, research output from Chinese and other East Asian countries currently eclipses that of Western nations.

4.3 Descriptive of the studies

The analytical results of this study consist of four parts: Effectiveness of Recruitment Strategy, Influence on Training Program, Integration in Performance Evaluation, and Green Governance Policy Scope and Sustainability Outcome Metrics as follows.

Figure 3. Research Volume by GHRM Focus Area

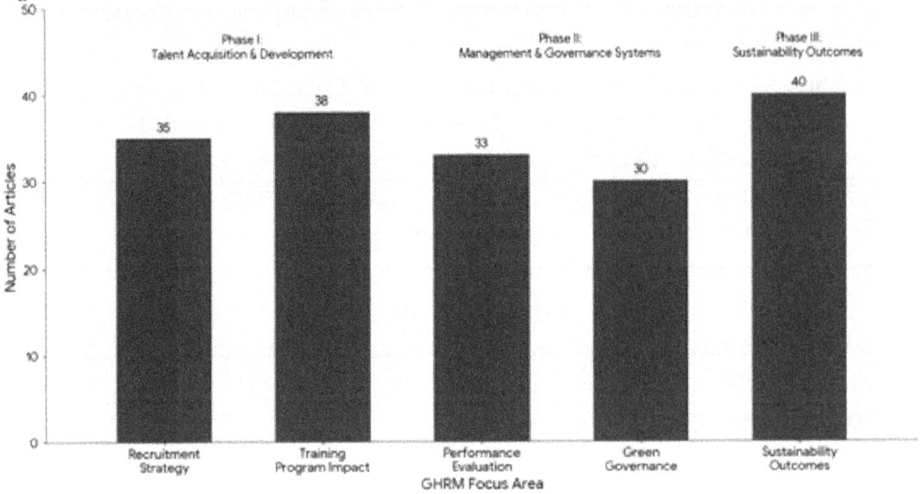

1. **Recruitment Strategy Effectiveness** The literature emphasizes the role of recruitment in Green HRM. Hiring greens has been proven to be effective, not only in drawing environment-friendly staff but also in impacting an organization's economic and environmental performance in 35 studies (Akbar, et al., 2025; Hakuduwal, & Kc, 2025; Hien, 2025). In addition, recruitment is positioned as a building block for mainstreaming values (i.e., sustainability) and a green organisational culture at the start of the employment and hiring process by (Roscoe, et al., 2019), Kumari & Sharma (2025) for greening the organization, and (Kumari, 2024). However, recruitment effectiveness is highly influenced by employees' motivation and the extent of their commitment to environmental goals (Pawar, 2025; Saeed, et al., 2019). However, there are practical challenges in implementing such systems, especially resource constraints and humanitarian bureaucracy inherent in both the public sectors and developing countries (Rahman, Saad, et al., 2025; Trang, 2025).
2. **Impact of Training Program** Training is a strong influencer of the employee's nascent environmental knowledge. Thirty-eight studies indicate that green training has a substantial positive impact on environmental knowledge, skills, and behavior (Rosyafah et al., 2025; Shah Ridwan Chowdhury, 2024; Veerasamy et al., 2023). The value of such initiatives to developing sustainability-oriented organizational culture and achieving sustainable innovation is well-recognized (Boominathan et al., 2023; Muhammad et al., 2025; Tu et al., 2024). However, the impact of training is also dependent on employee commitment and top

management support (Ahmad, et al., 2023; Memon, et al., 2024). Recognized barriers include the absence of expertise and resistance within an organisation (Rahman, Saad, et al., 2025; Khan, 2023).

3. **Evaluation Integration.** The incorporation of green metrics into performance management systems has yielded better sustainability outcomes. Additionally, 33 studies confirm that environmental performance appraisals encourage employee green behavior (Begum, & Arshi, 2020; Hien, 2025; Devi, et al., 2024). Leadership's attention and a supportive organizational culture are often referred to as the keys to this system's success (Lee, & Ahn, 2024; Roscoe, et al., 2019). However, technical constraints persist, particularly in the area of poor integration and the challenge of effectively measuring green performance (Rahman, Saad, et al., 2025; Yong et al., 2020).
4. **Green governance: Policy scope.** Green governance is a framework that integrates several HR practices. Thirty-three of the articles acknowledge 'holistic governance policies' comprising recruitment, training, performance management, rewards, and employee participation (Gaikwad, et al., 2025; Shahrulnizam, et al., 2024; Shaikh et al., 2025). The extent of governance also differs across spheres and locations, with developing countries confronting unique contextual challenges (Permata, et al., 2024; Yong et al., 2020). Leadership, Commitment, and Stakeholder Engagement. These are factors that contribute to effective governance (Kumari, & Sharma, 2025; Manao, & Senen, 2024; Roscoe, et al., 2019)While ineffective, these might pose as challenges such as Devastating Filling Gaps and Financial Resources Constraints. There should be financial resources dedicated to CSR investment. Lack of Clear Metrics Information; This will enhance the measurement body responsible, but the operational process remains unclear, particularly regarding its ability to serve internal drivers. Cultural Resistance. CSRs emphasis is based on a cultural context.
5. **Outcome of Sustainability Measures.** Finally, GHRM will lead to clearly measurable results. Positive changes in environmental, social, and economic performance resulting from GHRM are supported by 40 studies (Akbar et al., 2025; Hien, 2025; Javed et al., 2024). Tangible benefits include improved energy efficiency, reduced waste, and greater workforce participation and corporate image (Amini et al., 2024; Kumari & Sharma, 2025; Ojo & Raman, 2019). Other studies also emphasized the functions of green culture and employee commitment as mediators of sustainable development (Ahmad et al., 2023; Pawar, 2025; Roscoe. Also, GHRM shapes the firm's long-term resilience and competitive advantage & Senen, 2024; Roscoe et al., 2019), though quantitative measures of these are generally industry-specific and dependent on extant governance mechanisms (Hakuduwal & Kc, 2025; Lee.

4.4 Industrial Variation and Methodological Approach

A review of secondary data indicates that research on Net Zero HRM has expanded into more industry sectors than before. It might imply that formerly, the green transition was almost exclusive to heavy industries.

Figure 4. Industry Sector Coverage

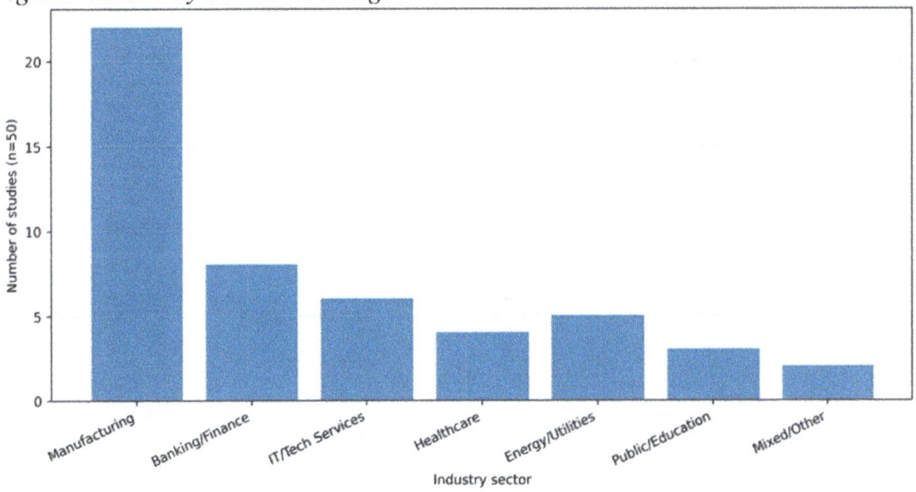

1. **Manufacturing Sector:** This sector remains the most researched, as its resource consumption is among the highest. Therefore, most researchers focus on technical training on energy consumption cuts.
2. **Service Sector:** This has emerged as a significant new growth area. The Banking, IT, and Healthcare subsectors appear to be the most suitable for further academic development.
3. **Methodological Approaches:** The approach to staffing, as emphasized herein, is green recruitment as an employer branding and employee performance management tool to satisfy ESG standards. Moreover, meta-analysis, as the primary methodological approach for testing scientific theories through SEM, is generally declining in favour of the more domain-specific systematic literature review and comparative case study, yielding detailed policy insights.

In summary, it might be emphasized that Net Zero HRM is not purely a theoretical subject but an academically empirical one, as the bodies involved originate from diverse parts of the globe.

4.5 Thematic Analysis: The Pillars of Net Zero HRM

The Pillars of Net Zero HRM. The review indicates that four thematic discourse directions form the foundation of HR policy implementation in pursuit of Net Zero objectives. The four themes demonstrate the strategic integration of HR procedures and environmental governance necessary for an organization to become a sustainable ecosystem.

Figure 5. Thematic Analysis: Pillars of Net Zero HRM

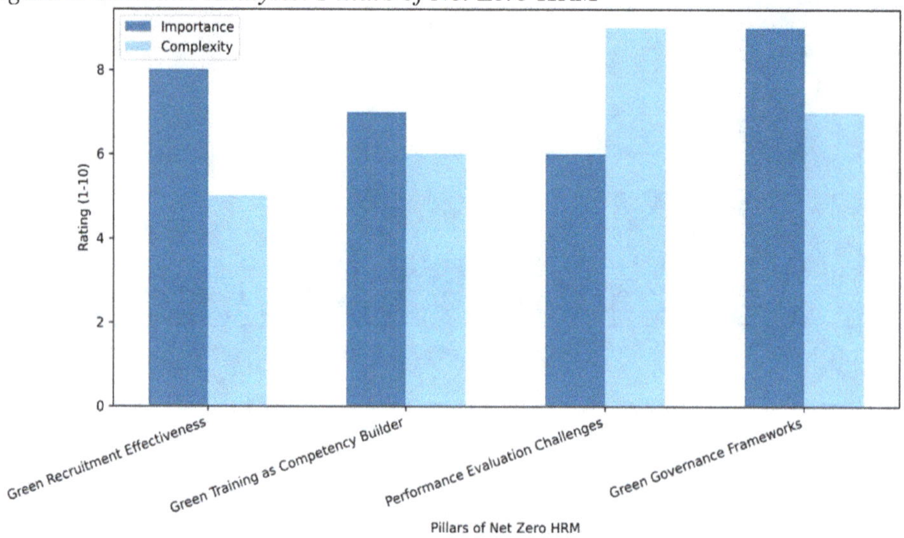

1. *Green Recruitment Effectiveness*

Green recruitment is treated as an introductory "engine" to advance new environmental policy directions. The literature demonstrates that the talent that adheres to environmental ethics prior to employment costs significantly less in the long run and is twice as productive.

- *Strategic fit,* successful green recruitment activity does not just focus on job descriptions, but that sustainability should be integrated into the "employer brand, which acts as a discriminator to attract environmentally conscious applicants (Akbar, et al., 2025).
- *Performance implications*, Green hiring is related to positive performance outcomes; hence, greener hires are more likely to engage in pro-environmen-

tal behaviors and be committed to the organization (Hakuduwal, & Kc, 2025). However, the effectiveness of this pillar is often assessed by the organization's legitimacy and perceived "green worthiness.

2. *Green Training as Competency Builder*

If recruitment is the inlet, green training serves as the "accumulation mechanism" for green human capital. The literature review indicates that training is the most common strategy to increase employees' knowledge and technical competencies for the Net Zero transition (Faisal, 2023; Veerasamy, et al., 2023).

- *Skill Diversification*, Training programs captured in the literature focused on particular technical skills such as energy efficiency, waste reduction, and eco-innovation (Shahzad, et al., 2024).
- *Beyond Awareness*, Thematic analysis points to the importance of training that goes beyond general environmental awareness and moves into an "empowerment-based" zone, where employees are empowered to apply green knowledge in their daily operational routines (Rosyafah, et al., 2025). This procedure metabolizes codified green policies into best practices, such as organizational know-how.

3. *Performance Evaluation Challenges*

Performance evaluation in the green sphere is the most complex and least standardised part of Net Zero HRM, but hardly less significant. There exists an enduring "*measurement gap*" between organizations' measures of and reward to green behaviour (Rahman et al., 2025).

- *Quantitative Issues*, The challenge lies in finding the objective and measurable KPIs for 'soft' green acts, including environmental volunteering.
- *The Accountability Gap*, Although some corporations state in their mission statements that they need to 'go green,' they have done little to link this to individual performance appraisals (Hien, 2025). Some of your employees may think that corporate" is "greenwashing" without a tangible link between their "green" behaviour and rewards.

4. *Green Governance Frameworks*

Green governance forms the foundation; it is the "strategic scaffolding" that supports the other three pillars. It sets the table for Net Zero HRM to succeed by offering formal rules, leadership commitment, and policy structures (Shaikh et al., 2025).

- *Policy Integration*, Studies show they are environmentally friendly because they rarely integrate recruitment, training, and rewards into a broad green governance policy, instead treating these as separate HR activities (Shahrulnizam, et al., 2024).
- *Leadership as the Engine*, Commitment to leadership is the cornerstone of effective governance. According to the thematic synthesis, top-down governance should be balanced with bottom-up public participation to avoid green policies being "on paper" problems rather than culturally embedded (Kumari, & Sharma, 2025). This alignment is important for connecting the mezzo sustainable goals with the micro, longer-term actions of employees.

5. DISCUSSIONS

5.1 Critical Synthesis: Consensus and Conflicts

A systematic review of 50 core studies demonstrates broad academic consensus on the strategic importance of Net Zero HRM. However, we also find a disparate literature regarding "strength of Influence" and implementation mechanisms. This section summarizes the results to distill where scholars converge and diverge, or what significant academic tensions (or "boundary conditions") exist.

1. Areas of Consensus: The Strategic Imperative

There is consensus among the reviewed papers that embedding ecological values into HR is no longer optional but imperative if organizational legitimacy and sustainability are to be achieved. Academics, including Akbar et al. (2025), Hien (2025), and Shaikh et al. (2025), consistently demonstrate that Green Recruitment serves not only as a hiring tool but also as a strong signal of corporate social responsibility. This signalling effect effectively creates a pull for "value-aligned" talent people who care about leading sustainable lifestyles. The argument is that this alignment with values and other purposes minimises friction at the psychological and turnover levels (as per above), while also improving organisational economic and ecological performance. More generally, Green Training is widely acknowledged to be the central "human capital accumulator". Applying the Resource-Based View (RBV), scholars believe that physical resources (e.g., green machinery) can be imitated, whereas organization-specific knowledge embedded in employees cannot. Thus, training is considered crucial for the development of eco-innovation needed to support a Net Zero transition (Faisal, 2023; Veerasamy et al., 2023). Without this ongoing investment in human capacity, tech net-zero targets are all pie-in-the-sky rather than operational realities.

2. Areas of Conflict: The Contextual and Operational Paradox

Although there is broad agreement regarding the strategic importance, academic debates centre on the operational effectiveness of practices per the Leadership Dependency Paradox. The significance was the "Leadership Dependency" of training programs. Most of the literature tends to regard training itself as the direct cause of change, although an important subset of studies provides an exception. Researchers such as Kwerawarni (2023) and Shahrulnizam et al. (2024) claim that "Green HRM may develop a policy-practice gap" or a decoupling effect when leadership support is lacking. They argue that, in the absence of strong leadership and an enabling culture, green training is likely to become no more than 'knowledge without action.' In such situations, employees can act green. However, they do not have the opportunity or resources provided by the organization to apply those abilities to green causes rather than transformation, which breeds cynicism.

The Sectoral Paradox. A further critical deviation concerns what has been termed the Sectoral Paradox, pointing out that the Net Zero HRM instrument is heavily dependent on context. A comparison of the manufacturing and service sectors, however, yields strong evidence. In manufacturing, attention is overwhelmingly focused on technical efficiency, adherence, and waste reduction via operation-based training (Akbar et al., 2025). Quantitative indicators determine productivity here, like a lower carbon footprint per unit. In the banking and service industry, the priority becomes green recruitment for employer branding and to gain psychological engagement (Hakuduwal & Kc, 2025). Here, the objective tends to be behavioral and reputational. This variation implies that a 'one-size-fits-all' approach to achieving Net Zero HRM is unlikely to be viable. Theories that hold for the shop floor in Vietnam may not be materialised in a banking hall of Pakistan; industry effects should be controlled more carefully in the future. These crucial tensions are summarized in the following table:

Table 2. Meshing Frame: Agreement and Disagreement in Net Zero HRM Sources

HRM Dimension	Areas of Agreement (Consensus)	Areas of Conflict (Divergence)	Key References
Green Recruitment	Great for attracting environmentally responsible candidates and increasing the "Employer of Choice" profile.	There is an ongoing debate regarding the direct, tangible benefit to long-term sustainability if the internal culture does not echo it.	(Hien, 2025; Muhammad et al., 2025; Yong et al., 2020)
Green Training	Cited as a crucial means to develop technical skills and environmental awareness.	Efficacy is very moderate: Training without rewards or autonomy will, most of the time, have no significant effect.	(Faisal, 2023; Kwerawarni, 2023; Veerasamy, et al., 2023)
Performance Evaluation	Key to establishing personal accountabilities with respect to the Net Zero targets.	A major problem with measurement: many organizations measure at the "symbolic" level rather than the "substantive" level. "	(Gaikwad et al., 2025; Hien, 2025; Rahman, Abidah Saad, et al., 2025)
Green Governance	The successful implementation of green HRM practices depends on holistic governance.	There is an enduring "policy-practice gap" between policies being written down and adequately operationalized.	(Roscoe, et al., 2019; Shahrulnizam, et al., 2024; Shaikh et al., 2025)
Sectoral Context	Green HRM is relevant to every sector, regardless of organization size.	Mechanisms of success vary: manufacturing around technical processes, and service around culture.	(Akbar, et al., 2025; Hakuduwal, & Kc, 2025; Permata, et al., 2024)

This integrative review suggests that the "gap" may lie in the literature, where practices are packaged together. The conflicting results in performance evaluation, for example, highlight the inherent complexity of consistently measuring employees' contributions towards macro-environmental objectives (Rahman et al., 2025). These conflicts indicate that the effectiveness of Net Zero HRM is largely dependent on the degree to which practices are congruent with an organization's wider green governance infrastructure.

Figure 6. Dimensions chart: Diverging Bar Chart representing the levels of Academic Consensus versus Divergence for Key Net-Zero HRM Dimensions.

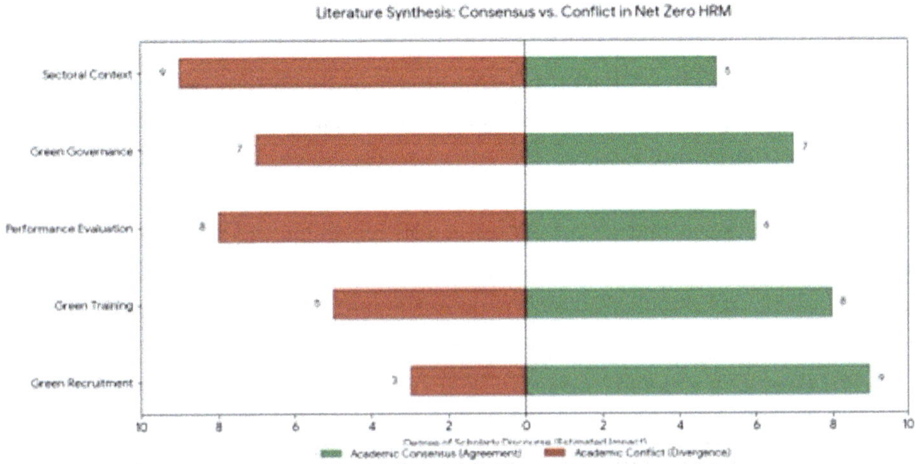

The Green Bars (Right Side - Consensus). These bars show a level of evidence: the longer the green bar, the greater the consistency and the support the existing literature provides for this dimension. The green bar for Green Recruitment is the longest, indicating that scholars agree irrefutably on its positive impact on talent attraction and retention and consider it a "mature" field. The Red Bars (Left side Divergence). These bars show the strength of discord or controversy in academia. A longer red bar represents a disagreement, conflicting results, or non-standardized methods in the literature. The widest columns are the ones for the red bars. This provides visual evidence that these counties are "competition terrain" with high inconsistency, and highlights them as major "black holes" of research interest that you will fill over the next 10-20 years.

5.2 Theoretical Implications

The results of this systematic review make considerable theoretical contributions by bridging and combining the two major schools in SHRM: The Resource-Based View (RBV) and the Ability-Motivation-Opportunity (AMO) theory. This goes beyond simply employing these theories as descriptions and instead proposes a more policy-informed and fluid synthesis.

1. Reconceptualizing RBV: The "Dynamic Green Resource."

According to the RBV, human capital is strategic because it satisfies the VRIN (Valuable, Rare, Inimitable, and Non-substitutable) criteria on a static basis. However, this paper pushes the theoretical envelope even further, by re-conceptualizing Green Human Capital not simply as a stock of competences but, "through activation", as a "dynamic capability" that can be actively orchestrated. The term strategic, in this context, refers to the latent quality of committed actors that, in the absence of appropriate system-bound constraint-action relations, would not be activated. Recruitment is a potential, whereas the application of green goals in training and performance management converts the potential into kinetic behavior (DEV, 2025; Shahzad, et al., 2024). Also, without this repetitive regeneration, the "value" of green talent devalues quickly because environmental technologies and regulatory protocols change over time

Additionally, this research significantly reshapes the origin of Inimitability in Net Zero. The review makes the case that a firm's green advantage does not lie in mobile, poachable individual talent but rather in its social complexity, its green-filled relationships, and its network of resource exchange. This is consistent with the theory of causal ambiguity; perhaps competitors can see some visible artefacts of Green HRM (e.g., Appointment policy) but cannot imitate the invisible, socially laden "soft infrastructure" beneath that allows these policies to work (Akbar, et al., 2025; Roscoe, et al., 2019). Accordingly, it shifts the focus of RBV analysis from the micro individual level to the organisational level, arguing that sustainable competitive advantage arises from the idiosyncratic legacy ecosystem of values and governance, which is historically contingent and interlocks human capital with environmental objectives.

2. Refining AMO: Identifying the "Governance-Opportunity" Link

While RC attempts to explain where differences between firms' performances come from, complementary to the resource-based view is this investigation into how variation in firm performance occurs within a firm through AMO-based work processes." A second key finding is what we might call "The AMO Imbalance. The synthesis uncovers that the antecedent variables: 'Ability' (by using green recruitment to attract value-aligned talent and technical training to develop decarbonization skill set among employees (Faisal, 2023), 'Motivation' (through performance metrics that institutionalize accountability (Devi, et al., 2024; Veerasamy, et al., 2023) have been successfully operationalized, but a significant gap remains in the dimension of 'Opportunity.' The situation behind this lack is established in the literature as a failure of Green Governance; in other words, without a participatory architecture that decentralizes decision making and resources, at which purely pro-environmental acts become bogged down," meaning that their office runs into (tedious) bureaucracy. Therefore, it creates the "Latent Potential Paradox": The Employees have Cognitive Competence (Ability) and Psychological Urges (Motivation) to Act, but there are Hierarchies that are inflexible and a lack of path (The authorization to participate, Gaikwad). We interpret this as suggesting that the blockage in Net Zero is not usually a failure of quality of human capital, but a failure to design governance that unlocks the platform (Opportunity) needed for individual intent to become systemic response.

3. Theoretical Synergy: A Holistic Net Zero HRM Model

Integrating the RBV with the AMO model, this article develops a systemic Net Zero HRM model that is not merely a bundle of disconnected practices but an interdependent system. In this synergistic dynamic, RBV provides the 'Strategic Why,' framing green human capital as an unshared and non-substitutable resource necessary to establish competitive parity and advantage. At the same time, AMO contributes the 'Operational How,' specifying the operational levers needed to mobilize these potential capabilities into actual eco-effective actions. What is genuinely new in this theoretical amalgamation, then, is the explicit nomination of Green Governance as a moderating filter conceptualized by way of metaphor as "the structural linchpin," which can help overcome and fill in the enduring gap between decarbonization goals at macro-levels of organizational activity and front-line employee behaviour (Rahman et al., 2025; Shahrulnizam, et al., 2024). By conceptualizing governance as the stabilizing agent that anchors HR incentive design to corporate sustainability goals, this integrated model provides a sturdier foundation for future empirical analysis with greater policy relevance: one in which sustainable human capital is no longer treated as an isolated phenomenon but rather as part and parcel of a system.

5.3 Managerial & Policy Implications

1. Implications for HR Managers and Organizational Leaders

Right from the administrative facilitator, HR professionals have to become "Green Strategic Partners."

a) Synchronizing Impact through "Green Bundling."

One of the most significant problems we have in corporate sustainability is that 'green' ends up in siloed projects or initiatives that never follow through. To counter this, HR managers need to resort to Green Bundling. Third, as stressed by

Zaid (2022), Green HRM practices are most effective when implemented in tandem rather than separately. For example, hiring environmentally motivated employees (Green Recruitment) is of little long-term benefit if the firm's reward system (Green Rewards) fails to encourage ongoing green innovation. Recruitment, training, and performance management, when "bundled," form a self-reinforcing loop. This alchemy will ensure that sustainability is baked into the employee life cycle, driving us towards Net Zero in a way that is powerful and impervious to inertia within the organization.

b) Cultivating Green Transformational Leadership

The problems of "leadership dependency" hit hardest in any green transition. According to (Kwerawarni, 2023)Green projects lose momentum if they are not top-down-led. In that sense, the role of HR is to develop Green Transformational Leaders who lead not by managing systems but by inspiring an environmental dream. Authenticity is the coin of the realm in this new order. It is HR's job to ensure that leadership development programmes focus on "valid green visions" and avoid engaging in greenwashing (Roscoe, et al., 2019). Greenwashing occurs when the gap between an organization's environmental rhetoric and its internal actions widens to an unacceptable degree. HR contributes to establishing organizational trust by creating pro-environmental behavioral leader prototypes used in daily decision-making, and ensures that a company's green commitment is considered a core value rather than a marketing strategy.

c) Digitization of Green HR

The backbone feature of Green HRM is digital transformation. They can directly reduce the HR department's carbon footprint by deploying Digital HR technologies. Moving to digital recruitment activities, cloud document management solutions, and AI-reliant resource platforms reduces the amount of traveling and paper waste they generate. Digitalization is not just an issue of efficiency; it also produces the internal coherence needed. There is little logic in a business that champions sustainability while pushing paper and taking hours to process approvals. Only with the help of digital tools can we achieve the data transparency needed to reliably record such green performance indicators. "Aligning the internal structures of HR function with the external green goals of the company communicates a strong message to employees on one hand about the company being sincere and on the other about its technological preparedness for a low-carbon future."

5.4 Implications for Policymakers and Regulators

What the report recommends for government departments and industry associations: The requirement for an enabling external environment to promote widespread change in organisations.

a) **Green HR Metrics Standardization:** A critical barrier in the Net Zero journey is the lack of standardization in measuring (Rahman, et al., 2025). Policymakers should engage with academics, researchers, and industry experts to establish 'traditional' benchmarks on Environment, Social, and Governance issues related to human capital resources. Policymakers need to encourage a multi-stakeholder partnership (i.e., universities, industry leaders, and researchers) specializing in the creation of Conventional ESG Standards on human capital. Finally, these standards might go beyond carbon or GHG emissions to measure factors such as employees' "Green Literacy Rates" or "Eco-Innovation Contributions" per staff member. By establishing a standard structure, regulators can enable Green HR data to be as transparent and reliable as financial data, thereby enhancing investor trust and genuine corporate responsibility
b) **Incentivizing SME Adoption:** Moreover, after some time and after some palpable evidence had been agreed upon by thousands of biochemists in hundreds of laboratories all over the world, they could not believe it until they satisfied themselves. The government can offer tax breaks, grants, and easy access to green certificates for Small & Medium Enterprises in developing countries as policy incentives to embrace Net Zero HRM (Permata, et al., 2024). Governments should address and bridge this "sustainability divide" by implementing targeted tools. That might mean financial incentives, like tax credits for businesses that offer green training courses, or direct subsidies for digital HR upgrades. In addition, simplified green certification schemes tailored to SMEs' levels of activity can provide them with the visibility and access to markets required to compete in a green economy. By reducing the cost of entry, they are inviting a Net Zero transition that is not elitist but all-inclusive.
c) **Sector Focused Regulatory Guidance:** Acknowledging the "sectoral paradox," it is essential for regulators to stop issuing universal instructions and guidelines tailored to specific sectors. For instance, manufacturing companies are suggested to prioritize technical skills audit and waste management, whereas service industries should consider worker advocacy and sustainable workplace culture (Hakuduwal, & Kc, 2025; Shaikh et al., 2025). As Lyne mentioned, the current regulation is failing at scale, and the "one-size fits all" approach overlooks the specifics of how those rules work in practice. "Sectoral Paradox" means that one-size-fits-all rules won't cut it; the environmental impact and HR needs of

a software company are very different from those of a chemical plant. To address this, regulators should transition toward more sector-specific regulation that offers more nuanced models. For example, in manufacturing and industry, policies would need to prioritize technical skill audits and safety-related waste management training for "hard" skills (circular economy-related). In contrast, guidance for services and corporations should always point to "soft", inclusive measures such as employee advocacy, virtual work policies, and sustaining workplace culture. To make GHRM practices feasible and practical, legislation can be adapted to meet the needs and effects of different sectors.

6. CONCLUSIONS

This chapter aims to synthesise conceptual progress in Net Zero Human Resource Management (HRM) published between 2019 and 2025. Moreover, it examines how green employment, training, and performance appraisal are integrated with upper-level green governance policies to facilitate low-carbon organizational management approaches. Fifty articles were selected from an initial total of 578 records according to the PRISMA guidelines. Citation chaining and query transformation were used to include recent empirical studies, with a particular focus on those from emerging economies. This review shows a positive relationship between green human capital and green recruitment and training. By contrast, the green performance appraisal process is far more problematic for a variety of reasons, not least ongoing measurement difficulties. In addition, green governance at the macro level is important for integrating sustainability policies and employee practices within companies. This paper offers a theory-driven conceptual framework that bridges partially disjointed HR practices with broader governance structures. Instead of focusing on standalone green initiatives, the model suggests that HR has an important role to play in integrating HR into decarbonization through a green bundling approach, particularly in the Global South. The chapter contributes to the Resource-Based View by conceptualizing green human capital as a resource that evolves. It further advances the Ability Motivation Opportunity framework and elevates governance to a high status as an enabler of pro-environmental behaviour. The findings are valuable for HR managers; they now know that green HRM should be practiced in an interconnected manner to generate synergies. Policymakers should note that ESG measurements need to be more transparent. The net-zero transition necessitates a comprehensive corporate strategy in which Green Human Resource Management

(GHRM) and strong governance policies act as essential facilitators for embedding sustainability into the fundamental business strategy and culture.

Net Zero HRM has transitioned from a niche administrative adjustment to the fundamental 'soft infrastructure' necessary for organizations as they steer through the global climate transition. Based on the assimilation and compilation of 50 core studies, this study further affirms that achieving Net Zero ambitions cannot be merely technology-driven; it must be anchored in the strategic re-sculpting of human capital. The synthesis shows that Green Recruitment and Green Training are the key means of developing "Green Human Capital," encompassing environmental values and the technical competencies needed for decarbonization. The review, however, points out that these practices are often hindered by a "measurement gap" in Green Performance Evaluation, which refers to difficulties in setting macro-scale sustainability goals at the organizational and personal levels. "One of the most important aspects resulting from this research is the importance Green Governance has played in enhancing. This is what allows this 'strategic scaffolding' on which the division is built, and from which it is later possible to overcome. Truly, in the event that there is nothing governance and management can do about this, then HRM becomes more symbolic than reality in its relationship to sustainability." Strategically, this review's contribution to RBV perspectives and AMO logic is to show that green competitive advantage is a dynamic capability that should be constantly revitalized through "green bundling" and underpinned by a participatory governance culture. In conclusion, to achieve Net Zero-resilient performance among firms in emerging markets, HR practices will need to move out of the silo and follow an integrated, governance-driven HRM approach where talent management is aligned with ecological imperatives.

7. LIMITATIONS AND RECOMMENDATIONS

7.1 Limitations

First, there is the bias of geography: while the rise in research from developing countries (e.g., Malaysia, Pakistan, and Vietnam) brings new perspectives on the Global South, those views seldom intersect with comparative studies that compare emerging findings with more mature Western markets. Second, the majority of the literature we reviewed has employed cross-sectional designs that provide a window into what Green HRM arrangements reflect, but are less convenient for establishing causality between HR actions and Net Zero outcomes. Third, the hegemony in the current literature of surveys that are vulnerable to social desirability bias (i.e., interviewees inflating their organizations' "greenness"). Moreover, the literature is still dominated by manufacturing firms, thus creating a dearth of knowledge about how Net Zero HRM works in complex, disaggregated service contexts or at gig economy platforms.

7.2 Recommendations for Further Studies

In the interest of moving the field forward to more robust and impactful findings, the following high-impact areas should be prioritized in future work:

1. **Longitudinal Impact Assessment.** Researchers should also broaden their focus from cross-sectional surveys to longitudinal studies and track the consequences of these Green HRM 'bundles' over a five- to ten-year period. This is important to assess whether these practices result in enduring decarbonization or fade out when organizational courses are adjusted. Watching these practices over an extended period, researchers can determine the "maturity levels" of green initiatives and whether they take hold within corporate DNA or remain skin-deep. This is important because it helps better understand the lifecycle of green commitment and provides empirical evidence that GHRM leads to permanent environmental change rather than symbolic or cosmetic change.
2. **Digital Green HRM & AI Integration.** One key research gap is a high-level investigation into how new technologies, i.e., Blockchain, Artificial Intelligence, and Big Data analytics, can be integrated with HR systems to monitor carbon footprints at the individual level and automatically reward green performance. One of the principal domains of scientific research is in the "Digital" and "Green" twin transitions. Such exploration should consider how AI could offer personalized "green nudges" to employees and how data analytics could help ensure a 'transparent audit trail' for sustainability reporting. By further examining this synergy, one may gain a more accurate picture of how technology enables precision-based GHRM by shifting the focus of organizations from general environmental targets to very specific personal responsibility through quantitative assessments.
3. **Green Governance and "Policy Practice" Gap.** Future research should explore psychological and structural barriers that hinder the translation of "top-down" green governance into micro-level employee behavior. Research focused on "Green Decoupling," where companies declare success on their Net Zero targets, but HR lags will be key to accountability at the institutional level. Future inquiries should pay particular attention to the "middle management bottleneck," where high-level sustainability policies are frequently disrupted. If researchers can unravel the dissonance between a company's public image, "green image," and how the company actually operates, they may help craft ways

to align institutional rhetoric more closely with genuine employee engagement, all but erasing the prospect of internal greenwashing.

4. **Cross-Cultural and Multi-Sectoral Synthesis.** In order to increase the generizability of the Net Zero HRM model, studies have been encouraged to compare nations other than Europe (politically stable developed countries, emerging economies, as well as comparing sectors within a non-one-size-fits-all approach (i.e., High-Tech: Heavy Industry), with implications for success factors that need tailoring to a particular context. This requires an understanding of how country-specific regulations, cultural norms, and economic challenges may shape the application of GHRM. Cross-sectoral synthesis will enable the development of a Contingency Framework and help organisations appreciate that what works for a carbon-neutral tech start-up organisation will need significant modification when applied to a traditional manufacturing organisation, thereby ensuring Net Zero HRM models are universalisable and applicable across the globe.

REFERENCES

Ahmad, J., Mamun, A. A., Masukujjaman, Md., Makhbul, Z. K. M., & Ali, K. A. M. (2023). Modeling the workplace pro-environmental behavior through green human resource management and organizational culture: Evidence from an emerging economy. *Heliyon*, *9*(9), e19134–e19134. DOI: 10.1016/j.heliyon.2023.e19134 PMID: 37681178

Akbar, K. F., Muhammad B., & Syed Kamran H. S. (2025, September 4). *The Relationship Between Green HRM and Environmental Performance in SMEs*. SciSpace - Paper. https://doi.org/DOI: 10.63056/acad.004.03.0707

Amini, M. Y., Tang, Z., & Besharat, A. (2024). Greening university practices: Empowering eco-conscious behavior, transforming sustainable culture, and shaping greener institutional awareness through strategic green HRM initiatives. *International Journal of Research In Business and Social Science*. https://doi.org/DOI: 10.20525/ijrbs.v13i1.3149

Begum, V., & Arshi, T. A. (2020). An Impact-Based Model of Green Human Resource Management: Evidence from UAE. *Journal of Security and Sustainability Issues*, *10*(2), 573–584. DOI: 10.9770/JSSI.2020.10.2(16) PMID: 39858411

Boominathan, V., Selvi, J. T., Dhilipan, C., Arasu, M. T., & Elamurugan, B. (2023). *Palanivel Rathinasabapathi Velmurugan*. Advances in Business Information Systems and Analytics Book Series. Empirical Study on the Impact of Select Green HRM Dimensions on Green Innovation Culture., DOI: 10.4018/979-8-3693-0049-7.ch026

Chowdhury, S. R. (2024). Enhancing Employee Performance Through Green Human Resource Management: A Comprehensive Literature Review and Suggestions for Potential Researchers. *American Journal of Theoretical and Applied Business*, *10*(2), 21–32. DOI: 10.11648/j.ajtab.20241002.11

DEV, A. (2025). ECO-HR: Integrating Sustainability into workforce management" "Reducing HR's Environmental Footprint. *International Journal For Multidisciplinary Research*, *7*(2). Advance online publication. DOI: 10.36948/ijfmr.2025.v07i02.41484

Devi, S. (2024). The impact of green human resource management practices on employee's job performance. *International Journal of Management*, *13*(3), 506–522.

Faisal, S. (2023). Green Human Resource Management—A Synthesis. *Sustainability*, *15*(3), 2259–2259. DOI: 10.3390/su15032259

Gaikwad, D., Amitabh P., Gaurav P., & Francis, A. T. (2025, March 5). *Impact of human resource practices on the dissemination of green HRM*. SciSpace - Paper; Informa. https://doi.org/DOI: 10.4324/9781003616306-12

Hadi, S., Purnasari, E., Hidayat, R., Lukitaningsih, A., Kirana, K. C., & Siripipatthanakul, S. (2025). Challenges and Management Practices in Protecting Employee Personal Data in the Digital Era. In *Privacy and Security Management Practices for Organizations* (pp. 89–116). IGI Global Scientific Publishing., DOI: 10.4018/979-8-3693-9720-6.ch004

Hadi, S., Sulistyawati, E., Kirana, K. C., Dianti, C. A. I., Purnasari, E., & Ashakah, F. O. (2025). Abusive leadership and turnover intention: How employee and work engagement act as shields. *Asian Management and Business Review*, 74–89. https://doi.org/DOI: 10.20885/AMBR.vol5.iss1.art6

Hadi, S., Tjahjono, H. K., & Palupi, M. (2020). *Systematic Review: Meta Sintesis Untuk Riset Perilaku Organisasional*. Viva Victory Abadi.

Hakuduwal, K., & Kc, S. (2025). Impact of Green Human Resource Management Practices on Perceived Environmental Sustainability: Employee Perspectives from Nepalese Commercial Banks. *Deleted Journal, 13*(1), 70–87. DOI: 10.3126/ljbe.v13i1.80257

Hien, P. T. T. (2025). Does Green Human Resources Management Enhance Sustainable Performance? An Empirical Study from Manufacturing Firms. *International Journal of Research and Innovation in Social Science*, *IX*(IV), 1161–1167. DOI: 10.47772/ijriss.2025.90400089

Javed, S., Zhu, Y., Kumar, N., & Meghwar, S. (2024). Green human resource management practices and organizational sustainability performance in Pakistan's healthcare sector. *International Journal of Research In Business and Social Science*. https://doi.org/DOI: 10.20525/ijrbs.v13i1.3091

Khan, T. I. (2023). Green Human Resource Management practices: A study on the banking sector of Pakistan. International Journal of Economics and Business Issues.

Kumari, D., & Sharma, A. (2025). Building A Greener Brand: The Role Of GHRM In Enhancing Corporate Image And Sustainable Commitment. *Journal of Informatics Education and Research*, *5*(3). Advance online publication. DOI: 10.52783/jier.v5i3.3480

Kumari, S. (2024). A study of green human resource management practices adopted by the new age organizations. *Indian Scientific Journal Of Research In Engineering And Management*, *08*(04), 1–5. DOI: 10.55041/ijsrem32694

Kwerawarni, D. T. (2023). The Conceptual Model of Green Human Resource Management in Business Organization: A Literature Review. *International Journal of Multicultural and Multireligious Understanding*, *10*(12), 452–463. DOI: 10.18415/ijmmu.v10i12.5447

Lee, D., & Ahn, J. S. (2024). Unlocking the potential of green human resource management in Korea: A comprehensive review. *Asia Pacific Business Review*, •••, 1–24. DOI: 10.1080/13602381.2024.2358945

Manao, A., & Senen, S. H. (2024). Sustainable Human Resource Management: Building an Adaptive and Inclusive Organizational Culture. [DIJEMSS]. *Dinasti International Journal of Education Management and Social Science*, *6*(2), 863–871. DOI: 10.38035/dijemss.v6i2.3673

Memon, H., Anwar, S., & Jamali, A. (2024). The Role of Green HRM Practices and Employee Empowerment in Boosting Environmental Sustainability in Higher Educational Institutions in Sindh, Pakistan. *Progressive Research Journal of Arts & Humanities*, *6*(2), 55–67. DOI: 10.51872/prjah.vol6.iss2.357

Muhammad, N., Laghari, A., Amin, D. H., Bangash, S. A., Ishfaq, U., & Nosh, N. (2025). The Green Innovation Engine: How GHRM Practices Influence Sustainable Performance in Manufacturing Industries. *Social Science Review Archives*, *3*(1), 417–435. DOI: 10.70670/sra.v3i1.321

Ojo, A. O., & Raman, M. (2019). Role of Green HRM Practices in Employees' Pro-environmental IT Practices. *SciSpace - Paper*, 678–688. https://doi.org/DOI: 10.1007/978-3-030-16181-1_64

Pawar, P. (2025). The relationship between green hrm and employee environmental commitment. *International Journal of Research in Commerce and Management Studies*, *07*(04), 481–485. DOI: 10.38193/ijrcms.2025.7436

Permata, A. K., Nur W., & Hariyadi, E. (2024). Analisis Penerapan Green HRM pada Kinerja Keberlanjutan Industri Keuangan di Indonesia. *Reviu Akuntansi Dan Bisnis Indonesia*, *8*(3), 371–382. DOI: 10.18196/rabin.v8i3.22777

Rahman, N. H. A., Saad, A., & Ismail, S. (2025). Green human resource management (ghrm): A review of implementation challenges. *ADVANCED INTERNATIONAL JOURNAL OF BUSINESS, ENTREPRENEURSHIP AND SME's*, *7*(24), 315–324. DOI: 10.35631/aijbes.724021

Roscoe, S., Subramanian, N., Jabbour, C. J. C., & Chong, T. (2019). Green human resource management and the enablers of green organisational culture: Enhancing a firm's environmental performance for sustainable development. *Business Strategy and the Environment*, *28*(5), 737–749. DOI: 10.1002/BSE.2277

Rosyafah, S., Hidayat, W. G. P. A., & Nitawati, E. Y. (2025). Sustainable HRM in the face of climate change: Building organizational resilience through green human capital practices. *Annals of Human Resource Management Research*, *5*(2), 323–333.

Saeed, B. B., Afsar, B., Hafeez, S., Khan, I., Tahir, M., & Afridi, M. A. (2019). Promoting employee's proenvironmental behavior through green human resource management practices. *Corporate Social Responsibility and Environmental Management*, *26*(2), 424–438. DOI: 10.1002/CSR.1694

Shahrulnizam, N. A. A., Hassan, M. A., Mohamad Azizie, N. A., Rashid, A. H. A., & Wahab, S. R. A. (2024). Roles of Ghrm in Sustainable Organization Practices: A Systematic Literature Review. *International Journal of Academic Research in Business & Social Sciences*, *14*(10).

Shahzad, M. A., Du, J., Junaid, M., & Shahzad, F. (2025). From green HRM practices to green innovation performance: A mediation-moderation model. *Journal of Environmental Planning and Management*, *68*(5), 1187–1212.

Shaikh, D. M. R., Ali, M. I., Tunio, M. M. S. M., & Naeem, M. M. (2025). Green Human Resource Management and Sustainability Performance: A Systematic Review Using PRISMA Methodology. *Journal of Management & Social Science*, *2*(3), 330–344. DOI: 10.63075/80rydk30

Siripipatthanakul, S., Rungsiyanont, S., & Siripipattanakul, S. (2025a). Sustainable Dentistry Using Artificial Intelligence (AI) for Digital Education and Business Among University Dental Hospitals. In *Sustainable Development for Hospitals and Clinics* (pp. 33-64). IGI Global Scientific Publishing.

Siripipatthanakul, S., Phuangsuwan, P., Siripipattanakul, S., & Jaipong, P. (2025, b). Sustainable Tourism Based on Challenges, Tourist Experiences, and Host Communities: A Case Study of Ayutthaya, Thailand. In Best Practices for Human Resources Management in Tourism and Hospitality (pp. 361-394). IGI Global Scientific Publishing.

Siripipatthanakul, S., Rakpathum, W., Siripipattanakul, S., Kaewnaknaew, C., Phuangsuwan, P., & Jaipong, P. (2025, c). The Future Artificial Intelligence (AI) for the Hotel and Hospitality Industry: The Case of Hotel's Human Resource Management (HRM). In *Best Practices for Human Resources Management in Tourism and Hospitality* (pp. 395-422). IGI Global Scientific Publishing.

Trang, T. T. H. (2025). The Impact of Green Human Resource Management on Employee Performance in Public Sector Units: A Survey in Vietnam. *Journal of Information Systems Engineering & Management, 10*(39s), 263–274. DOI: 10.52783/jisem.v10i39s.7149

Tu, Y., Xiao, Q., Liu, Z., & Yin, S. (2024). Driving Green Development in Manufacturing Enterprises: Green Human Resource Management. *Scholars Journal of Economics, Business and Management, 11*(12), 404–412. DOI: 10.36347/sjebm.2024.v11i12.004

Veerasamy, U., Joseph, M. S., & Parayitam, S. (2024). Green human resource management practices and employee green behavior. *Journal of Environmental Planning and Management, 67*(12), 2810–2836.

Yong, J. Y., Yusliza, M.-Y., Ramayah, T., Chiappetta Jabbour, C. J., Sehnem, S., & Mani, V. (2020). Pathways towards sustainability in manufacturing organizations: Empirical evidence on the role of green human resource management. *Business Strategy and the Environment, 29*(1), 212–228. DOI: 10.1002/bse.2359

Zaid, A. (2022, July 8). *The Impact of Green Human Resource Management Practices with Sustainable and Operational Performance: A Conceptual Model.* SciSpace - Paper; Springer eBooks. https://doi.org/DOI: 10.1007/978-3-031-08090-6_37

KEY TERMS AND DEFINITIONS

Governance Policies: Governance policies are systematic guidelines and regulations that govern an organization's conduct, decision-making, and accountability, ensuring ethical, legal, and efficient operations. Key areas encompass corporate governance, risk management, ethics, and transparency for both public entities (establishing laws) and private enterprises (developing frameworks for leadership and stakeholders). These rules establish frameworks for leadership, mitigate wrongdoing (such as corruption), safeguard interests (of shareholders, the public, and stakeholders), and promote sustainable value creation through defined roles, duties, and performance oversight.

Green governance: A comprehensive framework that integrates environmental sustainability across all tiers of decision-making, uniting governments, businesses, communities, and stakeholders to manage resources, mitigate pollution, and foster a harmonious relationship between human society and nature for enduring well-being. It entails systemic transformations such as sustainable procurement, climate budgeting, closed-loop systems, and the integration of economic development and ecological integrity, ensuring that policies foster a stable, sustainable future.

Green Human Resource Management: Green HRM (Human Resource Management) involves integrating environmental sustainability into all HR functions—such as recruitment, training, performance evaluation, and compensation—to cultivate a workforce that advocates eco-friendly practices, minimizes the company's environmental impact, and nurtures a green corporate culture. The objective is to cultivate environmental consciousness among employees to encourage their alignment with the organization's overarching ecological objectives, resulting in cost reductions, enhanced reputation, and greater sustainability.

The Net Zero Transition: The Net Zero Transition refers to the worldwide economic transformation aimed at reconciling greenhouse gas (GHG) emissions with their removal, especially to meet the Paris Agreement's objective of limiting warming to 1.5°C. This necessitates substantial investment in clean energy, decarbonization, and sustainable practices by approximately 2050. It necessitates substantial reductions in emissions (over 90%) and the mitigation of remaining emissions, requiring fundamental transformations in energy, transportation, industry, and land utilization, while simultaneously fostering economic opportunities and resolving fairness concerns.

Chapter 11
Corporate Sustainability and Green Governance Environmental Policy Effectiveness

Syamsul Hadi
https://orcid.org/0000-0001-7226-4978
Universitas Sarjanawiyata Tamansiswa, Indonesia

Lin Dar Ong
https://orcid.org/0000-0003-1182-4323
Faculty of Business and Economics, Universiti Malaya, Malaysia

Parichat Jaipong
https://orcid.org/0000-0002-9249-3169
GlobalNxt University, Malaysia

Kusuma Chandra Kirana
Universitas Sarjanawiyata Tamansiswa, Indonesia

Saizal Pinjaman
https://orcid.org/0000-0002-6790-2688
Universiti Malaysia Sabah, Malaysia

Ambar Lukitaningsih
https://orcid.org/0009-0002-9847-9242
Universitas Sarjanawiyata Tamansiswa, Indonesia

ABSTRACT

Corporate sustainability affects emissions and waste policy. To show how internal management affects regulatory inputs' environmental implications. An extensive SLR literature review. PRISMA synthesized 50 high-impact empirical research on productive purposes (2018–2025) from 388 records (280 database, 108 citation chaining).Coded governance, regulation, and environment are discussed. Analysis shows pervasive "institutional decoupling" where governance rarely affects operations and compliance is symbolic. Circular economy and markets may reduce industry emissions. Greenwashing may result from generic regulations neglecting high-emission industry. Epistemological Paradox: Self-reported data conceals

DOI: 10.4018/979-8-3373-6821-4.ch011

greenwashing hazards despite methodological breakthroughs. Stakeholder and Neo-Institutional theories contrast business goals and behavior. To improve performance prediction, future models should include internal governance and external limitations. Policymakers must advise sectors, not horizontally.

1. INTRODUCTION

The escalating climate crisis and environmental decay have driven the relationship between corporate ecological behaviours and policy performance to the forefront of academic discussions. Traditionally, the business response to environmental protection was largely voluntary and philanthropic. In fact, over the past few years, there has been a major paradigm shift: Regulatory-type CSR is what we are seeing! This evolution has been perceived as moving from cosmetic image-making to real progress in environmental improvement, stimulated by a growing number of investors and increasing regulatory attention (AlHares, 2025; Haque, & Ntim, 2018). As a result, comprehending how companies respond to these changing requirements is no longer optional but imperative for the firm's existence and legitimacy.

In addition to providing a practical rationale for the study, the research is supported by international efforts to reduce carbon emissions, in which the business sector is responsible for most of these global emissions, and the natural resources consumed (Chen, 2024; Chikhaoui, 2025). Environmental governance norms have become potent agents of systemic transformation. Evidence shows that the introduction of the Non-Financial Reporting Directive in the European Union reduced corporate emissions by an average of 22% (Ernst, 2025). In this sense, those figures not only depict the prospective force of government intervention for refitting unreal plus un-conventionalism under new parameters but also gesture towards an emergent consensus that good corporate governance, stakeholder engagement, and challenging regulatory environments are necessary to provide more traction for a sustainable shift (Qi, et al., 2023; Shang, 2025).

Notwithstanding the growing adoption of such policies, there are clear conceptual and empirical barriers to understanding exactly how corporate sustainability efforts lead to effective environmental policy outcomes, especially in reducing emissions and improving waste disposal. The present state of research is nothing if not fragmented and divergent. On the one hand, literature reports a positive association between well-governed firms and environmental regulation (AlHares, 2025; Haque, & Ntim, 2022). Conversely, the systemic challenges also persist to subvert policy intentions by means of carbon leakage and 'greenwashing' when firms merely create an eco-friendly image without taking material changes in their operations (Kim, 2025; Okon, & Marauche, 2025). In addition, specific regulations can stimulate innovation and industry costly fights for material firms, and trigger policymaker competitions (Chen, et al., 2020; Hadi, Faridiana, et al., 2024; Hadi, Kirana, et al., 2024; Li, et al., 2025). This means that not only is the adoption of eco-legislation important, but also its actual implementation.

This lack of understanding is exacerbated by a poor grasp of the operationalization of these mechanisms in diverse institutional contexts (Borowiec, et al., 2024; Li, et al., 2025). A great deal of extant literature treats the corporation as a metaphorical "black box," with most efforts concentrated on the 'input' (regulation) and 'output' (emission) aspects, while failing to consider how internal managerial and operational activities bring the two together. There are serious dangers associated with this lack of analytic depth, in terms of both suboptimal policy and missed chances for high-impact climate action (Lister, 2018). Building on institutional and stakeholder theories and focusing on the interdependent relationship among governance systems, regulatory stringency, and stakeholder pressures over time, this paper adds to the extant literature by investigating whether corporate sustainability programs affect the effectiveness of environmental regulation. In this study, corporate sustainability practices are defined as emission-reduction efforts, eco-innovation, and resource-efficiency endeavours (Haque, & Ntim, 2022). Meanwhile, such policy performance is numerically measured by the degree to which it leads to actual environmental gains in terms of emission and pollution curtailment through corporations' compliance (Jiang, et al., 2024; Li, et al., 2025; Shang, 2025). By consolidating empirical findings on emissions and waste, the review addresses gaps in current research by elucidating when and under what conditions corporate initiatives can contribute to government policy outcomes.

2. LITERATURE REVIEW

2.1 Stakeholder Theory

According to *Stakeholder Theory*, which was proposed by Freeman (2010) conclude that the state of the company does not exist solely to serve shareholders but also needs to maintain good relationships with stakeholders. From an environmental sustainability perspective, this theory provides an underpinning for why firms are motivated to go beyond legal compliance in emissions reduction and waste management. Stakeholders are typically split into two groups: primary stakeholders (such as regulators, investors, and employees) and secondary stakeholders (such as communities or NGOs). Haque & Ntim (2018) state that, in many cases, corporate sustainability practices are strategic responses to the diverse, if not opposing, appetites of these constituencies. This model proposes that environmental performance is a dependent variable influenced by 'stakeholder salience', the extent to which managers prioritize competing demands. Companies with superior corporate governance can more effectively reconcile operational strategy with government-mandated environmental goals, thereby enhancing the efficiency of policy (Shang, 2025). Weak stakeholder engagement, on the other hand, usually leads to a disconnect between government policies and real corporate behavior.

2.2 Neo-Institutional Theory

Neo-Institutional Theory. Whereas Stakeholder Theory specifies the causes of change, Neo-Institutional Theory by DiMaggio & Powell (1983) explains why and how organizations within a given institutional domain tend to adopt the same or similar firm-level practices, a phenomenon also referred to as isomorphism. The theory is essential for explaining the effectiveness of environmental policies, as it distinguishes successful measures from essentially symbolic ones, such as greenwashing. Neo-institutionalism distinguishes three arenas of pressure that influence corporate sustainability. First, Coercive Isomorphism is derived from formal rules and legal obligations (in this case, the environmental policy on emissions and waste), which constitute a coercive force that compels organizations to "have no other option than to fit" into that environment. However, Kim (2025) argue that very strict levels of stringency, lacking adequate backstop protection, can result in counterproductive outcomes, such as carbon leakage. If the accumulated burdens increase, firms may move elsewhere to avoid emissions reductions more conveniently. A different type of MIM occurs when a firm imitates the organizational patterns or routines of successful rivals across varying environments. Firms could also use the waste management technology, not for instrumental reasons alone, but in order to acquire an institutional legitimacy in the eyes of the public (Li, et al., 2025). Finally, compliance with Normative Isomorphism is driven by professional norms and connections. The emergent Environmental, Social and Governance (ESG) reporting schema has contributed to establishing a normative context in which sustainability is perceived as an institutionalized business norm (Ernst, 2025).

2.3 Theoretical Synthesis and Relevance

Theoretical Synthesis and Relevance. The significance of these theories lies in their (potential) ability to "push back" against shallow policy borrowing. In the public sector, epistemic justification theorists suggest that in the absence of a tight internal control system, environmental regulations can prompt "symbolic compliance," whereby organizations recalibrate reporting practices without altering productive processes and thereby nullify the intent of the policymakers (Okon, &

Marauche, 2025). A simple agency-theory interpretation is provided by Stakeholder Theory: businesses act in response to regulators and shareholders. In contrast, Neo-Institutional Theory is a structural argument: the regulatory context `sets the rules of the game'. The combination of these two theories accounts for the variation in policy effects. Policies are maximally effective when coercive regulatory powers can be effectively transformed into normative corporate values, moving firms from mere symbolic compliance (institutional) to a proactively responsive phase (Chen, et al., 2020; Qi, et al., 2023).

3. METHODOLOGY

The study uses an SLR approach and adheres to PRISMA (Preferred Reporting Items for Systematic Reviews and Meta-Analyses) guidelines. The use of the PRISMA model was also based on promoting transparency, replicability, and scientific rigor in the review process to reduce selection bias that is inherent in traditionally typified narrative-review styles (Hadi et al., 2020; Purnasari et al., 2025; Siripipatthanakul et al., a-c). The paper aggregates high-impact evidence on the convergence between corporate sustainability practices and the effectiveness of environmental policy. We performed the literature search in SciSpace Database 19 using its advanced semantic search features and query translation algorithm. This method enabled the extraction of context-related studies, not just exact keyword matches. The primary search question focused on the intersection of 'corporate sustainability practices' and policy effectiveness from an emission reduction and waste management perspective.

Screening and inclusion criteria. This was performed in accordance with PRISMA guidelines. A further extensive eligibility check was carried out on 388 records (280 from the first search and 108 via citation chaining) using inclusion criteria such as relevance, empirical quality, and English language. Another 383 articles were excluded at the screener level: they were not directly policy-relevant, did not meet methodological criteria, or were duplicates. Afterward, we included 50 high-quality studies in the final sample. A schematic presentation of the information flow at each stage of the systematic review process is shown in Table 1.

Table 1. Stage selection of articles

Stages Selection	Number of Articles	Description
Initial Identification	280	Records were identified through SciSpace semantic search.
Citation Chaining	108	More records from backward/forward citations.
Absolute	388	Candidate papers combined together (280 + 108).
Screening & Eligibility	(338)	Excluded by relevance, language, or document type.
Included Core Studies	50	Excluded Records (did not meet inclusion criteria).

A thematic analysis was performed on the 50 included articles. Three major dimensions were identified for data extraction: governance mechanism, regulatory compliance strategy, and measurable environmental performance. This synthesis facilitated the development of a comprehensive story on "agreement and divergence" across extant work, detailed in the remainder of this review.

4. RESULTS

4.1 Chronological Review of Literature

10 years of academic discussion on corporate sustainability and the effectiveness of environmental policy: A perspective based on bibliometric results. With the increase in environmental problems, these works have shifted from an emphasis on voluntary compliance to more intricate and multidimensional governance arrangements. This evolution can be distinguished into four necessary steps as illustrated in the diagram in Figure 1, below:

Figure 1. Chronological Evolution of Research Publication (2018-2025)

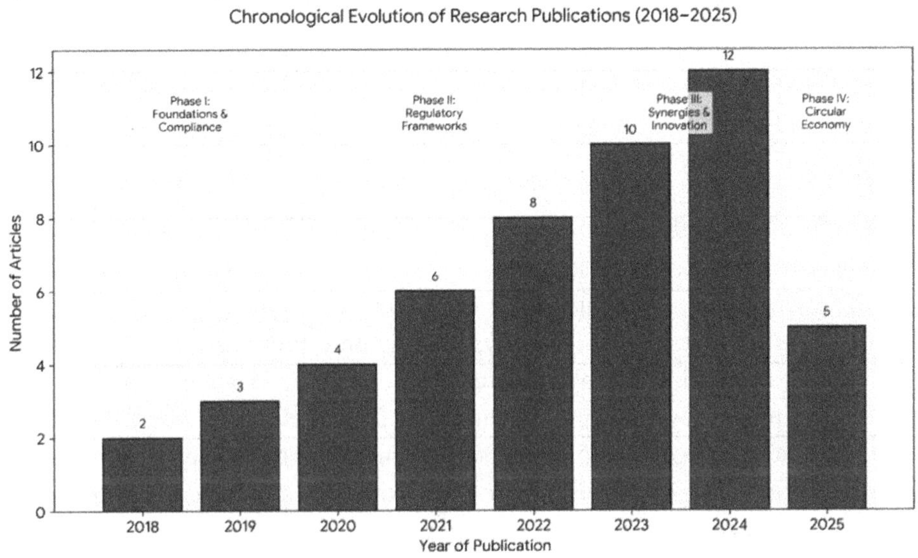

a. *2017–2019: Core to Corporate Sustainability and Environmental Compliance.* Phase 1 research was grounded in an initial exploration of the effects of corporate resources and governance on regulatory compliance. During that time, academics were very interested in the notion of organizational managers' responsibility and in CSR disclosures, particularly in light of emerging non-financial reporting requirements. The virtue of voluntary business action versus required policy edicts was a key argument during this period. Findings from these studies sometimes suggested that voluntary sustainability practices, without rigorous enforcement, do not lead to any real improvements in environmental performance.

b. *2020–2022: Development of Regulatory Policy and Market Participants.* The publications have shifted towards an evidence-based scrutiny of particular regulatory regimes. Interest in studying the effectiveness of Cleaner Production standards and a carbon trading system soared. In contrast to the previous phase, research in this period focused more on examining the interaction between "command and control" legislation and mbi. Moreover, this period saw the initial inclusion of Environmental, Social, and Governance (ESG) factors in waste and emissions management planning. More importantly, researchers began to detect content and regional heterogeneity, finding that policies did not have the same effect across industries or the overall economy.

c. *2023–2024: Going Beyond the Trial and Error Pool, Innovation, Synergistic Policies, and Stakeholder Engagement.* The research scope in 2023 and 2024 was further expanded to investigate the integrated power of multifarious environmental institutions, green finance, and public participation mechanisms. In such a circumstance, the source In1 of corporate green transformation was reinforced by moulding managers' "green perception" and prompting eco-innovation. Rising is now the realization in literature that it is necessary for companies to employ digital technology and governance innovations as enabling conditions to handle complex policy settings.
d. *2025-Deep Interconnection of Circular Economy, Informationized Power, and All-Domain Governance.* Recently, the literature presents a mature stage of discussion, stressing the role of Circular Economy (CE) practices in achieving resource efficiency and carbon neutrality. "Digital empowerment," especially the use of new tools to more precisely monitor and report emissions, and the introduction of synergy policy mixes, are two often highlighted 1. Introduction of models currently being researched (Acosta Llano et al., 2024; Ochoa et al., 2025). There is a sharp focus on good governance quality as a means of countering "greenwashing" (the publication of highly positive accounts without an accompanying methodology) and to ensure that sustainability reports present actual environmental performance. The belief in this step underscores the importance of specific, enabling policies that support corporate actions in aligning with dynamic global climate targets and regulatory environments

4.2 Thematic Review of Literature

A systematic review of the selected 50 studies presents two significant findings. First, the discourses on corporate sustainability and environmental policy effectiveness are not monolithic but are anchored in several separate yet related themes. As shown in Table 4, FAQ A8 focuses on internal governance structures across the examined articles, and FAQ A9 on external regulatory frameworks. FAQ A10 focuses on the transition to a circular economy and emergent synergistic governance mechanisms. The literature on EIP includes a range of aspects, such as self-perceived control.

Figure 2. Frequency of Dominant Themes in Selected Literature

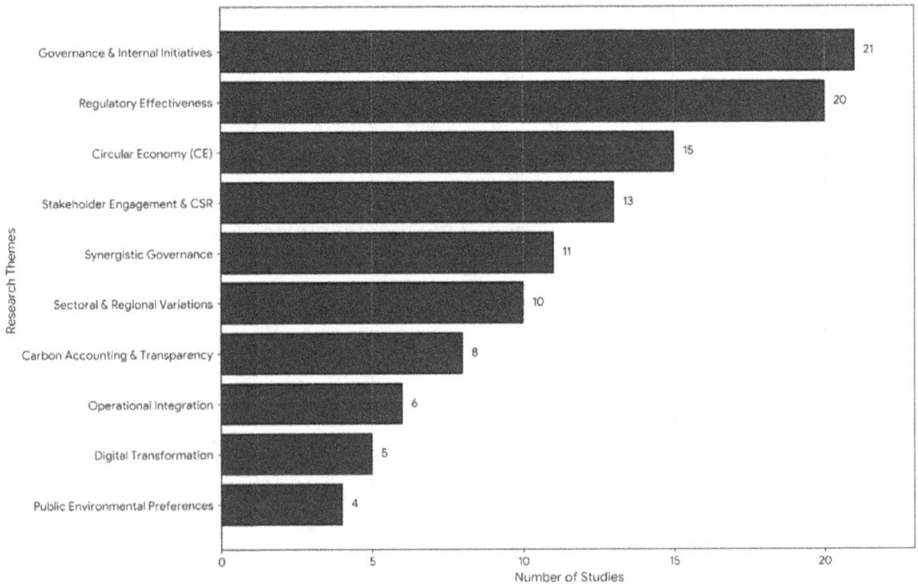

a. *Governance and internal initiatives come first.* The most common theme, found in 21 studies, is that corporate sustainability can be reduced to a governance issue. To find evidence of this, the literature suggests that emission mitigation policies and resource-efficient practices are more sustainable when implemented within the bounds of strong corporate governance. Research by AlHares (2025) and Haque & Ntim (2022) provides empirical evidence that overly rigorous board monitoring and the existence of dedicated sustainability committees are indispensable tools for combating "green washing." Plenary 4: Public-Private Accountability & Commitment: Lack of internal commitment. Without this internal promise, sustainability initiatives are likely to lack credibility. Moreover, managerial accountability can serve as an impetus for sectors to coordinate their daily operational mechanisms toward the broader decarbonization target (Shang, 2025).

b. *Regulatory Frameworks:* Market vs. Command-and-Control. The second most dominant theme (20 studies) related to the study of "rules of the game." A second type of study analyzes the efficiency of alternative policy instruments. There is agreement that "command-and-control " laws establish a baseline, but also, market incentives (e.g., carbon trading) often have an even stronger effect on company behavior due to their flexibility (Li, et al., 2025; Liu, 2024).

However, Ernst (2025) cautions that these policies are crucially dependent on the stringency of regulation and the extent of enforcement gaps.

c. *The Circular Economy Frontier.* An emerging theme in the literature, this subset focuses on more complex 'system-level' Circular Economy (CE) approaches to differentiate from reductionist 'end-of-pipe' pollution control (15 articles). Kaur (2024) and Zahran (2024) suggest that incorporating CE principles, e.g., green supply chain management and closed-loop recycling, can reduce emissions and promote resource conservation. This theme, however, identifies persistent barriers to progress, such as the availability of finance and the preparedness for new digital waste-monitoring technologies (Albitar, et al., 2024).
d. *The Mediating Position of Stakeholders and CSR.* Spanning internal strategy and external policy are Stakeholder Engagement and CSR (13 studies). The main conclusion here is that CSR mediates the way firms perceive and respond to environmental regulations (Chen, et al., 2020; Zaim, et al., 2023), and Qi et al. (2023). 80 secondary stakeholders (communities, NGOs) often increase the effectiveness of government policy, and in 2023, they are demanding corporate resistance to progress: Organizations must evolve from minimal compliance to strategic innovation.
e. *Synergistic Governance and Innovation.* Emerging powerfully over the last few years (11 studies), this theme stands in opposition to policy isolation. Jiang et al. (2024) and Xu et al. (2023)"Synergistic governance," the incorporation of green finance, digital empowerment, and pollution control policies, provides a more cost-effective achievement with respect to environmental policy than any single policy instrument in isolation. This theme highlights the importance of digital transformation as a key enabler of contemporary ecological governance (Peng, et al., 2025).
f. *Carbon Accounting and Transparency Challenges.* One of the main operational themes that emerged from 8 studies is "addressing the measurement problem of corporate sustainability. Haque & Ntim (2018), Okon & Marauche (2025) argue that managerial values, as well as regulatory pressure, are critical to a stronger influence on robust carbon accounting embodiment. There are, however, significant differences: institutional capacity constraints weigh especially heavily on SMEs. This is a message I have seen repeatedly and one that underscores the risk of poor reporting quality and greenwashing, pointing to the need for an easy-to-use reporting framework that links corporate intention to real climate action (To & Tran, 2025).
g. *The tension of integration:* Operationalizing sustainability Peter Dobers & Staffan Furusten How are we to make sense of the tightening spiral between economic and ecological unsustainability, which is building up? From strategy to implementation, six studies focus on integrating sustainability into core business

operations, including production lines and the supply chain. It is not surprising that companies with a higher operational dynamic used their resources more efficiently, as indicated by either Zia (2024) or Fu & Su (2020). Nonetheless, there is a certain conflict in this direction; Guo et al. (2019) conclude the sustained trade-off faced by managers between short-term operations and long-term environmental objectives, such a tension calling for new approaches to strategic management.

h. *The Role of Digital Transformation in Facilitating Sustainability.* The impact of technology themes. It is observed that technology is a discrete factor in 5 of the studies. In the literature, digital technologies, in particular IoT, AI, and cloud platforms, are acknowledged to play a key role in the accurate tracking of emissions in order to have sustainable waste management (Balasubramanian, 2025). Peng et al. (2025) argue that digitalization not only increases the efficiency of resources but also offers transparency, which is essential for ESG reporting, improves stakeholder engagement, and helps the implementation of complex policy environments (Faieq, & Çek, 2024).

i. *Collaboration and Co-Regulation:* Corporate actors, government agencies, and local/regional governments need to collaborate to generate synergistic environmental effects. The research suggests that co-regulation should be central to policy formulation, with shared incentives and joint action among the public and private sectors. Policy makers need to develop mechanisms for joint innovation, through which companies will not only enforce environmental policy but also actively adopt and enhance it. It helps achieve synergistic environmental benefits and activities by linking corporate sustainability efforts to larger governmental sustainability objectives (Chen, et al., 2020; Jiang, et al., 2024; Leffel, et al., 2024).

5. DISCUSSIONS

5.1 Critical Analysis and Synthesis

Drawing on a detailed analysis of 50 studies, the paper rigorously examines how corporate sustainability practices interact with effective environmental policy. Instead, the review suggests that although the literature agrees that regulatory enforcement pressure and internal controls are critical to achieving reduced emissions, a more critical synthesis reveals epistemological frictions, institutional decoupling, and geopolitical asymmetries. The reflection on the seven key issues raised by this review at least leads us to conclude that, although the field shows methodological maturity, it also identifies structural gaps that challenge the sustainability paradigm in place.

a. The Epistemological Paradox: Strict Techniques vs. Expedients information

An underlying tension that emerges from this review concerns the relationship between the complexity of analysis and the quality of the input data. It is a maturing field, with work such as AlHares (2025), Ernst (2025), and data by Li et al. (2025) using sophisticated econometric methods/estimators, such as Difference-in-Differences (DID) and fixed-effects models, to filter for causal effects. Moreover, the addition of facility-level data and ESG indicators in Kim (2025) further increases the granularity of the analysis. Nevertheless, there is an "epistemological fragility" because it is based on self-reported corporate data. Soulful states that, without independent verification, these data sets could be exposed to "greenwashing" and the foibles of management. This includes the hazard that econometric models perform well with the data, even when the data itself is somewhat flawed. Further, there is a dominance of empirical studies with a focus on big enterprises and developed economies, a known research gap in flipping the talking point to SMEs for emerging markets; limitations of generalisability results across settings (Okon, & Marauche, 2025; To, & Tran, 2025).

b. Institutional Decoupling: Structure and Practice of Governance

The literature depicts an alarming form of (institutional) decrepitude where institutions matter only at the `sociotechnical MP'-stage. Despite finding corporate governance as an important moderating factor in reducing greenwashing (AlHares, 2025; Haque & Ntim, 2018) and stakeholder engagement as promoting accountability (Balasubramanian, 2025; Shang, 2025), uncertainty persists. Okon & Marauche (2025) observe a common "symbolic compliance," in which management with sustainability committees is more concerned with making their company look good than with effecting real change. This separation is most noticeable in the examination of effects: while measures have a positive association with carbon performance in polluting industries (Su, 2020; Haque & Ntim, 2022), their translation into absolute reductions in emissions remains mixed (Hannouf). Kim (2025) and Molinaro et al. (2025) argue that the latter challenge of causal extra-boundary emissions, which are not covered by conventional frameworks, is exacerbated by the global outsourcing of carbon-intensive activities, making it even more difficult to determine true responsibility.

c. The Boundaries of Regulatory Convergence and Spatial Externalities

A damning convergence of regulations shows what "one-size-fits-all" has wrought. A combination of complementary market-based instruments and policy mixes is usually the most effective way to promote green transition (Chen, 2024; Jiang, et al., 2024; Li, et al., 2025; Liu, 2024)but their effectiveness can vary widely depending on the concrete context. Ernst (2025), Su (2024), and Tang et al. (2023) found that horizontal governance is left to be ineffective in high-emission sectors like Energy and Materials, which calls for sector-specific adjustments. Moreover, the literature reveals interesting spatial paradoxes. There is extensive regional variation in enforcement and market creation, which in turn shapes outcomes (Handoyo, et al., 2024; Liu, 2024). Zhu et al. (n.d.) and Yu et al. (2024) anticipate spatial 'spillover effects' and "pollution refuges ": regional policies increase environmental loadings at one regulatory jurisdiction (which has weak enforcement), outweighing total burden reductions. Added to this is a fragmentation of institutional power and Authority in developing countries, which erodes regulatory firmness (To, & Tran, 2025).

d. The 'Innovation Divide' and the Difficulties of Collaboration

The power of technology is playing out as a double-edged sword. Similar to Behera (2023), Zia (2024), and Albitar et al. (2024), the use of digital tools and eco-innovation increases competition and improves waste management, but their use remains irregular. (Balasubramanian, 2025) emphasizes that, as with anything related to e-inclusion, there will also be a sustainability divide around digitization, as financial and technical obstacles keep SMEs off this digital train. Lastly, there is a "collaborative dilemma" between corporate and governmental duties. Collaborative governance has potential value (Chen et al., 2020) and is supported by the public's environmental preference. However, Qi et al. (2023) present a counter story of crowding-out effects, in which government diktats might lead to a diminution of voluntary corporate altruism. This points to a complex, nonlinear relationship in which policy makers must tread carefully between mandatory provisions and incentives for voluntary innovation.

5.2 Agreement and Divergence Across Studies

A synthesising comparison of the selected literature reveals a multifaceted knowledge domain, with broad agreement on the driving forces of sustainability alongside wide divergence in outcomes and means. This dichotomy is explained in five key axes.

a. Corporate Sustainability Impact: Designed and Implemented in Two Different Ways

There is widely shared optimism about corporate projects. Research in C Agreement finds that environmental management processes, including emission-reduction legislation and eco-innovation, lead to a considerable decrease in GHG emissions and better waste management outcomes. It is a well-documented fact that CE practices are acknowledged for playing the dual role in waste reduction and carbon mitigation (Kaur, 2024; Molinaro, et al., 2025). However, the Divergent results point to a "commitment-performance gap." Researchers present evidence of symbolic compliance, or "greenwash, in which sustainability is given lip service that does not necessarily lead to substantial emission reductions by corporations (Hannouf, & Assefa, 2017; Haque, & Ntim, 2018). Still others note that total emissions are on the rise even within some companies' programs, revealing an inability to handle Scope 3 emissions (Leffel, et al., 2024; Okon, & Marauche, 2025). This discrepancy is primarily due to differences in measurement (intention vs. behavior) and to the existence of "green blushing" tactics.

b. Regulatory Regime Impact: Effectiveness and Leakage

The consensus in the literature is that policy intervention works. MBIs (such as carbon trading) are widely acknowledged to spur green innovation, with command-and-control regulations serving as a necessary floor (Chen, 2024; Hu et al., 2022; Li et al., 2025; Liu, 2024). Conversely, divergence appears regarding uniformity. Even though the efficiency of these instruments has already been demonstrated, it remains uneven and, in other fields, is much less effective than in high-emission sectors, which are constrained by narrow regulatory frameworks (Ernst, 2025; Kim, 2025). In addition, some studies find "diminishing returns" or nonlinear effects of economic incentives (24). One key reason for this disconnect is regulatory design: sector-specific policies work where "one-size-fits-all" policies do not.

c. Governance and the Paradox of Credibility

It is the opinion of all that governance is a credibility. It is understood that better corporate governance and stakeholder engagement can act to improve the performance of sustainability programs and reduce greenwashing potential (Al-Hares, 2025; Balasubramanian, 2025; Shang, 2025). But divergence occurs at the governance-to-impact level. Some evidence has shown that good governance does not automatically lead to positive environmental performance, attributed to the disparities in managers' perceptions and values (Chen, et al., 2020; Guo, et al., 2019;

To, & Tran, 2025). This position seems inconsistent with the heterogeneous power of stakeholders across regions or between SMEs and MNCs.

d. Sectoral and regional diversity: Spill-ups and differences

The evidence is consistent with the idea that policy impacts are sector- and region-specific. More reactive are for polluting industries or in tightly enforced areas (Ernst, 2025; Handoyo, et al., 2024). However, significant range-insensitive effects are found. Other studies identify "pollution leakage" effects, in which the reduction of emissions in regulated areas is compensated for by elevating the level of activities elsewhere, even under stronger enforcement (Kim, 2025; Yu, et al., 2024; Zhu, et al., n.d.). And, corporate readiness disparity is also product-specific between developing and developed countries (Borowiec, et al., 2024; Chikhaoui, 2025).

e. Modes of interaction: synergy versus crowding-out

Most of all, there is agreement on the effectiveness mediated by mechanisms such as green innovation and financial incentives. Joint vision-language models are all about stronger transfigurations (Chen, 2024; Jiang, et al., 2024; Xu, et al., 2023). Formal vs informal rules. On the other hand, there is much debate about the distinction between formals and informals. There are some indications of 'crowding-out' whereby forceful government mandates can crowd out or mitigate voluntary corporate commitments (Lyu, et al., 2024; Qi, et al., 2023). Also, the effects of digital technology are divergent, as uneven adoption gaps affect the findings (Balasubramanian, 2025; Peng, et al., 2025).

5.3 Theoretical and Practical Implications

5.3.1 Theoretical Implications

a. **Corporate Governance and Stakeholder Engagement:** The research cements the pivotal contributions of corporate governance and stakeholder engagement in improving the performance of corporate sustainability programmes, especially in emissions reduction and waste treatment. In addition, strong governance will guard against risks such as greenwashing and boost the credibility of corporate climate pledges. This is in line with the neo-institutional theory, which argues that organizations adjust to external pressures by embracing institutionalized actions (AlHares, 2025; Haque, & Ntim, 2018, 2022). The results indicate that a firm's internal governance has a significant Influence on its environmental performance and are consistent with calls to organize theories to consider the interplay between institutional and rational determinants of corporate responses to regulatory imperatives.

b. **Environmental Regulations:** The research, while shedding light on the interaction effects between various forms of ecological regulations, command-and-control, market-based, and public participation, is less detailed. Nonlinear and heterogeneous causal effects defy simplistic assumptions about regulatory impact and require more sophisticated theoretical models. These models must include both internal governance structures and external regulation, whose impacts are often sector-specific. Such an implication calls for the generalization of the theories of VP inclusion under different regulatory settings confronting firms (Li, et al., 2025; Liu, 2024; Lyu, et al., 2024).

c. **Circular Economy and Eco-Innovation:** The incorporation of circular economy principles and eco-innovation into corporate sustainability strategies is arguably a theoretical development. This result also aligns with carbon reduction and the relationship between resource efficiency and waste minimisation, which are in line with both co-benefit theory and legitimacy theory, and further expands the current framework of corporate environmental performance. In many cases, circularity in sustainability isn't simply about connecting the compliance dots; it requires a more proactive stance to minimize its ecological footprint. This facilitates a shift of sustainability frameworks from simple regulation-driven compliance to an innovation-driven transition, shifting the focus of sustainability in corporate strategies (Albitar, et al., 2024; Kaur, 2024; Molinaro, et al., 2025).

d. **Market Mechanism:** The potential of a market mechanism (e.g., Carbon trading systems, markets) as a motivating force for corporate environmental innovations and emissions reduction is recognized. However, obstacles like carbon leakage and regulatory complexity remain and should prevent us from implementing these mechanisms without further consideration. This aligns with economic incentive theories, but it also underscores the need for a more sophisticated governance structure to mitigate unintended consequences. These results call for theoretical models of market-based mechanisms that account for the economic and regulatory challenges embedded in them, such that they actually deliver on environmental objectives (Chen, 2024; Kim, 2025; Li et al., 2025).

e. **Multi-Level Governance and Co-Regulation:** This study has confirmed the significance of multi-level governance for the effectiveness of environmental policies. When companies assume ecological responsibility in line with state requirements, this may lead to a theoretical shift from voluntarism to co-regulation and collaboration in governance models. These models present a more accurate reflection of how the effectiveness of environmental policy operates, and that future theoretical measures should center around governance modes where government regulation is combined with firm responsibility in local to national as well as international contexts to fix this environmental problem (Chen, et al., 2020; Jiang, et al., 2024; Lister, 2018).

f. **Digital Transformation and Innovation:** Value, Sustainable Innovation. The significance of digital transformation (DT) as a mediator or moderator in the relationship between companies' sustainable development activities and environmental returns is increasing5.3.. This points to the theoretical extension of digital governance and innovation diffusion theories. Operational efficiency, reporting quality, and transparency can be enhanced with digital tools, which are essential to improving environmental performance. The results suggest that digital governance should be integrated into sustainability frameworks to ensure that technological innovation contributes to sustainable development goals (Balasubramanian, 2025; Peng, et al., 2025; Zia, 2024).

5.3.2 Practical Implications

a. **Targeted and Industry-Specific Measures:** Sector-specific environmental rules, tailored to industry characteristics and regional conditions, should be developed by the government to increase the impact of regulation. The research suggests that a uniform approach, such as broad non-financial reporting mandates, has differential sector-level impacts. "There may be stronger or more specific regulations needed in high-emitting sectors, for example. Therefore, practical policies should be tailored to the industry's specific requirements and bottlenecks to ensure that rules achieve their intended environmental outcomes (Ernst, 2025; Su, 2024; Tang, et al., 2023).

b. **Enhancing Corporate Governance:** Strengthening corporate governance systems and improving stakeholder engagement are realistic levers to enhance the credibility and effectiveness of corporate sustainability initiatives. According to the results, it may be necessary to include governance quality measures in policymakers' or investors' assessments of environmental performance. Good governance is essential if the promises of sustainability are to be translated into practice, and regulators should accord special attention to governance quality when measuring corporate environmental performance (AlHares, 2025; Okon & Marauche, 2025; Shang, 2025).

c. **Circular Economy and Eco-Innovation:** Adopting circular economy strategies and eco-innovation are practical avenues for firms to reduce emissions and minimize waste. Potential also means business resilience and competitiveness. Operationally, companies must embed the circular economy into their business models by adopting cleaner technologies, minimizing waste, and maximizing resource use through innovation. This necessitates investment in technology and partnerships through collaboration for industries to achieve long-term sustainable aspirations (Kaur, 2024; Zahran, 2024).

d. **Further refinement of market-based instruments:** With market-based instruments, such as carbon emission trading schemes and green finance incentives, having demonstrated practical effectiveness, the study underscores the need to fine-tune them continuously. Concerns, such as carbon leakage and design complexity, dilute their efficacy, suggesting that these mechanisms need to be more flexible to be more inclusive of different types and sizes of firms. Policymakers should also strive to make their design even more effective in providing incentives for environmental improvement and fair across sectors and regions (Chen, 2024; Kim, 2025; Li, et al., 2025).
e. **Digitization:** Incorporating digital technologies and ESG standards into environmental corporate practices can improve transparency, reporting quality, and operational efficiency. Companies need to develop digital capabilities to enhance their environmental performance and more closely meet regulatory requirements. The takeaway: Businesses must invest in digital tools that improve monitoring, reporting, and the optimization of their environmental performance. Additionally, governments/regulatory authorities should encourage inter-sectoral collaboration to foster the application of digital technologies to sustainability activities (Balasubramanian, 2025; Guo, et al., 2019; Peng, et al., 2025).
f. **Collaboration and Co-Regulation:** Corporate actors, government agencies, and local/regional governments need to collaborate to generate synergistic environmental effects. The research suggests co-regulation should be central to policy formulation, with shared incentives and joint action among public and private sectors. Policy makers need to develop mechanisms for joint innovation, through which companies will not only enforce environmental policy but also actively adopt and enhance it. It helps achieve synergistic environmental benefits and activities by linking corporate sustainability efforts to larger governmental sustainability objectives (Chen, et al., 2020; Jiang, et al., 2024; Leffel, et al., 2024).

6. CONCLUSIONS

The purpose of the study is to examine the extent to which corporate environmental sustainability activities influence the effectiveness of environmental policy, specifically in terms of emissions and waste. It is intended to fill the gap in understanding of how internal management activities structure the association between regulatory input factors and environmental outputs. SLR Systematic Literature

Review. A total of 388 records (280 from the database, 108 from citation chaining) were obtained according to PRISMA guidelines, and 50 high-impact empirical articles on productive goals (2018–2025) were synthesized. Coding of board discussion combined governance instruments, regulatory activities, and measurable environmental performance. Analysis demonstrates such a pattern of widespread" institutional decoupling"; the governance does not often penetrate operations, and practice is symbolic compliance. Circular economy approaches and market levers are associated with (but do not guarantee) emissions reductions – the outcomes will depend on industry-specific applications. Generic "one-size-fits-all" regulations do not address high-emission sectors, particularly regarding the risk of greenwashing. The review found an "Epistemological Paradox, which is that despite rigorous methodological advancement, reliance on self-reported data conceals greenwashing risks. It integrates Stakeholder and Neo-Institutional theories to account for the gap between corporations' goals and their observable behavior. Building on Neo-Institutional Theory, this study shows that unaligned coercive pressures lead to "pollution bypassing. Internal governance quality should be incorporated into future models to improve performance prediction, alongside external regulatory factors. Policy makers need to shift their focus from horizontal to sectoral guidance. To address the "commitment-performance gap" and be sustainable, board monitoring and the organization's digital carbon accounting need to improve.

This systematic study underscores the significance of corporate sustainability activities for enhancing the economy, particularly in carbon minimization and waste management. The evidence accumulated seems to point toward a positive relationship between firm-level action, including circular economy strategies and efficiency in resource use, and actual progress on environmental outcomes, particularly for high-polluting firms and ESG leaders. And there remain pitfalls, such as symbolic compliance and greenwash, that call for transparent corporate governance and effective stakeholder management to underpin the credibility and performance of sustainability promises. At the core of environmental policy is the interaction between market-based instruments, such as carbon trading, and regulatory requirements that induce investment in research and development. This research suggests that tailored regulations are much more effective than one-size-fits-all policies, especially in high-emissions sectors and developing countries. Institutional readiness and enforcement capacity are also noted as necessary for successfully implementing such measures, particularly with small and medium-sized enterprises (SMEs), which are commonly challenged by a lack of resources and expertise within their organizations.

Finally, governance quality and digital technology adoption are the main antecedents of the inculcation of a culture of sustainability within organizations, enhancing corporate responsibility. While technological advancements can enable advanced systems for emissions monitoring and waste tracking, a range of hurdles, including high implementation costs and limited technical knowledge, are preventing adoption at scale. Tourism research should prioritize linking the rhetoric of commitment to the reality of environmental impacts, favoring transparent adaptation and transformation in sustainable practices through integrated multi-level governance frameworks that accommodate sectoral and regional heterogeneity.

7. LIMITATIONS OF THE LITERATURE

a. The academic literature on corporate sustainability and the effectiveness of environmental policy has several significant limitations that need to be addressed if it is to have greater practical relevance and impact. The first limitation is geographic bias, with the majority of studies heavily concentrated in certain areas, such as China, Europe, or the US. This small geographic concentration limits the generalizability of findings to other regions and may exclude region-specific dynamics that influence corporate sustainability practices and public policy outcomes. The external validity and generalizability of conclusions to different parts of the world would be increased if research were extended beyond a few geographic settings (AlHares, 2025; Chen, et al., 2020; Ernst, 2025; Hu, et al., 2022; Li, et al., 2025).
b. Another significant limitation is sectoral coverage since the majority of studies focus on high-emission or manufacturing sectors rather than other industries. This sectoral predisposition reduces the ability to draw comprehensive conclusions regarding corporate sustainability impacts and has implications for understanding impacts across various economic activities. To achieve a more comprehensive vision of corporate sustainability, it is essential to consider a wide variety of industries, including even those with less pollution and different process characteristics (Ernst, 2025; Fu, & Su, 2020; Handoyo, et al., 2024; Haque, & Ntim, 2018; Molinaro, et al., 2025).
c. Issues of greenwashing and reporting bias are also prevalent in the literature. Some studies highlight the risks posed by symbolic compliance, whereby organizations report on sustainability activities without implementing those that benefit the environment. This poses a problem not only because it would compromise the trustworthiness of self-reports and policy assessments but also because it would give a misleading impression of corporate sustainable activities and their real environmental implications (AlHares, 2025; Haque & Ntim, 2018; Okon & Marauche, 2025). This shortcoming highlights a need for stronger validation of the company's sustainability claims and for more standardized reporting frameworks.

d. Lastly, the lack of studies on scope three emissions should be emphasized. These scopes also fail to consider scope three emissions, indirect emissions generated in the production of goods and services, and thus cause a partial understatement of companies' full environmental burden. Closing this blind spot is critical for better evaluating corporate sustainability and the effectiveness of policies, since disregarding Scope 3 GHG emissions results in an incomplete understanding of a company's environmental impact (Ernst, 2025; Kim, 2025; Molinaro, et al., 2025).

Gaps and Future Research Directions

a. The literature on corporate sustainability and green policies has gaps that need to be addressed to support the quality of existing practices. One of the most glaring such gaps is sector-specific policy efficacy. Existing environmental regulations typically deliver broad horizontal policies that do not make a substantial dent in emissions from high-emitting sectors, including Energy, Materials, and Utilities. This gap demands that sector-specific environmental policy and Climate Transition Plans be formulated and empirically tested, as they address the specific challenges posed by carbon emissions in these high-emitting industrial sectors. Due to the variety of industries, blanket policies are not very effective, and targeted interventions are essential for any serious attempt at emission regulations (Ernst, 2025; Su, 2024).
b. Another critical gap is the corporate mechanisms for governance and greenwashing. High governance quality is associated with improved emission outcomes, but the mechanisms by which governance reduces greenwashing and leads to real environmental performance remain unclear. Future research is required to conduct qualitative and quantitative analyses using governance structure, board roles, and accountability mechanisms as themes to unpack governance structures that are instrumental in ensuring that symbolic compliance gives way to actual reductions in emissions. This evidence is crucial in addressing the chasm between organisational sustainability rhetoric and empirical environmental effects (AlHares, 2025; Haque, & Ntim, 2018).
c. Furthermore, the literature reveals a disproportionate involvement of small and medium-sized enterprises (SMEs) in sustainability studies. The majority of these studies were conducted with large companies, and little attention has been paid to SMEs, despite their economic and environmental importance. Future research should also explore the obstacles and facilitators to SME s taking up sustainability activities, carbon accounting, and compliance. Custom frameworks are necessary to help SMEs overcome their specific resource and knowledge

limitations. These initiatives are of utmost importance for the design of effective policies to help SMEs engage in sustainability (Okon & Marauche, 2025; To & Tran, 2025).

d. Finally, Scope 3 emissions, which reflect indirect value chain emissions, are frequently omitted from corporate sustainability strategies. In general, emissions are poorly studied, undermining the overall effectiveness of Reduction strategies. In the future, efforts should be made to develop methodologies and case studies for incorporating. Scope 3 into corporate sustainability, Approaches, and legislation. It is a necessary part of full corporate environmental responsibility and to prevent Carbon leakage, so that companies take into account their ecological impact (Kim, 2025; Molinaro, et al., 2025).

REFERENCES

Acosta Llano, E., Hurmelinna-Laukkanen, P., & Haapanen, L. (2024). Blockchain for the circular economy, implications for public governance. *International Journal of Public Sector Management, 38*(1), 30–52. DOI: 10.1108/IJPSM-12-2023-0365

Albitar, K., Nasrallah, N., Hussainey, K., Wang, Y., Nasrallah, N., Hussainey, K., Wang, Y., & Wang, Y. (2024). Eco-innovation and corporate waste management: The moderating role of ESG performance. *Review of Quantitative Finance and Accounting*. Advance online publication. DOI: 10.1007/s11156-024-01281-5

AlHares, A. (2025). Evaluating Emission Reduction Policies and the Influence of Corporate Governance. *Sustainability, 17*(18), 8204–8204. DOI: 10.3390/su17188204

Balasubramanian, A. (2025). Business Strategies For Sustainable Waste Management: Integrating Esg And Environmental Stewardship. *International Journal of Environmental Sciences*, •••, 3366–3374. DOI: 10.64252/93h8n420

Behera, D. K. (2023). Promoting Sustainable Development Through Environmental Policy, Green Technologies, and Effective Waste Management: A Comprehensive Review. *Journal of Multidisciplinary Sciences, 1*(2), 179–198. DOI: 10.58578/mikailalsys.v1i2.1675

Borowiec, J., Papież, M., & Śmiech, S. (2024). The impact of environmental regulations on carbon emissions in countries with different levels of emissions. *Environmental Science and Pollution Research International*. Advance online publication. DOI: 10.1007/s11356-024-35702-8

Chen, X., Chen, X., Zhang, J., & Zeng, H. (2020). Is corporate environmental responsibility synergistic with governmental environmental responsibility? Evidence from China. *Business Strategy and the Environment, 29*(8), 3669–3686. DOI: 10.1002/BSE.2603

Chen, Z. (2024). Path Analysis of The Carbon Market's Impact on Corporate Carbon Emission Reduction. *Highlights in Business. Economics and Management, 43*, 297–302. DOI: 10.54097/a7jeqk60

Chikhaoui, M. (2025). Do Environmental Management Policies Decrease Carbon Emissions? A Probabilistic Analysis of G7 Countries. *International Journal of Energy Economics and Policy, 15*(4), 388–400. DOI: 10.32479/ijeep.19664

DiMaggio, P. J., & Powell, W. W. (1983). The Iron Cage Revisited: Institutional Isomorphism and Collective Rationality in Organizational Fields. *American Sociological Review, 48*(2), 147–160. DOI: 10.2307/2095101

Ernst, E. (2025). From Mandatory CSR Reporting to Environmental Impact: Assessing the Effectiveness of the EU NFRD in Reducing Corporate GHG Emissions. *European Company Law*, *22*(4/5), 184–192. DOI: 10.54648/eucl2025027

Faieq, H. T., & Çek, K. (2024). Enhancing Kurdistan's manufacturing companies' sustainable waste management: A norm activation approach to green accounting, CSR, and environmental auditing oversight. *Heliyon*, *10*(12), e32725–e32725. DOI: 10.1016/j.heliyon.2024.e32725

Freeman, R. E. (2010). *Strategic Management: A Stakeholder Approach*. Cambridge University Press., DOI: 10.1017/CBO9781139192675

Fu, W., & Su, H.-C. (2020). Take actions or outsource? An empirical examination of strategic environmental options on greenhouse gas emissions. *International Journal of Operations & Production Management*, *40*(6), 753–776. DOI: 10.1108/IJOPM-02-2020-0088

Guo, L., Xu, Y., Liu, G., & Wang, T. (2019). Understanding Firm Performance on Green Sustainable Practices through Managers' Ascribed Responsibility and Waste Management: Green Self-Efficacy as Moderator. *Sustainability*, *11*(18), 4976. DOI: 10.3390/SU11184976

Hadi, S., Faridiana, A., Kirana, K. C., Lukitaningsih, A., Rahmawati, C. H. T., Purnasari, E., & Wahyuningtyas, N. (2024). Political Skill And Transactional Leadership On Employee Performance: The Mediating Effect Of Organizational Work Culture. [Ikonomicheski Izsledvania]. *Economic Studies Journal*, *33*(4), 112–129.

Hadi, S., Kirana, K. C., Setiawati, L., Lada, S. B., & Rahmawati, C. H. T. (2024). The Effect of Digital Leadership and Organizational Support on Innovative Work Behavior: The Mediating Role of Emotional Intelligence. *Quality - Access to Success*, *25*(199). Advance online publication. DOI: 10.47750/QAS/25.199.09

Hadi, S., Tjahjono, H. K., & Palupi, M. (2020). *Systematic Review: Meta Sintesis Untuk Riset Perilaku Organisasional*. Viva Victory Abadi.

Handoyo, S., Yudianto, I., & Dahlan, M. (2024). Exploring firm and country's specific factors affecting carbon emission reduction performance: Study on selected ASEAN countries. *Heliyon*, *10*(17), e37036–e37036. DOI: 10.1016/j.heliyon.2024.e37036

Hannouf, M., & Assefa, G. (2017). The Role of Sustainability Resources of Large Greenhouse Gas Emitters: The Case of Corporations in Alberta, Canada. *Sustainability*, *9*(2), 182. DOI: 10.3390/SU9020182

Haque, F., & Ntim, C. G. (2018). Environmental Policy, Sustainable Development, Governance Mechanisms and Environmental Performance. *Business Strategy and the Environment, 27*(3), 415–435. DOI: 10.1002/BSE.2007

Haque, F., & Ntim, C. G. (2022). Do corporate sustainability initiatives improve corporate carbon performance? Evidence from European firms. *Business Strategy and the Environment, 31*(7), 3318–3334. DOI: 10.1002/bse.3078

Hu, J., Wu, H., & Ying, S. X. (2022). Environmental regulation, market forces, and corporate environmental responsibility: Evidence from the implementation of cleaner production standards in China. *Journal of Business Research, 150*, 606–622. DOI: 10.1016/j.jbusres.2022.06.049

Jiang, Y., An, W., & Yang, J. (2024). A Configurational Approach to the Driving Forces for Synergistic Reduction in Pollution and Carbon Emissions. *Sustainability, 16*(24), 10940–10940. DOI: 10.3390/su162410940

Kaur, H. (2024). Towards sustainable business practices: Circular economy principles for resource optimisation and wastereduction. *Global Journal for Research Analysis*. Advance online publication. DOI: 10.36106/paripex/2506575

Kim, J. (2025). <scp>ESG</scp> and Carbon Leakage: How Cap-and-Trade Regulations Influence Firms' Emission Allocation Strategies. *Corporate Social Responsibility and Environmental Management*. Advance online publication. DOI: 10.1002/csr.70142

Leffel, B., Lyon, T. P., & Newell, J. P. (2024, March 1). *Filling the climate governance gap: Do corporate decarbonization initiatives matter as much as state and local government policy?* SciSpace - Paper. DOI: 10.1016/j.erss.2023.103376

Li, N., Zhang, H., Zhang, X., & Xie, X. (2025). Does Market-Based Environmental Regulatory Policy Improve Corporate Environmental Performance? Evidence from Carbon Emission Trading in China. *Sustainability, 17*(2), 623–623. DOI: 10.3390/su17020623

Lister, J. (2018). The Policy Role of Corporate Carbon Management: Co-regulating Ecological Effectiveness. *Global Policy, 9*(4), 538–548. DOI: 10.1111/1758-5899.12618

Liu, K. W. (2024). Impact of Heterogeneous Environmental Regulation on Carbon Emissions: Firm-Level Evidence from China's Manufacturing Industry. *International Journal of Economics and Finance, 16*(12), 44–44. DOI: 10.5539/ijef.v16n12p44

Lyu, M., Chen, Y., & Chen, S. (2024). Prevention or ex-post forcing? The impact of dual environmental regulation on corporate environmental behavior. *Heliyon*. Advance online publication. DOI: 10.1016/j.heliyon.2024.e30850

Molinaro, M., Orzes, G., & Sarkis, J. (2025). Unpacking Circular Economy Practices and Carbon Emissions Relationships: Co-benefits and Legitimacy Perspectives. *Business Strategy and the Environment*. Advance online publication. DOI: 10.1002/bse.70157

Ochoa, W. A. A., Arce, A. G. T., Neto, A. I., Munaro, M. R., Calabokis, O. P., Ballesteros-Ballesteros, V. A., Ochoa, W. A. A., Arce, A. G. T., Neto, A. I., Munaro, M. R., Calabokis, O. P., & Ballesteros-Ballesteros, V. A. (2025). Interlinking Urban Sustainability, Circular Economy and Complexity: A Systematic Literature Review. *Sustainability*, *17*(15). Advance online publication. DOI: 10.3390/su17157118

Okon, D. U. O., & Marauche, O. M. O. (2025, July 1). *The Drivers and Barriers of Corporate Carbon Accounting Adoption: A Systematic Review of Managerial, Organizational, and Policy Factors*. SciSpace - Paper. DOI: 10.35629/5252-0707717727

Peng, J., Wang, H., & Wen, L. (2025). Editorial: Advancing carbon reduction and pollution control policies management: theoretical, application, and future impacts. *Frontiers in Environmental Science*, *13*. Advance online publication. DOI: 10.3389/fenvs.2025.1687660

Purnasari, E., Hermuningsih, S., Hidayat, R., Hadi, S., & Kumar, A. (2025). *Antecedents and Consequences of Financial Literacy: A Comprehensive Systematic Literature Review of the Past Decade*. 3–14. DOI: 10.2991/978-94-6463-698-7_2

Qi, Y., Xie, X., & Chen, X. (2023). Effects of environmental regulation and corporate environmental commitment: Complementary or alternative? Evidence from China. *Journal of Cleaner Production*. Advance online publication. DOI: 10.1016/j.jclepro.2023.138641

Shang, K. (2025). Balancing Corporate Social Responsibility, Environmental Policies, and Stakeholder Expectations for Sustainable Development: A Stakeholder-Centric Analysis. *Sustainable Development*. Advance online publication. DOI: 10.1002/sd.70204

Siripipatthanakul, S., Phuangsuwan, P., Siripipattanakul, S., & Jaipong, P. (2025, a). Sustainable Tourism Based on Challenges, Tourist Experiences, and Host Communities: A Case Study of Ayutthaya, Thailand. In *Best Practices for Human Resources Management in Tourism and Hospitality* (pp. 361-394). IGI Global Scientific Publishing.

Siripipatthanakul, S., Phuangsuwan, P., Siripipattanakul, S., & Jaipong, P. (2025, b). Sustainable Tourism Based on Challenges, Tourist Experiences, and Host Communities: A Case Study of Ayutthaya, Thailand. In Best Practices for Human Resources Management in Tourism and Hospitality (pp. 361-394). IGI Global Scientific Publishing.

Siripipatthanakul, S., Rakpathum, W., Siripipattanakul, S., Kaewnaknaew, C., Phuangsuwan, P., & Jaipong, P. (2025, c). The Future Artificial Intelligence (AI) for the Hotel and Hospitality Industry: The Case of Hotel's Human Resource Management (HRM). In *Best Practices for Human Resources Management in Tourism and Hospitality* (pp. 395-422). IGI Global Scientific Publishing.

Su, L. (2024). Environmental regulation and corporate green innovation: Evidence from the implementation of the total energy consumption target in China. *Journal of Business Economics*. Advance online publication. DOI: 10.1007/s11573-024-01207-6

Tang, Y., Ye, Q., Zhou, F., & Liu, H. (2023, November 28). *A Study on the Differential Effects of Heterogeneous Environmental Regulations on the Green Transformation of Chinese Manufacturing Enterprises under 'Double Carbon' Target*. SciSpace - Paper; *Research Square*. DOI: 10.21203/rs.3.rs-3667338/v1

To, M. H., & Tran, B. M. (2025). Environmental regulatory stringency and corporate reporting quality: Evidence from waste, emissions, water, and energy regulations in Vietnamese firms. *Environment and Ecology*, *16*(3), 67–81. DOI: 10.21511/ee.16(3).2025.05

Xu, H., Xu, J., Wang, J., & Hou, X. (2023). Reduce production or increase efficiency? Hazardous air pollutants regulation, energy use, and the synergistic effect on industrial enterprises' carbon emission. *Energy Economics*. Advance online publication. DOI: 10.1016/j.eneco.2023.107027

Yu, W., Li, Z., & Hu, C. (2024). Carbon reduction and corporate sustainability: Evidence from low-carbon city pilot policy. *Heliyon*. Advance online publication. DOI: 10.1016/j.heliyon.2024.e28992

Zahran, S. (2024). Investigating the Nexus between Green Supply Chain Practices and Sustainable Waste Management in Advancing Circular Economy. *Sustainability*. Advance online publication. DOI: 10.3390/su16093566

Zaim, R. L., Lukman, S., & Syafrizal. (2023). Pengaruh Environmental Regulation terhadap Environmental Performance Melalui Mediasi Corporate Social Responsibility. *Jurnal Informatika Ekonomi Bisnis*, 331–338. DOI: 10.37034/infeb.v5i2.257

Zhu, F., Hunjra, A. I., Roubaud, D., Zhao, S., & Grebinevych, O. (n.d.). *Assessment of Synergistic Governance of Pollution and Carbon Reduction*. SciSpace - Paper., DOI: 10.1016/j.jenvman.2025.124226

Zia, F. (2024). The Interaction between Corporate Operations and Environmental Management. *Global Regional Review*, *IX*(I), 206–217. DOI: 10.31703/grr.2024(ix-i).18

KEY TERMS AND DEFINITIONS

Corporate Sustainability: Corporate Sustainability incorporates environmental, social, and governance (ESG) factors into a company's fundamental strategy and operations to generate enduring value for all stakeholders, not solely shareholders, emphasizing responsible resource utilization, ethical conduct, and beneficial community effects for sustainable growth rather than mere short-term profits. Principal initiatives encompass emission reduction, diversity promotion, equitable labor practices, transparent reporting, and investment in green research and development, advancing from mere compliance to comprehensive, sustainable development.

Green Governance: Green governance is a comprehensive approach to decision-making that incorporates environmental sustainability across all tiers of policy and management, aligning governments, businesses, and society to reconcile economic development with ecological preservation for enduring well-being, emphasizing pollution reduction, resource conservation, and climate change mitigation through systemic transformations and collaborative engagement among multiple stakeholders. It necessitates integrating environmental factors into conventional frameworks, advocating for evidence-based decision-making, and transitioning to circular economies.

Neo-Institutional Theory (NIT): Neo-Institutional Theory (NIT) elucidates the phenomenon of organizational isomorphism, whereby entities converge in form and practice due to the influence of external cultural regulations, norms, and stakeholder expectations, rather than only via considerations of efficiency. It perceives institutions as socially constructed ideas that influence behavior, rather than as mere logical models, and has developed to investigate how organizations actively transform these institutions through language, rhetoric, and social engagement, transcending mere conformity.

Stakeholder Theory: Stakeholder Theory is a business philosophy that posits organizations should cater to all parties impacted by their operations (stakeholders), rather than solely shareholders, by balancing the interests of employees, consumers, suppliers, communities, and investors for mutual advantage and sustainable success. Proposed by R. Edward Freeman, it asserts that a company's success depends on robust interactions across its entire ecosystem, shifting the emphasis from shareholder primacy to comprehensive corporate responsibility and ethical governance.

The PRISMA METHOD: The PRISMA method (Preferred Reporting Items for Systematic Reviews and Meta-Analyses) is an evidence-based guideline that features a checklist and flow diagram to assist researchers in transparently and comprehensively reporting systematic reviews and meta-analyses, particularly in the health sciences, thereby enhancing quality, reproducibility, and clarity in research outcomes. It offers a consistent framework for documenting the entire process, from the identification and selection of studies (using the renowned flow diagram) to data extraction and reporting of results, thereby enhancing credibility for authors, reviewers, and readers.

Compilation of References

Abdessadak, A., Ghennioui, H., Thirion-Moreau, N., Elbhiri, B., Abraim, M., & Merzouk, S. (2025). Digital twin technology and artificial intelligence in energy transition: A comprehensive systematic review of applications. *Energy Reports, 13*, 5196–5218. DOI: 10.1016/j.egyr.2025.04.060

Abdulmuhsin, A. A., & Hussein, H. D., AL-Abrrow, H., Masa'deh, R. E., & Alkhwaldi, A. F. (2025). Impact of artificial intelligence and knowledge management on proactive green innovation: The moderating role of trust and sustainability. *Asia-Pacific Journal of Business Administration, 17*(3), 765–795. DOI: 10.1108/APJBA-05-2024-0301

Abu-Mahfouz, S., Halim, M. S. A., Bahkia, A. S., & Noryati Alias, P. D. in M., & Tambi, A. M. (2023). Sustainable human resource management practices in organizational performance: The mediating impacts of knowledge management and work engagement. *Journal of Entrepreneurship. Management and Innovation, 19*(2), 57–97. DOI: 10.7341/20231922

Acosta Llano, E., Hurmelinna-Laukkanen, P., & Haapanen, L. (2024). Blockchain for the circular economy, implications for public governance. *International Journal of Public Sector Management, 38*(1), 30–52. DOI: 10.1108/IJPSM-12-2023-0365

Adebayo, T. S., Genç, S. Y., Castanho, R. A., & Kirikkaleli, D. (2021). Do public–private partnership investment in energy and technological innovation matter for environmental sustainability in the East Asia and Pacific region? An application of a frequency domain causality test. *Sustainability, 13*(6), 3039. Advance online publication. DOI: 10.3390/su13063039

Adewusi, A. O., Asuzu, O. F., Olorunsogo, T., Iwuanyanwu, C., Adaga, E., & Daraojimba, D. O. (2024). AI in precision agriculture: A review of technologies for sustainable farming practices. *World Journal of Advanced Research and Reviews, 21*(1), 2276–2285. DOI: 10.30574/wjarr.2024.21.1.0314

Adger, W. N. (2010). Climate Change, Human Well-Being, and Insecurity. *New Political Economy, 15*(2), 275–292. DOI: 10.1080/13563460903290912

Advancing Climate Justice at the International Court of Justice. (2025, November). https://www.ciel.org/project-update/advancing-climate-justice-at-the-icj/

Agrawal, N., Beriwal, K., & Daga, N. (2024). Sustainability Through Human Resource Management: A Conceptual Framework. In R. Sharma, I. Bhardwaj, S. Grima, T. Sachdeva, K. Sood, & E. Özen (Eds), *Sustainable Development Goals: The Impact of Sustainability Measures on Wellbeing* (Vol. 113B, p. 0). Emerald Publishing Limited. https://doi.org/DOI: 10.1108/S1569-37592024000113B003

Agrawal, R., Majumdar, A., Kumar, A., & Luthra, S. (2023). Integration of artificial intelligence in sustainable manufacturing: Current status and future opportunities. *Operations Management Research, 16*(4), 1720–1741. DOI: 10.1007/s12063-023-00383-y

Agudelo-Vera, C. M., Mels, A. R., Keesman, K. J., & Rijnaarts, H. H. (2011). Resource management as a key factor for sustainable urban planning. *Journal of Environmental Management, 92*(10), 2295–2303. DOI: 10.1016/j.jenvman.2011.05.016

Ahmadin, M. (2022). Social research methods: Qualitative and quantitative approaches. *Jurnal Kajian Sosial Dan Budaya: Tebar Science, 6*(1), 104-113. http://ejournal.tebarscience.com/index.php/JKSB/article/view/103

Ahmad, J., Mamun, A. A., Masukujjaman, Md., Makhbul, Z. K. M., & Ali, K. A. M. (2023). Modeling the workplace pro-environmental behavior through green human resource management and organizational culture: Evidence from an emerging economy. *Heliyon, 9*(9), e19134–e19134. DOI: 10.1016/j.heliyon.2023.e19134 PMID: 37681178

Ahmad, R., Nawaz, M. R., Ishaq, M. I., Khan, M. M., & Ashraf, H. A. (2023). Social exchange theory: Systematic review and future directions. *Frontiers in Psychology, 13*, 1015921. DOI: 10.3389/fpsyg.2022.1015921

Ahmed, M. U., Farhad, F., & Hussain, M. G. (2025). *The Effect of Sustainable Practices on Organizational Success and Worker Well-Being*. Business, Economics and Management. https://doi.org/DOI: 10.20944/preprints202507.1755.v1

Aijaz, N., Lan, H., Raza, T., Yaqub, M., Iqbal, R., & Pathan, M. S. (2025). Artificial intelligence in agriculture: Advancing crop productivity and sustainability. *Journal of Agriculture and Food Research, 101762*. Advance online publication. DOI: 10.1016/j.jafr.2025.101762

Ai-min, Q., Shao, G., & Zheng, W. (2018). Assessing China's Cybersecurity Law. *Computer Law & Security Review*, *34*(6), 1342. DOI: 10.1016/j.clsr.2018.08.007

Akbar, K. F., Muhammad B., & Syed Kamran H. S. (2025, September 4). *The Relationship Between Green HRM and Environmental Performance in SMEs*. SciSpace - Paper. https://doi.org/DOI: 10.63056/acad.004.03.0707

Akbar, A., Ahmad, S., Khalid, M., Aslam, M. F., & Bhatti, M. A. A. (2024). Analyzing the Effect of Green HRM on Organizational Performance. [BBE]. *Bulletin of Business and Economics*, *13*(2), 864–869. DOI: 10.61506/01.00402

Akhter, R., & Sofi, S. A. (2022). Precision agriculture using IoT data analytics and machine learning. *Journal of King Saud University. Computer and Information Sciences*, *34*(8), 5602–5618. DOI: 10.1016/j.jksuci.2021.05.013

Akoijam, A. S., & Krishna, V. V. (2017). Exploring the Jawaharlal Nehru National Solar Mission (JNNSM): Impact on innovation ecosystem in India. *African Journal of Science, Technology, Innovation and Development*, *9*(5), 573. DOI: 10.1080/20421338.2017.1359466

Alamandi, M. (2025). Sustainable innovation management: Balancing economic growth and environmental responsibility. *Sustainability*, *17*(10), 4362. DOI: 10.3390/su17104362

Alam, M. N., Campbell, N., Das, S., Hashim, F., Ur Rehman, I. H., & Iqbal, J. (2023). Green Training and Development Revolutionizing Organizational Performance: The Moderating Role of Green Employee Involvement in the Bangladeshi Pharmaceutical Industry. *International Business Research*, *16*(9), 36. DOI: 10.5539/ibr.v16n9p36

Albino, V., Berardi, U., & Dangelico, R. M. (2015). Smart cities: Definitions, dimensions, performance, and initiatives. *Journal of Urban Technology*, *22*(1), 3–21. DOI: 10.1080/10630732.2014.942092

Albitar, K., Nasrallah, N., Hussainey, K., Wang, Y., Nasrallah, N., Hussainey, K., Wang, Y., & Wang, Y. (2024). Eco-innovation and corporate waste management: The moderating role of ESG performance. *Review of Quantitative Finance and Accounting*. Advance online publication. DOI: 10.1007/s11156-024-01281-5

Aldy, J., Pizer, W., Tavoni, M., Reis, L. A., Akimoto, J., Blanford, G., Carraro, C., Clarke, L. E., Edmonds, J., Iyer, G. C., & McJeon, H. C. (2016). Economic Tools to Promote Transparency and Comparability in the Paris Agreement. *Nature Climate Change*, *6*(11), 1000–1004. https://www.nature.com/articles/nclimate3106

Alegbesogie, A. I. (2023). The Impact of Green Human Resource Management Practices on Organisational Performance. *Vilnius University Proceedings*, *37*, 6–11. DOI: 10.15388/VGISC.2023.1

Alenezi, M., & Akour, M. (2025). AI-driven innovations in software engineering: A review of current practices and future directions. *Applied Sciences (Basel, Switzerland)*, *15*(3), 1344. DOI: 10.3390/app15031344

Algburi, S., Kareem, S. S. A., Sapaev, I., Mukhitdinov, O., Hassan, Q., Khalaf, D. H., & Jabbar, F. I. (2025). The role of artificial intelligence in accelerating renewable energy adoption for global energy transformation. *Unconventional Resources*, *100229*. Advance online publication. DOI: 10.1016/j.uncres.2025.100229

AlHares, A. (2025). Evaluating Emission Reduction Policies and the Influence of Corporate Governance. *Sustainability*, *17*(18), 8204–8204. DOI: 10.3390/su17188204

ALICE. (2024). Leveraging the digital transition for efficient and zero-emissions logistics – ALICE Alliance for Logistics Innovation through Collaboration in Europe. https://www.etp-logistics.eu/leveraging-the-digital-transition-for-efficient-and-zero-emissions-logistics/

Al-kateeb, Z. N., & Abdullah, D. B. (2024). Unlocking the potential: Synergizing IoT, cloud computing, and big data for a bright future. *Iraqi Journal for Computer Science and Mathematics*, *5*(3), 25. DOI: 10.52866/ijcsm.2024.05.03.001

AlKetbi, A., & Rice, J. (2024). The Impact of Green Human Resource Management Practices on Employees, Clients, and Organizational Performance: A Literature Review. *Administrative Sciences*, *14*(4), 78. DOI: 10.3390/admsci14040078

Alkhalaf, T., & Al-Tabbaa, O. (2024). The effect of ability, motivation and opportunity (AMO) on SMEs' innovation performance. *Creativity and Innovation Management*, *33*(1), 21–38. DOI: 10.1111/caim.12578

Allen, R., & Bergin, A. (2009). Ocean Acidification: An Emerging Australian Environmental Security Challenge. *Australian Journal of Maritime and Ocean Affairs*, *1*(2), 49–56. DOI: 10.1080/18366503.2009.10815637

Almulhim, A. I., & Yigitcanlar, T. (2025). Understanding smart governance of sustainable cities: A review and multidimensional framework. *Smart Cities*, *8*(4), 113. DOI: 10.3390/smartcities8040113

Alnaqbi, M., & Alami, A. (2023). Sustainability and Renewable Energy in the UAE: A Case Study of Sharjah. *Energies*, *16*(20), 7034. Advance online publication. DOI: 10.3390/en16207034

Alshaflut, A. (2024). Enhancing smart city infrastructure: An IoT-integrated system approach for real-time urban management. *Indian Journal of Science and Technology, 17*, 3012–3017. DOI: 10.17485/IJST/v17i29.1070

Al-Sulaiti, K. (2024). Sustainability and Sustainable HRM Case Study: Sustainable HRM in Light of Ministry of Culture, Qatar. *WSB Journal of Business and Finance, 58*(1), 19–28. DOI: 10.2478/wsbjbf-2024-0003

Alvarez, D. del B., Sasakawa, A., Yamaguchi, K., & Wang, J. (2022). Strategic priorities for regional power connectivity in Asia in the overlapping of subregional initiatives and China's global energy interconnection. *Asia Pacific Business Review, 30*(2), 376. DOI: 10.1080/13602381.2022.2093528

Alzahrani, A. I., Chauhdary, S. H., & Alshdadi, A. A. (2023). Internet of Things (IoT)-based wastewater management in smart cities. *Electronics (Basel), 12*(12), 2590. DOI: 10.3390/electronics12122590

AlZubi, A. A., & Galyna, K. (2023). Artificial intelligence and internet of things for sustainable farming and smart agriculture. *IEEE Access : Practical Innovations, Open Solutions, 11*, 78686–78692. DOI: 10.1109/ACCESS.2023.3298215

Amini, M. Y., Tang, Z., & Besharat, A. (2024). Greening university practices: Empowering eco-conscious behavior, transforming sustainable culture, and shaping greener institutional awareness through strategic green HRM initiatives. *International Journal of Research In Business and Social Science*. https://doi.org/DOI: 10.20525/ijrbs.v13i1.3149

Amirkhizi, P. J., Pedrammehr, S., Pakzad, S., & Shahhoseini, A. (2025). Generative artificial intelligence in adaptive social manufacturing: A pathway to achieving industry 5.0 sustainability goals. *Processes (Basel, Switzerland), 13*(4), 1174. DOI: 10.3390/pr13041174

Amjad, M. A., Khoso, R. A., Soomro, M. A., & Khan, A. A. (2025). Engaging Minds and Greening Workspaces: Investigating the Mediating Role of Employee Engagement in Strengthening the Relationship Between Green HRM Practices and Organizational Sustainability Outcomes. *Journal of Management & Social Science, 2*(3), 186–208. DOI: 10.63075/9j2zwv02

Andrade, A. C. C., & Vasquez, M. C. S. (2024). The compatibility between SDGs and the EU regulatory framework of AI. *Journal of Ethics and Legal Technologies, 6*(1), 111–144. DOI: 10.14658/pupj-JELT-2024-1-6

Andriani, A. D., & Prasetyo, B. (2024). Building digital sustainability awareness through organizational communication in higher education. *Journal of Environmental Communication Studies*, *12*(3), 201–219.

Anshima, A., Bhardwaj, B., Sharma, D., & Chand, M. (2025). Ensuring sustainability through green HRM practices: A review, synthesis and research avenues. *Journal of Organizational Effectiveness: People and Performance*, *12*(1), 57–90. DOI: 10.1108/JOEPP-10-2023-0493

Antonopoulos, I., Robu, V., Couraud, B., Kirli, D., Norbu, S., Kiprakis, A., Flynn, D., Elizondo-Gonzalez, S., & Wattam, S. (2020). Artificial intelligence and machine learning approaches to energy demand-side response: A systematic review. *Renewable & Sustainable Energy Reviews*, *130*, 109899. DOI: 10.1016/j.rser.2020.109899

Apple Inc. (2024). *Environmental progress report 2024.* https://www.apple.com/environment/pdf/Apple_Environmental_Progress_Report_2024.pdf

Article 6 of the Paris Agreement. (2016). https://unfccc.int/process-and-meetings/the-paris-agreement/article6

ASEAN. (2021). *ASEAN digital masterplan 2025.* Association of Southeast Asian Nations. https://asean.org

Asrifan, A., Siripipatthanakul, S., Alaqad, M.H., Saodi, S., Sadaruddin, S. (2025). Privacy and Security Management Practices for Organizations. Open source preview, 2025, pp. 151–182

Atiku, S. O., Jeremiah, A., Semente, E., & Boateng, F. (Eds.). (2024). *Sustainable development in industry and society 5.0: Governance, management, and financial implications*. IGI Global Scientific Publishing., DOI: 10.4018/979-8-3693-7322-4

Avagyan, A. (2025). AI-Integrated smart traffic systems for Carbon-Neutral cities. Trends Research & Advisory. https://trendsresearch.org/insight/ai-integrated-smart-traffic-systems-for-carbon-neutral-cities/?srsltid=AfmBOoqtne30QOrZYW5fgC34fGubBBMSspz8h-ppLJXBPtvXBWBIseqj

Awwad Al-Shammari, A. S., Alshammrei, S., Nawaz, N., & Tayyab, M. (2022). Green Human Resource Management and Sustainable Performance With the Mediating Role of Green Innovation: A Perspective of New Technological Era. *Frontiers in Environmental Science*, *10*. Advance online publication. DOI: 10.3389/fenvs.2022.901235

Axelrod, R. S. (1996). The European Energy Charter Treaty. *Energy Policy*, *24*(6), 497. DOI: 10.1016/0301-4215(96)00022-5

Babatunde, O. B., Okonji, C. T., Olanihun, Z. S., & Daniel, H. A. (2025). Public-private partnerships in advancing AI-based logistics. *NIU Journal of Humanities*, *10*(2), 87–100. DOI: 10.58709/niujhu.v10i2.2199

Bahamazava, K. (2025). *AI-driven scenarios for urban mobility: Quantifying the role of ODE models and scenario planning in reducing traffic congestion*. Transport Economics and Management., DOI: 10.1016/j.team.2025.02.002

Bahaw, P., Forgenie, D., Sadiq, G., & Sookhai, S. (2025). Generative AI for business sustainability: Examining usability, usefulness, and triple bottom line impacts in small and medium enterprises. *Sustainable Futures : An Applied Journal of Technology, Environment and Society*, *100815*. Advance online publication. DOI: 10.1016/j.sftr.2025.100815

Bai, X., Lu, C., Dong, X., & Li, Y. (2024). Role of collaborative governance in unlocking private investment in sustainable projects. *Humanities & Social Sciences Communications*, *11*, 699. Advance online publication. DOI: 10.1057/s41599-024-03175-2

Balasubramanian, A. (2025). Business Strategies For Sustainable Waste Management: Integrating Esg And Environmental Stewardship. *International Journal of Environmental Sciences*, •••, 3366–3374. DOI: 10.64252/93h8n420

Baldé, C. P., Forti, V., Gray, V., Kuehr, R., & Stegmann, P. (2024). *The Global E-Waste Monitor 2024*. United Nations Institute for Training and Research (UNITAR), International Telecommunication Union (ITU), and International Solid Waste Association (ISWA). https://ewastemonitor.info

Banerjee, G. (2025). The impact of Employee Engagement and Organizational Environmental Performance: A Literature Review. *Journal of Information Systems Engineering & Management*, *10*(35s), 991–1001. DOI: 10.52783/jisem.v10i35s.6179

Bansal, N., & Choudhary, H. (2024). Fostering digital equity: Evaluating impact of digital literacy training on internet outcomes in rural marginalised communities in India. *International Journal of Lifelong Education*, *43*(5), 473–493. DOI: 10.1080/02601370.2024.2347327

Bansal, S., Garg, I., & Sharma, G. D. (2019). Social entrepreneurship as a path for social change and driver of sustainable development: A systematic review and research agenda. *Sustainability*, *11*(4), 1091. DOI: 10.3390/su11041091

Barakat, S., Alneyadi, R., Alderei, M., Bani Malek, S., Alahbabi, F., & Jasim, M. (2024). The Impact Of Sustainable Hrm Practices On Job Performance In Uae Organizations. *International Journal of Management Trends: Key Concepts and Research, 3*(1), 29–38. DOI: 10.58898/ijmt.v3i1.29-38

Barnett, J. (2009). Environmental Security. In Kitchin, R., & Thrift, N. (Eds.), *International Encyclopedia of Human Geography* (pp. 553–557). Elsevier., DOI: 10.1016/B978-008044910-4.00774-4

Barney, J. (1991). Firm resources and sustained competitive advantage. *Journal of Management, 17*(1), 99–120. DOI: 10.1177/014920639101700108

Barritt, E., & Sediti, B. (2019). The Symbolic Value of Leghari v Federation of Pakistan: Climate Change Adjudication in the Global South. *King's Law Journal : KLJ, 30*(2), 203. DOI: 10.1080/09615768.2019.1648370

Barton, J. H. (2009, March 18). *Patenting and Access to Clean Energy Technologies in Developing Countries.*

Batty, M., Axhausen, K. W., Giannotti, F., Pozdnoukhov, A., Bazzani, A., Wachowicz, M., Ouzounis, G., & Portugali, Y. (2012). Smart cities of the future. *The European Physical Journal. Special Topics, 214*(1), 481–518. DOI: 10.1140/epjst/e2012-01703-3

Beck, U. (1992). *Risk society: Towards a new modernity.* SAGE Publications.

Begum, V., & Arshi, T. A. (2020). An Impact-Based Model of Green Human Resource Management: Evidence from UAE. *Journal of Security and Sustainability Issues, 10*(2), 573–584. DOI: 10.9770/JSSI.2020.10.2(16) PMID: 39858411

Behera, D. K. (2023). Promoting Sustainable Development Through Environmental Policy, Green Technologies, and Effective Waste Management: A Comprehensive Review. *Journal of Multidisciplinary Sciences, 1*(2), 179–198. DOI: 10.58578/mikailalsys.v1i2.1675

Beloglazov, A., Abawajy, J., & Buyya, R. (2012). Energy-aware resource allocation heuristics for efficient management of data centers for cloud computing. *Future Generation Computer Systems, 28*(5), 755–768. DOI: 10.1016/j.future.2011.04.017

Benítez-Núñez, C., Díaz-Díaz, N. L., Ballesteros-Rodríguez, J. L., & de Saá-Pérez, P. (2024). Explaining academic researchers' performance from the Ability–Motivation–Opportunity (AMO) perspective. *Studies in Higher Education, 49*(12), 2423–2440. DOI: 10.1080/03075079.2024.2307965

Bestuzheva, K., Besançon, M., Chen, W. K., Chmiela, A., Donkiewicz, T., Van Doornmalen, J., & Witzig, J. (2023). Enabling research through the SCIP optimization suite 8.0. *ACM Transactions on Mathematical Software*, *49*(2), 1–21. https://dl.acm.org/doi/abs/10.1145/3585516

Bharati, S., & Podder, P. (2022). Machine and deep learning for iot security and privacy: Applications, challenges, and future directions. *Security and Communication Networks*, *2022*(1), 8951961. DOI: 10.1155/2022/8951961

Bibri, S. E., Alexandre, A., Sharifi, A., & Krogstie, J. (2023). Environmentally sustainable smart cities and their converging AI, IoT, and big data technologies and solutions: An integrated approach to an extensive literature review. *Energy Informatics*, *6*, 9. Advance online publication. DOI: 10.1186/s42162-023-00259-2 PMID: 37032812

Bibri, S. E., Huang, J., Jagatheesaperumal, S. K., & Krogstie, J. (2024). The synergistic interplay of artificial intelligence and digital twin in environmentally planning sustainable smart cities: A comprehensive systematic review. *Environmental Science and Ecotechnology*, *20*, 100433. DOI: 10.1016/j.ese.2024.100433

Blau, P. M. (2017). *Exchange and Power in Social Life* (1st ed.). Routledge., DOI: 10.4324/9780203792643

Blokhin, A. (2025, February 3). *The Five Countries that Produce the Most Carbon Dioxide (CO_2)*. Investopedia. https://www.investopedia.com/articles/investing/092915/5-countries-produce-most-carbon-dioxide-co2.asp#

Boateng, E. Y., & Abaye, D. A. (2019). A review of the logistic regression model with emphasis on medical research. *Journal of Data Analysis and Information Processing*, *7*(4), 190–207. DOI: 10.4236/jdaip.2019.74012

Bocken, N. M. P., Short, S. W., Rana, P., & Evans, S. (2022). Sustainable business models and circular economy transitions. *Journal of Cleaner Production*, *372*, 133764.

Bolón-Canedo, V., Morán-Fernández, L., Cancela, B., & Alonso-Betanzos, A. (2024). A review of green artificial intelligence: Towards a more sustainable future. *Neurocomputing*, *599*, 128096. DOI: 10.1016/j.neucom.2024.128096

Boominathan, V., Selvi, J. T., Dhilipan, C., Arasu, M. T., & Elamurugan, B. (2023). *Palanivel Rathinasabapathi Velmurugan*. Advances in Business Information Systems and Analytics Book Series. Empirical Study on the Impact of Select Green HRM Dimensions on Green Innovation Culture., DOI: 10.4018/979-8-3693-0049-7.ch026

Borowiec, J., Papież, M., & Śmiech, S. (2024). The impact of environmental regulations on carbon emissions in countries with different levels of emissions. *Environmental Science and Pollution Research International*. Advance online publication. DOI: 10.1007/s11356-024-35702-8

Boumaiza, A., & Maher, K. (2024). Leveraging blockchain technology to enhance transparency and efficiency in carbon trading markets. *International Journal of Electrical Power & Energy Systems*, *162*, 110225. DOI: 10.1016/j.ijepes.2024.110225

Bouquet, P., Jackson, I., Nick, M., & Kaboli, A. (2023). AI-based forecasting for optimised solar energy management and smart grid efficiency. *International Journal of Production Research*, *62*(13), 4623. DOI: 10.1080/00207543.2023.2269565

Bouyahrouzi, E. M., Benmimoun, R., El Kihel, Y., & Bajjou, M. S. (2025). Integrating industry 4.0 technologies and maintenance 4.0 for sustainable manufacturing: A systematic literature review. *International Journal of Advanced Manufacturing Technology*. Advance online publication. DOI: 10.1007/s00170-025-16194-3

Brandenburg, M., Govindan, K., Sarkis, J., & Seuring, S. (2014). Quantitative models for sustainable supply chain management: Developments and directions. *European Journal of Operational Research*, *233*(2), 299–312. DOI: 10.1016/j.ejor.2013.09.032

Bratton, A., Renwick, D. W. S., Paillé, P., Matthews, B., Leidner, S., Karamali, E., & Holland, P. (2025). Guest editorial: Sustainable human resource management and organizational performance: new definitions, navigating tensions, and global insights. *Journal of Organizational Effectiveness: People and Performance*, *12*(1), 1–13. DOI: 10.1108/JOEPP-03-2025-569

Broadhead, L.-A. (2001). Canada as a Rogue State: Its Shameful Performance on Climate Change. *International Journal (Toronto, Ont.)*, *56*(3), 261–280. DOI: 10.1177/002070200105600305

Broadhead, L.-A. (2019). Environmental Insecurity: Another Case for Concept Change. *Peace and Conflict Studies*, *26*(1), 1. https://nsuworks.nova.edu/pcs/vol26/iss1/1

Buczynski, W., Steffek, F., Cuzzolin, F., & Jamnik, M., & SAHAKIAN, B. (2022). Hard law and soft law regulations of artificial intelligence in investment management. *Cambridge Yearbook of European Legal Studies*, *24*, 262–293. DOI: 10.1017/cel.2022.10

Bulut, H., & Samuel, R. (2025). Public support for carbon taxes varies across countries and policy design must consider the national context. *Communications Earth & Environment*, *6*(1). Advance online publication. DOI: 10.1038/s43247-025-02562-0

Burns, R. (2023, June 9). *Rising Sea Levels: Causes, Consequences, and Impact on India*. Give.do. https://give.do/blog/rising-sea-levels-causes-consequences-and-impact-on-india/#

Caggiano, A. (2018). Cloud-based manufacturing process monitoring for smart diagnosis services. *International Journal of Computer Integrated Manufacturing*, *31*(7), 612–623. DOI: 10.1080/0951192X.2018.1425552

Camilleri, M. A. (2024). Artificial intelligence governance: Ethical considerations and implications for social responsibility. *Expert Systems: International Journal of Knowledge Engineering and Neural Networks*, *41*(7), e13406. DOI: 10.1111/exsy.13406

Caputo, A., Tron, A., Sassetti, S., Giakoumelou, A., & Colantoni, F. (2025). Managing for resilience through social sustainable human resource approach: The effect on organisational performances. *Management Decision*. Advance online publication. DOI: 10.1108/MD-05-2024-1144

Chatterjee, S., Chaudhuri, R., & Vrontis, D. (2025). Creating organizational value and sustainability through green HR practices: An innovative approach with the moderating role of top management support. *Business Ethics, the Environment & Responsibility*, *34*(1), 17–31. DOI: 10.1111/beer.12569

Chauhan, M., & Sahoo, D. R. (2024). Towards a greener tomorrow: Exploring the potential of AI, blockchain, and IoT in sustainable development. *Nature Environment and Pollution Technology*, *23*(2), 1105–1113. DOI: 10.46488/NEPT.2024.v23i02.044

Chaves, R., & Monzón, J. L. (2012). Beyond the crisis: The social economy, prop of a new model of sustainable economic development. *Service Business*, *6*, 5–26. DOI: 10.1007/s11628-011-0125-7

Chawla, M., Sharma, P., Gautam, S., & Popescu, C. R. Gh. (2025). Sustainability in Human Capital Management: Shaping the Future With Technology. In T. Kapoor, A. K. Singh, K. Sahay, H. Kaur, & R. Verma (Eds), *Sustainable Management Practices for Employee Retention and Recruitment* (pp. 265–290). IGI Global. https://doi.org/DOI: 10.4018/979-8-3693-8679-8.ch013

Cheema, S. A. (2012, April 29). *India's Approach to Sanctions on Iran*. https://www.e-ir.info/2012/04/29/indias-approach-to-sanctions-on-iran/

Cheng, W. (2022, November 15). *International Clean Technology Diffusion: Pathways and Prospects*. https://www.southcentre.int/wp-content/uploads/2022/11/SV243_221115.pdf

Chennai, Manimalar, Dr. R. Pavithra, S., & B, Preethi. (2025). Green HRM and Inclusive Leadership: A Dual Strategy for Sustainable Growth. *International Scientific Journal of Engineering and Management, 04*(09), 1–9. DOI: 10.55041/ISJEM.ESEH025

Chen, X. H., Tee, K., Elnahass, M., & Ahmed, R. (2023). Assessing the Environmental Impacts of Renewable Energy Sources: A Case Study on Air Pollution and Carbon Emissions in China. *Journal of Environmental Management, 345*, 118525. DOI: 10.1016/j.jenvman.2023.118525

Chen, X. M., Sharma, A., & Liu, H. (2023). The Impact of Climate Change on Environmental Sustainability and Human Mortality. *Environments, 10*(10), 165. https://www.mdpi.com/2076-3298/10/10/165

Chen, X., Chen, X., Zhang, J., & Zeng, H. (2020). Is corporate environmental responsibility synergistic with governmental environmental responsibility? Evidence from China. *Business Strategy and the Environment, 29*(8), 3669–3686. DOI: 10.1002/BSE.2603

Chen, Y., Zhou, J., Ge, Y., & Dong, J. (2024). Uncovering the rapid expansion of photovoltaic power plants in China from 2010 to 2022 using satellite data and deep learning. *Remote Sensing of Environment, 305*, 114100. DOI: 10.1016/j.rse.2024.114100

Chen, Z. (2024). Path Analysis of The Carbon Market's Impact on Corporate Carbon Emission Reduction. *Highlights in Business. Economics and Management, 43*, 297–302. DOI: 10.54097/a7jeqk60

Cheong, B. C. (2024). Transparency and accountability in AI systems: Safeguarding wellbeing in the age of algorithmic decision-making. *Frontiers in Human Dynamics, 6*, 1421273. DOI: 10.3389/fhumd.2024.1421273

Chikhaoui, M. (2025). Do Environmental Management Policies Decrease Carbon Emissions? A Probabilistic Analysis of G7 Countries. *International Journal of Energy Economics and Policy, 15*(4), 388–400. DOI: 10.32479/ijeep.19664

Chowdhury, N. (2019). Constitutionally Shackled. *The Story of Environmental Jurisprudence in India, 159*. Advance online publication. DOI: 10.1007/978-981-13-9065-4_14

Chowdhury, S. R. (2024). Enhancing Employee Performance Through Green Human Resource Management: A Comprehensive Literature Review and Suggestions for Potential Researchers. *American Journal of Theoretical and Applied Business, 10*(2), 21–32. DOI: 10.11648/j.ajtab.20241002.11

Christensen, L. T., & Langer, R. (2020). Corporate communication and sustainability: An organizational legitimacy perspective. *Journal of Business Ethics*, *162*(2), 415–428.

Church, J. A., Monselesan, D., Gregory, J. M., & Marzeion, B. (2013). Evaluating the Ability of Process-Based Models to Project Sea-Level Change. *Environmental Research Letters*, *8*(1), 014051. https://iopscience.iop.org/article/10.1088/1748-9326/8/1/014051

Cornelissen, J. (2023). *Corporate communication: A guide to theory and practice* (7th ed.). SAGE Publications.

Corrêa do Lago, A. A. (2025, September 25). *COP30: An Integrated Vision for Climate Change and Trade*. https://tessforum.org/latest/cop30-an-integrated-vision-for-climate-change-and-trade *Country Analysis Brief: India*. (2025). [U.S. Department of Energy]. U.S. Department of Energy. https://www.eia.gov/international/content/analysis/countries_long/India/pdf/India.pdf

Couldry, N., MacDonald, R., Stephansen, H., Clark, W., Dickens, L., & Fotopoulou, A. (2015). Constructing a digital story circle: Digital infrastructure and mutual recognition. *International Journal of Cultural Studies*, *18*(5), 501–517. DOI: 10.1177/1367877913519313

Creemers, R. (2022). China's emerging data protection framework. *Journal of Cybersecurity*, *8*(1). Advance online publication. DOI: 10.1093/cybsec/tyac011

Culot, G., Podrecca, M., & Nassimbeni, G. (2024). Artificial intelligence in supply chain management: A systematic literature review of empirical studies and research directions. *Computers in Industry*, *162*, 104132. DOI: 10.1016/j.compind.2024.104132

Danish, M. S. S., & Senjyu, T. (2023). Shaping the future of sustainable energy through AI-enabled circular economy policies. *Circular Economy*, *2*(2), 100040. DOI: 10.1016/j.cec.2023.100040

Das, A., & Swain, P. K. (2024). Navigating the Sea Level Rise: Exploring the Interplay of Climate Change, Sea Level Rise, and Coastal Communities in India. *Environmental Monitoring and Assessment*, *196*, 1010. https://link.springer.com/article/10.1007/s10661-024-13191-z

De Wilde, J. (2008). Environmental Security Deconstructed. In Brauch, H. G. (Ed.), *Globalization and Environmental Challenges: Reconceptualizing Security in the 21st Century* (pp. 595–602). Springer.

Deb, K., & Kailash, S. (2023, September 5). *Mainstreaming the Energy Transition in India's Policy Framework*.

Del-Aguila-Arcentales, S., Alvarez-Risco, A., Jaramillo-Arévalo, M., De-la-Cruz-Diaz, M., & de las Mercedes Anderson-Seminario, M. (2022). Influence of social, environmental and economic sustainable development goals (SDGs) over continuation of entrepreneurship and competitiveness. *Journal of Open Innovation*, *8*(2), 73. DOI: 10.3390/joitmc8020073

Dell Technologies. (2023). *Sustainability and impact report 2023.* https://www.dell.com/learn/us/en/uscorp1/sustainability

Dev, A. (2025). ECO-HR: Integrating Sustainability into workforce management" "Reducing HR's Environmental Footprint. *International Journal For Multidisciplinary Research*, *7*(2), 41484. DOI: 10.36948/ijfmr.2025.v07i02.41484

Devi, S. (2024). The impact of green human resource management practices on employee's job performance. *International Journal of Management*, *13*(3), 506–522.

Dhanya, D., Kumar, S. S., Thilagavathy, A., Prasad, D. V., & Boopathi, S. (2023). Data analytics and artificial intelligence in the circular economy: Case studies. In Mishra, B. (Ed.), *Intelligent Engineering Applications and Applied Sciences for Sustainability* (pp. 40–58). IGI Global Scientific Publishing., DOI: 10.4018/979-8-3693-0044-2.ch003

DiMaggio, P. J., & Powell, W. W. (1983). The Iron Cage Revisited: Institutional Isomorphism and Collective Rationality in Organizational Fields. *American Sociological Review*, *48*(2), 147–160. DOI: 10.2307/2095101

Din, A. U., Yang, Y., Khan, K., Din, Q. M. U., Golo, M. M. A., & Khan, I. U. (2023). Role of employee commitment and individual value in green HRM practices and sustainable firm performance. *Second International Conference on Sustainable Technology and Management (ICSTM 2023), 12804*, 404–413. https://doi.org/DOI: 10.1117/12.3004029

Dira, A. F., Noor, G. P., Bangun, M. F. A., Winardi, N. M. A., Kamal, F., & Utomo, K. P. (2024). The Role of Employee Engagement in Green HRM to Create Sustainable Humanist Performance. *EKOMBIS REVIEW: Jurnal Ilmiah Ekonomi Dan Bisnis*, *12*(1). Advance online publication. DOI: 10.37676/ekombis.v12i1.4856

Dutz, M. A., & Sharma, S. (2012). *Green growth, technology, and innovation* (World Bank Policy Research Working Paper No. 5932). https://papers.ssrn.com/sol3/papers.cfm?abstract_id=1980586

Du, Z. Y., & Wang, Q. (2024). Digital infrastructure and innovation: Digital divide or digital dividend? *Journal of Innovation & Knowledge*, *9*(3), 100542. DOI: 10.1016/j.jik.2024.100542

Dyageti, S. R., & Kiran, K. C. U. (2025). THE ROLE OF HR IN DRIVING ORGANIZATIONAL SUSTAINABILITY THROUGH EMPLOYEE WELL BEING PROGRAMS. *International Journal of Advances in Business and Management Research, 03*(Supplementary), 78–85. DOI: 10.62674/ijabmr.2025.v3isupp1.010

Dyer, H. (2001). Theoretical Aspects of Environmental Security. In Petzold-Bradley, E., Carius, A., & Vincze, A. (Eds.), *Responding to Environmental Conflicts: Implications for Theory and Practice* (pp. 67–81). Kluwer.

Dyllick, T., & Muff, K. (2016). Clarifying the meaning of sustainable business: Introducing a typology from business-as-usual to true business sustainability. *Organization & Environment, 29*(2), 156–174.

Easterly, W. (2015). The Trouble with the Sustainable Development Goals. *Current History (New York, N.Y.), 114*(775), 322–324. DOI: 10.1525/curh.2015.114.775.322

Eccles, R. G., Ioannou, I., & Serafeim, G. (2014). The impact of corporate sustainability on organizational processes and performance. *Management Science, 60*(11), 2835–2857.

Eighth India-EU Cyber Dialogue (March 20, 2025). (2025, March 21). https://www.mea.gov.in/pressreleases.htm?dtl/39250/Eighth+IndiaEU+Cyber+Dialogue+March+20+2025

Ejiaku, S. A. (2014). Technology adoption: Issues and challenges in information technology adoption in emerging economies. *Journal of International Technology and Information Management, 23*(2), 5. DOI: 10.58729/1941-6679.1071

Ekhator, E. (2025). Environmental Justice in the Global South. *Palgrave Studies in Sustainable Futures, 595.* Advance online publication. DOI: 10.1007/978-3-031-55387-5_30

Elkington, J. (1997). *Cannibals with forks: The triple bottom line of 21st-century business.* Capstone Publishing.

Elkington, J. (1999). *Cannibals with Forks: The Triple Bottom Line of 21st Century Business.* Oxford University Press.

Emerson, R. M. (1976). Social Exchange Theory. *Annual Review of Sociology,* •••, 335–362.

Energising Asean: power grid deal set for Oct 2025 signing, Sarawak to play key role. (2025, May 25). https://www.scoop.my/news/258075/energising-asean-power-grid-deal-set-for-oct-2025-signing-sarawak-to-play-key-role/

Ernst, E. (2025). From Mandatory CSR Reporting to Environmental Impact: Assessing the Effectiveness of the EU NFRD in Reducing Corporate GHG Emissions. *European Company Law*, *22*(4/5), 184–192. DOI: 10.54648/eucl2025027

Erturk, A., Özşahin, M., & Simovic, V. (Eds.). (2025). *Human-centric, sustainable, and resilient organizations in the digital age*. IGI Global Scientific Publishing., DOI: 10.4018/979-8-3693-8181-6

European Commission. (2020). *A new circular economy action plan for a cleaner and more competitive Europe*. Publications Office of the European Union. https://eur-lex.europa.eu

European Commission. (2023). *E-governance and sustainability in the European digital agenda*. European Union Publications.

Eurostat. (2025). Quarterly greenhouse gas emissions in the EU. Eurostat. https://ec.europa.eu/eurostat/statistics-explained/index.php?title=Quarterly_greenhouse_gas_emissions_in_the_EU

Evans, S. (2021, October 5). *Analysis: Which Countries are Historically Responsible for Climate Change?* CarbonBrief. https://www.carbonbrief.org/analysis-which-countries-are-historically-responsible-for-climate-change/#

Ezeafulukwe, C., Okatta, C. G., & Ayanponle, L. (2022). Frameworks for sustainable human resource management: Integrating ethics, CSR, and Data-Driven Insights. *World Journal of Advanced Research and Reviews*, *13*(3), 583–592. DOI: 10.30574/wjarr.2022.13.3.0151

Faieq, H. T., & Çek, K. (2024). Enhancing Kurdistan's manufacturing companies' sustainable waste management: A norm activation approach to green accounting, CSR, and environmental auditing oversight. *Heliyon*, *10*(12), e32725–e32725. DOI: 10.1016/j.heliyon.2024.e32725

Faisal, S. (2023). Green Human Resource Management—A Synthesis. *Sustainability*, *15*(3), 2259. DOI: 10.3390/su15032259

Fatimah, Y. A., Govindan, K., Murniningsih, R., & Setiawan, A. (2020). Industry 4.0 based sustainable circular economy approach for smart waste management system to achieve sustainable development goals: A case study of Indonesia. *Journal of Cleaner Production*, *269*, 122263. DOI: 10.1016/j.jclepro.2020.122263

Feroz, A. K., Zo, H., & Chiravuri, A. (2021). Digital transformation and environmental sustainability: A review and research agenda. *Sustainability*, *13*(3), 1530. DOI: 10.3390/su13031530

Ficili, I., Giacobbe, M., Tricomi, G., & Puliafito, A. (2025). From sensors to data intelligence: Leveraging IoT, cloud, and edge computing with AI. *Sensors (Basel), 25*(6), 1763. DOI: 10.3390/s25061763

Fish: Indian Ocean - IORA Ministerial Blue Economy Conference. (2015). *Africa Research Bulletin Economic, Financial, and Technical Series, 52*(8). DOI: 10.1111/j.1467-6346.2015.06637.x

Folke, C., Carpenter, S. R., Walker, B., Scheffer, M., Chapin, T., & Rockström, J. (2010). Resilience thinking: Integrating resilience, adaptability and transformability. *Ecology and Society, 15*(4), 20.

Forti, V., Baldé, C. P., & Kuehr, R. (2020). *E-waste statistics: Guidelines on classification, reporting and indicators*. United Nations University.

Forti, V., Baldé, C. P., & Kuehr, R. (2024). Artificial intelligence and e-waste management: The next frontier of digital sustainability. *Journal of Cleaner Production, 441*, 141294.

Freeman, R. E. (2010). *Strategic Management: A Stakeholder Approach*. Cambridge University Press., DOI: 10.1017/CBO9781139192675

Fu, C., Lu, L., & Pirabi, M. (2023). Advancing green finance: A review of sustainable development. *Digital Economy and Sustainable Development, 1*, 20. Advance online publication. DOI: 10.1007/s44265-023-00020-3

Fu, W., & Su, H.-C. (2020). Take actions or outsource? An empirical examination of strategic environmental options on greenhouse gas emissions. *International Journal of Operations & Production Management, 40*(6), 753–776. DOI: 10.1108/IJOPM-02-2020-0088

Gaikwad, D., Amitabh P., Gaurav P., & Francis, A. T. (2025, March 5). *Impact of human resource practices on the dissemination of green HRM*. SciSpace - Paper; Informa. https://doi.org/DOI: 10.4324/9781003616306-12

Gamidullaeva, L., Shmeleva, N., Mityakov, E., Tolstykh, T., & Vasin, S. (2025). Strategic Tools for the Formation of Cluster Capital to Implement Technological Innovations. *Systems, 13*(4), 270. DOI: 10.3390/systems13040270

Ganin, A. A., Mersky, A. C., Jin, A. S., Kitsak, M., Keisler, J. M., & Linkov, I. (2019). Resilience in intelligent transportation systems (ITS). *Transportation Research Part C, Emerging Technologies, 100*, 318–329. DOI: 10.1016/j.trc.2019.01.014

Garay-Rondero, C. L., Martínez-Flores, J. L., Smith, N. R., Caballero Morales, S. O., & Aldrette-Malacara, A. (2020). Digital supply chain model in Industry 4.0. *Journal of Manufacturing Technology Management, 31*(5), 887–933. DOI: 10.1108/JMTM-08-2018-0280

Garcia, A. G. (2025). The impact of sustainable practices on employee well-being and organizational success. *Brazilian Journal of Development, 11*(3), e78599. DOI: 10.34117/bjdv11n3-054

Gautam, B. P. S., Qureshi, A., Gwasikoti, A., Gondwal, M., Kumar, V., & Gondwal, M. (2024). Global Scenario of Plastic Production, Consumption, and Waste Generation and Their Impacts on Environment and Human Health. In Soni, R., Debbarma, P., Suyal, D. C., & Goel, R. (Eds.), *Advanced Strategies for Biodegradation of Plastic Polymers* (pp. 1–34). Springer., https://link.springer.com/chapter/10.1007/978-3-031-55661-6_1#DOI

Gazi, Md. A. I., Masud, A. A., Emon, Md., Ibrahim, Md., & Senathirajah, A. R. bin S. (2025). The triadic relationship between green HRM, innovation, and pro-environmental behaviour: A study of their interactions and impacts on employee productivity and organizational sustainability. *Environmental Research Communications, 7*(1), 015016. DOI: 10.1088/2515-7620/ada676

Geissdoerfer, M., Savaget, P., Bocken, N. M., & Hultink, E. J. (2017). The circular economy: A new sustainability paradigm? *Journal of Cleaner Production, 143*, 757–768.

Ghisellini, P., Cialani, C., & Ulgiati, S. (2016). A review on circular economy: The expected transition to a balanced interplay of environmental and economic systems. *Journal of Cleaner Production, 114*, 11–32.

Ghobakhloo, M., & Hong, T. S. (2015). The business value of information system-enabled e-collaboration capabilities. [IJeC]. *International Journal of e-Collaboration, 11*(1), 22–56. DOI: 10.4018/ijec.2015010103

Ghobakhloo, M., Iranmanesh, M., Grybauskas, A., Vilkas, M., & Petraitė, M. (2021). Industry 4.0, innovation, and sustainable development: A systematic review and a roadmap to sustainable innovation. *Business Strategy and the Environment, 30*(8), 4237–4257. DOI: 10.1002/bse.2867

Ghosh, A., & Haque, S. (2024). The impact of GHRM and EGB on organization's financial and operational performance – an empirical analysis on Indian ICT sector. *Journal of Organizational Effectiveness: People and Performance, 12*(1), 105–126. DOI: 10.1108/JOEPP-08-2023-0359

Ghosh, A., Saini, A., & Barad, H. (2025). Artificial intelligence in governance: Recent trends, risks, challenges, innovative frameworks and future directions. *AI & Society*. Advance online publication. DOI: 10.1007/s00146-025-02312-y

Gilal, F. G., Nawaz, A., Riaz, A., Channa, K. A., & Jahanzeb, A. (2024). Achieving sustainability through green human resource management practices and environmental orientation fit: A multilevel modeling. *Evidence-Based HRM: A Global Forum for Empirical Scholarship, 13*(2), 299–318. https://doi.org/DOI: 10.1108/EBHRM-09-2022-0223

Gleeson, N., & Walden, I. (2016). Placing the state in the cloud: Issues of data governance and public procurement. *Computer Law & Security Review*, *32*(5), 683–695. DOI: 10.1016/j.clsr.2016.07.004

Glende, E., & Wolter, M. (2022). Extra-long distance ultra-high voltage direct current. *2022 IEEE Sustainable Power and Energy Conference (iSPEC)*, 1. DOI: 10.1109/ispec54162.2022.10033043

Golpeo, J. M., Andres, G. L., Faronal, I. N., & Casuat, C. D. (2025). Sustainable E-Waste Management Through a Circular Economy Framework: Integrating Smart Technologies for Resource Recovery in Urban Environments. In *E3S Web of Conferences* (Vol. 643, p. 01005). EDP Sciences., DOI: 10.1051/e3sconf/202564301005

Gomila, R. (2021). Logistic or Linear? Estimating causal effects of experimental treatments on binary outcomes using regression analysis. *Journal of Experimental Psychology. General*, *150*(4), 700–709. DOI: 10.1037/xge0000920 PMID: 32969684

Gooderham, P. N., Pedersen, T., Sandvik, A. M., Dasí, À., Elter, F., & Hildrum, J. (2022). Contextualizing AMO Explanations of Knowledge Sharing in MNEs: The Role of Organizational and National Culture. *Management International Review*, *62*(6), 859–884. DOI: 10.1007/s11575-022-00483-0

Grandhi, S. P., Dagwar, P. P., & Dutta, D. (2024). Policy pathways to sustainable E-waste management: A global review. *Journal of Hazardous Materials Advances*, *16*, 100473. DOI: 10.1016/j.hazadv.2024.100473

Greif, L., Röckel, F., Kimmig, A., & Ovtcharova, J. (2025). A systematic review of current AI techniques used in the context of the SDGs. *International Journal of Environmental Research*, *19*(1), 1–36. DOI: 10.1007/s41742-024-00668-5

Guerrero, M., Liñán, F., & Cáceres-Carrasco, F. R. (2021). The influence of ecosystems on the entrepreneurship process: A comparison across developed and developing economies. *Small Business Economics*, *57*(4), 1733–1759. DOI: 10.1007/s11187-020-00392-2

Gunawan, W. B., & Mikhail, L. (2025). Strategic and sustainable human resource management: Twin weapon for achieving competitive advantage in organization. *Priviet Social Sciences Journal, 5*(6), 45–58. DOI: 10.55942/pssj.v5i6.401

Guo, L., Xu, Y., Liu, G., & Wang, T. (2019). Understanding Firm Performance on Green Sustainable Practices through Managers' Ascribed Responsibility and Waste Management: Green Self-Efficacy as Moderator. *Sustainability, 11*(18), 4976. DOI: 10.3390/SU11184976

Gupta, S., Kartha, R. S., Ramirez-Asis, E., Flores-Albornoz, J., Vílchez-Vásquez, R., & Singh, S. K. (2022). A novel traffic surveillance system using an uncalibrated camera. *Journal of Advanced Transportation, 2022*, 1–10. DOI: 10.1155/2022/5867524

Gyensare, M. A., Adomako, S., & Amankwah-Amoah, J. (2024). Green HRM practices, employee well-being, and sustainable work behavior: Examining the moderating role of resource commitment. *Business Strategy and the Environment, 33*(4), 3129–3141. DOI: 10.1002/bse.3642

H., B., Altaf Q., Nabab, A., M., & Asmae, B. (2025). Smart Grid Security and Protection. *Lecture Notes in Electrical Engineering*. DOI: 10.1007/978-981-96-0824-9

Hadi, S., Lukitaningsih, A., Kirana, K.C., ... Hidayat, R., Siripipatthanakul, S. (2025).Privacy and Security Management Practices for Organizations. Open source preview, 2025, pp. 89–116

Hadi, S., Sulistyawati, E., Kirana, K. C., Dianti, C. A. I., Purnasari, E., & Ashakah, F. O. (2025). Abusive leadership and turnover intention: How employee and work engagement act as shields. *Asian Management and Business Review*, 74–89. https://doi.org/DOI: 10.20885/AMBR.vol5.iss1.art6

Hadi, N. U., Almessabi, B., & Khan, M. I. (2025). Leveraging industry 4.0 and circular open innovation for digital sustainability: The role of circular ambidexterity. *Journal of Open Innovation, 11*(2), 100545. DOI: 10.1016/j.joitmc.2025.100545

Hadi, S., Faridiana, A., Kirana, K. C., Lukitaningsih, A., Rahmawati, C. H. T., Purnasari, E., & Wahyuningtyas, N. (2024). Political Skill And Transactional Leadership On Employee Performance: The Mediating Effect Of Organizational Work Culture. [Ikonomicheski Izsledvania]. *Economic Studies Journal, 33*(4), 112–129.

Hadi, S., Kirana, K. C., Setiawati, L., Lada, S. B., & Rahmawati, C. H. T. (2024). The Effect of Digital Leadership and Organizational Support on Innovative Work Behavior: The Mediating Role of Emotional Intelligence. *Quality - Access to Success, 25*(199). Advance online publication. DOI: 10.47750/QAS/25.199.09

Hadi, S., Purnasari, E., Hidayat, R., Lukitaningsih, A., Kirana, K. C., & Siripipatthanakul, S. (2025). Challenges and Management Practices in Protecting Employee Personal Data in the Digital Era. In *Privacy and Security Management Practices for Organizations* (pp. 89–116). IGI Global Scientific Publishing., DOI: 10.4018/979-8-3693-9720-6.ch004

Hadi, S., Tjahjono, H. K., & Palupi, M. (2020). *Systematic Review: Meta Sintesis Untuk Riset Perilaku Organisasional.* Viva Victory Abadi.

Hakuduwal, K., & Kc, S. (2025). Impact of Green Human Resource Management Practices on Perceived Environmental Sustainability: Employee Perspectives from Nepalese Commercial Banks. *Deleted Journal, 13*(1), 70–87. DOI: 10.3126/ljbe.v13i1.80257

Handoyo, S., Yudianto, I., & Dahlan, M. (2024). Exploring firm and country's specific factors affecting carbon emission reduction performance: Study on selected ASEAN countries. *Heliyon, 10*(17), e37036–e37036. DOI: 10.1016/j.heliyon.2024.e37036

Hangl, J., Krause, S., & Behrens, V. J. (2023). Drivers, barriers and social considerations for AI adoption in SCM. *Technology in Society, 74*, 102299. DOI: 10.1016/j.techsoc.2023.102299

Hannouf, M., & Assefa, G. (2017). The Role of Sustainability Resources of Large Greenhouse Gas Emitters: The Case of Corporations in Alberta, Canada. *Sustainability, 9*(2), 182. DOI: 10.3390/SU9020182

Haque, F., & Ntim, C. G. (2018). Environmental Policy, Sustainable Development, Governance Mechanisms and Environmental Performance. *Business Strategy and the Environment, 27*(3), 415–435. DOI: 10.1002/BSE.2007

Haque, F., & Ntim, C. G. (2022). Do corporate sustainability initiatives improve corporate carbon performance? Evidence from European firms. *Business Strategy and the Environment, 31*(7), 3318–3334. DOI: 10.1002/bse.3078

Hariyani, D., Hariyani, P., Mishra, S., & Sharma, M. K. (2024). A literature review on green supply chain management for sustainable sourcing and distribution. *Waste Management Bulletin, 2*(4), 231–248. DOI: 10.1016/j.wmb.2024.11.009

Hashmi, S., & Ghai, R. K. (2025). *Applying Artificial Intelligence in Sustainable HR Practices: A Pathway to Organizational Endurance*. https://doi.org/DOI: 10.20944/preprints202507.1204.v1

Heath, R. L., & Palenchar, M. J. (2021). *Strategic issues management: Organizations and public policy challenges* (3rd ed.). SAGE Publications.

Heggelund, G. (2021). China's climate and energy policy: At a turning point? *International Environmental Agreement: Politics, Law and Economics*, *21*(1), 9. DOI: 10.1007/s10784-021-09528-5

Heldeweg, M. A., Sanders, M., & Harmsen, M. (2015). Public-private or private-private energy partnerships? Toward good energy governance in regional and local green gas projects. *Energy, Sustainability and Society*, *5*(1), 9.

Hemdan, E. E. D., El-Shafai, W., & Sayed, A. (2023). Integrating digital twins with IoT-based blockchain: Concept, architecture, challenges, and future scope. *Wireless Personal Communications*, *131*, 2193–2216. DOI: 10.1007/s11277-023-10538-6

Herlina, M. G., & Iskandar, K. (2025). Integrating Sustainable HRM, AI, and Employee Well-Being to Enhance Engagement in Greater Jakarta: An SDG 3 Perspective. *E3S Web of Conferences*, *601*, 00020. https://doi.org/DOI: 10.1051/e3sconf/202560100020

Hernandez, J. C. R., Villa-Enciso, E., Cardona-Acevedo, S., Valencia, J., & Velasquez Salas, S. (2025). Smart innovation for a circular economy: A systematic review of emerging trends and the future of AI in the sustainable economy. *Sustainability*, *17*(13), 5793. DOI: 10.3390/su17135793

Hickel, J. (2020). The Sustainable Development Index: Measuring the Ecological Efficiency of Human Development in the Anthropocene. *Ecological Economics*, *167*, 106331. DOI: 10.1016/j.ecolecon.2019.05.011

Hien, P. T. T. (2025). Does Green Human Resources Management Enhance Sustainable Performance? An Empirical Study from Manufacturing Firms. *International Journal of Research and Innovation in Social Science*, *IX*(IV), 1161–1167. DOI: 10.47772/ijriss.2025.90400089

Hofmann, M., Choi, S., & Binder, C. (2022). Collaborative governance for sustainability transitions: A systematic literature review. *Journal of Cleaner Production*, *338*, 130606. Advance online publication. DOI: 10.1016/j.jclepro.2022.130606

Hofmann, M., & Lindberg, K. B. (2023). Evidence of households' demand flexibility in response to variable hourly electricity prices – Results from a comprehensive field experiment in Norway. *Energy Policy*, *184*, 113821. DOI: 10.1016/j.enpol.2023.113821

Horowitz, C. A. (2016). Paris Agreement. *International Legal Materials*, *55*(4), 740. DOI: 10.1017/s0020782900004253

Hossain, I., Haque, A. M., & Ullah, S. A. (Eds.). (2025). *Intersecting environmental governance with technological advancements*. IGI Global Scientific Publishing., DOI: 10.4018/979-8-3693-7001-8

Housheya, N., & Atikbay, T. (2025). The Role of Green HRM in Promoting Green Innovation: Mediating Effects of Corporate Environmental Strategy and Green Work Climate, and the Moderating Role of Artificial Intelligence. *Sustainability*, *17*(16), 7238. DOI: 10.3390/su17167238

Huang, J., Lin, C., Zhou, H., Xu, Z., & Lin, C. (2019). Research on key technologies of deduction of multinational power trading in the context of Global Energy Interconnection. *Global Energy Interconnection*, *2*(6), 560. DOI: 10.1016/j.gloei.2020.01.010

Hu, J., Wu, H., & Ying, S. X. (2022). Environmental regulation, market forces, and corporate environmental responsibility: Evidence from the implementation of cleaner production standards in China. *Journal of Business Research*, *150*, 606–622. DOI: 10.1016/j.jbusres.2022.06.049

Huo, J., & Peng, C. (2023). Depletion of Natural Resources and Environmental Quality: Prospects of Energy Use, Energy Imports, and Economic Growth Hindrances. *Resources Policy, 86*(A), 104049. DOI: 10.1016/j.resourpol.2023.104049

Hussein, J. B., Workneh, T. S., Kassim, A., Ntsowe, K., Melesse, S. F., El-Mesery, H. S., & Zixheng, H. (2025). A review of the use of artificial intelligence in renewable energy for food processing and preservation process optimisation, challenges, and future prospects. *Renewable & Sustainable Energy Reviews*, *223*, 116076. DOI: 10.1016/j.rser.2025.116076

Husted, B. W., & de Sousa-Filho, J. M. (2017). The impact of sustainability governance, country stakeholder orientation, and country risk on environmental, social, and governance performance. *Journal of Cleaner Production*, *155*, 93–102.

Huzzat, A., Anpalagan, A., Khwaja, A. S., Woungang, I., Alnoman, A. A., & Pillai, A. S. (2025). A comprehensive review of digital twin technologies in smart cities. *Digital Engineering*, *100040*. Advance online publication. DOI: 10.1016/j.dte.2025.100040

Ibanez, A. A. E., Khalefa, M., & Aldulaimi, M. (2024). Green Human Resource Management Practices and Employee Engagement: A Case Study of Leading Sustainable Organizations. *International Journal of Green Management and Business Studies*, *4*(2). Advance online publication. DOI: 10.56830/IJGMBS12202401

Ikpe, E. O. (2024). Adoption and implementation of artificial intelligence in small businesses in selected developing countries. *Journal of Health. Applied Sciences and Management*, *8*(1), 26–34. DOI: 10.4314/johasam.v8i1.3

India inks agreement to establish Bimstec Energy Centre in Bengaluru. (2024, November 10).

India Reaffirms Commitment to Equity, Climate Justice and Multilateralism at CoP30, in Belém, Brazil. (2025, November 11). https://www.pib.gov.in/PressReleasePage.aspx?PRID=2189018

International Investment Agreements Navigator. (2025, May 27). https://investmentpolicy.unctad.org/international-investment-agreements/countries/98/iran-islamic-republic-of

Işık, C., Ongan, S., & Islam, H. (2025). Driving energy transition through artificial intelligence: Integrating economic, environmental, social, and governance (ECON-ESG) factors in OECD countries. *Journal of the Knowledge Economy*. Advance online publication. DOI: 10.1007/s13132-025-02687-9

Ivanov, D., Dolgui, A., & Sokolov, B. (2019). The impact of digital technology and Industry 4.0 on the ripple effect and supply chain risk analytics. *International Journal of Production Research*, *57*(3), 829–846. DOI: 10.1080/00207543.2018.1488086

Jain, A. D. S., Mehta, I., Mitra, J., & Agrawal, S. (2017). Application of big data in supply chain management. *Materials Today: Proceedings*, *4*(2), 1106–1115. DOI: 10.1016/j.matpr.2017.01.126

Jain, A. K. (2023). Sustainable Energy: One Sun-One World-One Grid. *Green Energy and Technology*, *91*. Advance online publication. DOI: 10.1007/978-981-99-0216-3_6

Jain, R., Singh, S. K., Palaniappan, D., Parmar, K., & Premavathi, T. (2025). Integrating smart cyber-physical systems into urban planning. In Poonia, R. C., Upreti, K., & Khan, M. S. (Eds.), *Smart cyber physical systems: Innovations and real-world implications* (pp. 82–106). CRC Press., DOI: 10.1201/9781003542513-5

Jaipong, P., & Vui, C. N. (2025). Sustainable Private Hospitals Based on Prices, Services, and Facilities. In *Sustainable Development for Hospitals and Clinics* (pp. 97–126). IGI Global Scientific Publishing., DOI: 10.4018/979-8-3693-9800-5.ch004

Jaiswal, S., & Pandey, S. (2024). Exploring generative AI as a catalyst for sustainability: Strategies for waste and energy reduction. *Management Insight*, *20*(1), 14–26. DOI: 10.21844/mijia.20.1.2

Jancar, B. (1993). The Environmental Attractor in the Former USSR: Ecology and Regional Change. In Lipschutz, R. D., & Conca, K. (Eds.), *The State and Social Power in Global Environmental Politics*. Columbia University Press.

Jangjarat, K., Kraiwanit, T., Satityapong, N., Sonsuphap, R., & Phaksipaeng, I. (2023). The social economy in the digital era: A perspective on community enterprises in a developing economy. *Journal of Social Economics Research, 10*(1), 13–21. DOI: 10.18488/35.v10i1.3317

Järlström, M., Saru, E., & Pekkarinen, A. (2023). Practices of Sustainable Human Resource Management in Three Finnish Companies: Comparative Case Study. *South Asian Journal of Business and Management Cases, 12*(1), 31–51. DOI: 10.1177/22779779231154656

Jaskeviciute, V., Stankeviciene, A., Diskiene, D., & Savicke, J. (2021). The relationship between employee well-being and organizational trust in the context of sustainable human resource management. *Problems and Perspectives in Management, 19*(2), 118–131. DOI: 10.21511/ppm.19(2).2021.10

Javed, S., Zhu, Y., Kumar, N., & Meghwar, S. (2024). Green human resource management practices and organizational sustainability performance in Pakistan's healthcare sector. *International Journal of Research In Business and Social Science*. https://doi.org/DOI: 10.20525/ijrbs.v13i1.3091

Jha, V. (2023). *International Solar Alliance: Bridging the Gap* [Center for Social and Economic Progress]. Center for Social and Economic Progress. https://csep.org/reports/international-solar-alliance-bridging-the-gap

Jiang, Y., An, W., & Yang, J. (2024). A Configurational Approach to the Driving Forces for Synergistic Reduction in Pollution and Carbon Emissions. *Sustainability, 16*(24), 10940–10940. DOI: 10.3390/su162410940

Jiatong, W., Li, C., Murad, M., Shahzad, F., & Ashraf, S. F. (2021). Impact of social entrepreneurial factors on sustainable enterprise development: Mediating role of social network and moderating effect of government regulations. *SAGE Open, 11*(3), 21582440211030636. DOI: 10.1177/21582440211030636

Jørgensen, B. N., & Grace, Z. M. A. (2025). Impact of EU regulations on AI adoption in smart city Solutions: A review of regulatory barriers, technological challenges, and societal benefits. *Information (Basel), 16*(7), 568. DOI: 10.3390/info16070568

Jørgensen, B. N., Gunasekaran, S. S., & Ma, Z. G. (2025). Impact of EU laws on AI adoption in smart grids: A review of regulatory barriers, technological challenges, and stakeholder benefits. *Energies, 18*(12), 3002. DOI: 10.3390/en18123002

Just Transition of the Workforce, and the Creation of Decent Work and Quality Jobs. (2020). [UNFCCC]. UNFCCC. https://unfccc.int/sites/default/files/resource/Just%20transition.pdf

Kaaria, A. G. (2024). Essential Human Resource Metrics and Analytics for Sustainable Work Environments: Literature Mapping and Conceptual Synthesis. *East African Journal of Business and Economics*, *7*(1), 241–262. DOI: 10.37284/eajbe.7.1.1976

Kabir, M., Habiba, U., & Farooqi, Z.-U.-R. (2021). Current Scenario of Environmental Pollution & Its Solution by Plantation in Changing Climate. *Journal of Sustainability Perspective*, 288-295. https://ejournal2.undip.ac.id/index.php/jsp/article/view/12016/6168

Kamboj, J., & A, E. (2024). Mapping the green human resource management practices: A systematic scoping review and its implications for employees' well-being. *Human Systems Management*, *43*(6), 932–970. DOI: 10.3233/HSM-230183

Kandpal, V., Jaswal, A., Gonzalez, E. D. R. S., & Agarwal, N. (Eds.). (2024). *Sustainable energy transition: Circular economy and sustainable financing for environmental, social and governance (ESG) practices.* Springer., DOI: 10.1007/978-3-031-52943-6

Kanellopoulou, D., Giannakopoulos, G., & Karkaletsis, V. (2025). Accelerating AI-powered digital innovation through "APSS": A novel methodology for sustainable business AI transformation. *Journal of Open Innovation*, *11*(1), 100495. DOI: 10.1016/j.joitmc.2025.100495

Kannan, H., Rodriguez, R. V., Paprika, Z. Z., & Ade-Ibijola, A. (Eds.). (2024). *Exploring ethical dimensions of environmental sustainability and use of AI.* IGI Global Scientific Publishing., DOI: 10.4018/979-8-3693-0892-9

Kannan, S., & Gambetta, N. (2025). Technology-driven sustainability in small and medium-sized enterprises: A systematic literature review. *Journal of Small Business Strategy*, *35*(1), 129–157. DOI: 10.53703/001c.126636

Kaptein, M. (2022). Understanding unethical behavior by unraveling ethical culture. *Human Relations*, *75*(2), 191–216.

Kar, S., Kar, A. K., & Gupta, M. P. (2021). Modeling drivers and barriers of artificial intelligence adoption: Insights from a strategic management perspective. *Intelligent Systems in Accounting, Finance & Management*, *28*(4), 217–238. DOI: 10.1002/isaf.1503

Kaur, H. (2024). Towards sustainable business practices: Circular economy principles for resource optimisation and wastereduction. *Global Journal for Research Analysis*. Advance online publication. DOI: 10.36106/paripex/2506575

Khagram, S., Clark, W., & Raad, D. F. (2003). From the Environment and Human Security to Sustainable Security and Development. *Journal of Human Development*, *4*(2), 289–313. DOI: 10.1080/1464988032000087604

Khan, T. I. (2023). Green Human Resource Management practices: A study on the banking sector of Pakistan. International Journal of Economics and Business Issues.

Khan, I. S., Ahmad, M. O., & Majava, J. (2021). Industry 4.0 and sustainable development: A systematic mapping of triple bottom line, circular economy and sustainable business models perspectives. *Journal of Cleaner Production*, *297*, 126655. DOI: 10.1016/j.jclepro.2021.126655

Khan, M. I., Yasmeen, T., Khan, M., Hadi, N. U., Asif, M., Farooq, M., & Al-Ghamdi, S. G. (2025). Integrating industry 4.0 for enhanced sustainability: Pathways and prospects. *Sustainable Production and Consumption*, *54*, 149–189. DOI: 10.1016/j.spc.2024.12.012

Khan, R. Z., Razak, L. A., & Premaratne, G. (2025). Green growth and sustainability: A systematic literature review on theories, measures and future directions. *Cleaner and Responsible Consumption*, *17*, 100274. DOI: 10.1016/j.clrc.2025.100274

Kharel, S., Chhetri, A. D., & Sharma, S. (2025). Advancing Organizational Sustainability Through Green Hrm: The Mediating Roles of Green Innovation and Employee Environmental Performance. *Journal of Balkumari College*, *14*(1), 1–9. DOI: 10.3126/jbkc.v14i1.80752

Khiawnoi, P., Nakudom, S. S., Noobutr, P., & Kaewzang, U. (2025). Empowering communities: The strategic role of community enterprise entrepreneurs in sustainable development. *Sustainability*, *17*(12), 5483. DOI: 10.3390/su17125483

Khokhar, M. (Ed.). (2025). *AI-driven business model innovation*. IGI Global Scientific Publishing., DOI: 10.4018/979-8-3693-9571-4

Kilroy, J., Dundon, T., & Townsend, K. (2023). Embedding reciprocity in human resource management: A social exchange theory of the role of frontline managers. *Human Resource Management Journal*, *33*(2), 511–531. DOI: 10.1111/1748-8583.12468

Kim, D., & Lim, U. (2017). Social enterprise as a catalyst for sustainable local and regional development. *Sustainability*, *9*(8), 1427. DOI: 10.3390/su9081427

Kim, J. (2025). <scp>ESG</scp> and Carbon Leakage: How Cap-and-Trade Regulations Influence Firms' Emission Allocation Strategies. *Corporate Social Responsibility and Environmental Management*. Advance online publication. DOI: 10.1002/csr.70142

King, L. C., & van den Bergh, J. C. J. M. (2021). Potential Carbon Leakage Under the Paris Agreement. *Climatic Change, 165*, 52. https://link.springer.com/article/10.1007/s10584-021-03082-4

Kingsbury, D., Remenyi, J., McKay, J., & Hunt, J. (2004). *Key Issues in Development*. Palgrave Macmillan.

Kirchherr, J., Matthews, N., Charles, K., & Walton, M. J. (2017). "Learning it the Hard Way": Social safeguards norms in Chinese-led dam projects in Myanmar, Laos and Cambodia. *Energy Policy, 102*, 529. DOI: 10.1016/j.enpol.2016.12.058

Kitchin, R., & Moore-Cherry, N. (2021). Fragmented governance, the urban data ecosystem, and smart city-regions: The case of Metropolitan Boston. *Regional Studies, 55*(12), 1913–1923. DOI: 10.1080/00343404.2020.1735627

Koirala, B., Cai, H., Khayatian, F., Munoz, E., An, J., Mutschler, R., Sulzer, M., De Wolf, C., & Orehounig, K. (2024). Digitalization of Urban Multi-Energy Systems – Advances in Digital Twin Applications across Life-Cycle Phases. *Advances in Applied Energy, 100196*. Advance online publication. DOI: 10.1016/j.adapen.2024.100196

Kojima, M., & Yoshida, A. (2022). E-waste management in East and Southeast Asia: Current policy and future perspectives. *Journal of Environmental Management, 310*, 114754.

Kok, S. L. S., & Sriboonruang, P. (2025). Healthcare Digitization and Sustainability of Penang Healthcare Providers: A Case of Medical Tourism. In *Sustainable Development for Hospitals and Clinics* (pp. 127-162). IGI Global Scientific Publishing. DOI: DOI: 10.4018/979-8-3693-9800-5.ch005

Kolawole, A. S., & Iyiola, A. O. (2023). Environmental Pollution: Threats, Impact on Biodiversity, and Protection Strategies. In Izah, S. C., & Ogwu, M. C. (Eds.), *Sustainable Utilization and Conservation of Africa's Biological Resources and Environment* (pp. 377–409). Springer., https://link.springer.com/chapter/10.1007/978-981-19-6974-4_14

Kong, B. (2009). *China's International Petroleum Policy*. DOI: 10.5040/9798400625749

Koppenjan, J., & Enserink, B. (2009). Public–private partnerships in urban infrastructure: Reconciling private sector efficiency and public accountability. *Public Administration Review, 69*(2), 284–296. DOI: 10.1111/j.1540-6210.2008.01974.x

Kosasih, K. (2024). The Role of Human Resource Management in Sustainable Development: A Systematic Literature Review. [DEMAND]. *Economics Studies and Banking Journal, 1*(2), 61–74. DOI: 10.62207/q4jya069

Krauss, J. E. (2022). Unpacking SDG 15, Its Targets and Indicators: Tracing Ideas of Conservation. *Globalizations, 19*(8), 1179–1194. DOI: 10.1080/14747731.2022.2035480

Kuhlmann, A., Moore, J., & Mehlum, E. (2021). Harnessing artificial intelligence to accelerate the energy transition. In World Economic Forum, BloombergNEF, & Deutsche Energie-Agentur (dena), W H I T E P a P E R [Report]. https://www3.weforum.org/docs/WEF_Harnessing_AI_to_accelerate_the_Energy_Transition_2021.pdf

Kuhlmann, S., Stegmaier, P., & Konrad, K. (2019). The tentative governance of emerging science and technology—A conceptual introduction. *Research Policy, 48*(5), 1091–1097.

Kulothungan, V., Mohan, P. R., & Gupta, D. (2025, May). AI regulation and capitalist growth: Balancing innovation, ethics, and global governance. In *2025 IEEE 11th Conference on Big Data Security on Cloud (BigDataSecurity)* (pp. 39-45). IEEE. https://doi.org/DOI: 10.1109/BigDataSecurity66063.2025.00020

Kumar, D., & Suthar, N. (2025). Balancing innovation and regulation: Ethical frameworks for AI policy and governance. In Youssef, A., & Arslan, A. (Eds.), *Navigating the Intersection of AI Policy, Technology, and Governance* (pp. 171–200). IGI Global Scientific Publishing., DOI: 10.4018/979-8-3373-1210-1.ch007

Kumari, D., & Sharma, A. (2025). Building A Greener Brand: The Role Of GHRM In Enhancing Corporate Image And Sustainable Commitment. *Journal of Informatics Education and Research, 5*(3). Advance online publication. DOI: 10.52783/jier.v5i3.3480

Kumari, S. (2024). A study of green human resource management practices adopted by the new age organizations. *Indian Scientific Journal Of Research In Engineering And Management, 08*(04), 1–5. DOI: 10.55041/ijsrem32694

Kutaula, S., Chowdhury, S., Gillani, A., Budhwar, P. S., & Dey, P. K. (2025). Linking HRM with Sustainability Performance Through Sustainability Practices: Unlocking the Black Box. *British Journal of Management, 36*(2), 615–632. DOI: 10.1111/1467-8551.12861

Kwerawarni, D. T. (2023). The Conceptual Model of Green Human Resource Management in Business Organization: A Literature Review. *International Journal of Multicultural and Multireligious Understanding*, *10*(12), 452–463. DOI: 10.18415/ijmmu.v10i12.5447

Lamb, W. F., Mattioli, G., Levi, S., Roberts, T. J., Capstick, S., Creutzig, F., Minx, J. C., Müller-Hansen, F., Culhane, T., & Steinberger, J. K. (2020). Discourses of Climate Delay. *Global Sustainability*, *3*, e17. DOI: 10.1017/sus.2020.13

Lanzalonga, F., Marseglia, R., Irace, A., & Biancone, P. P. (2024). The application of artificial intelligence in waste management: Understanding the potential of data-driven approaches for the circular economy paradigm. *Management Decision*. Advance online publication. DOI: 10.1108/MD-10-2023-1733

Laurenza, E., Quintano, M., Schiavone, F., & Vrontis, D. (2018). The effect of digital technologies adoption in healthcare industry: A case based analysis. *Business Process Management Journal*, *24*(5), 1124–1144. DOI: 10.1108/BPMJ-04-2017-0084

Lee, D., & Ahn, J. S. (2024). Unlocking the potential of green human resource management in Korea: A comprehensive review. *Asia Pacific Business Review*, •••, 1–24. DOI: 10.1080/13602381.2024.2358945

Lee, I., & Lee, K. (2015). The Internet of Things (IoT): Applications, investments, and challenges for enterprises. *Business Horizons*, *58*(4), 431–440. DOI: 10.1016/j.bushor.2015.03.008

Lee, S. U., Perera, H., Liu, Y., Xia, B., Lu, Q., Zhu, L., Cairns, J., & Nottage, M. (2025). Integrating ESG and AI: A comprehensive responsible AI assessment framework. *AI and Ethics*, •••, 1–28. DOI: 10.1007/s43681-025-00741-5

Lee, W.-J., Juskenaite, I., & Mwebaza, R. (2021). Public–Private Partnerships for Climate Technology Transfer and Innovation: Lessons from the Climate Technology Centre and Network. *Sustainability*, *13*(6), 3185. DOI: 10.3390/su13063185

Leffel, B., Lyon, T. P., & Newell, J. P. (2024, March 1). *Filling the climate governance gap: Do corporate decarbonization initiatives matter as much as state and local government policy?* SciSpace - Paper. DOI: 10.1016/j.erss.2023.103376

Lele, S. (2017). Sustainable Development Goal 6: Watering Down the Justice Concerns. *Wiley Interdisciplinary Reviews Water*, *4*(4), e1224. https://wires.onlinelibrary.wiley.com/doi/10.1002/wat2.1224

Lepawsky, J. (2018). *Reassembling rubbish: Worlding electronic waste*. MIT Press.

Le, T. T. (2022). How do corporate social responsibility and green innovation transform corporate green strategy into sustainable firm performance? *Journal of Cleaner Production*, *362*, 132228. DOI: 10.1016/j.jclepro.2022.132228

Liakos, K. G., Busato, P., Moshou, D., Pearson, S., & Bochtis, D. (2018). Machine learning in agriculture: A review. *Sensors (Basel)*, *18*(8), 2674. DOI: 10.3390/s18082674

Lieser, A. (2024, September 11). *Bilateral Investment Treaties as a Tool for Global Climate Governance?* https://voelkerrechtsblog.org/bilateral-investment-treaties-as-a-tool-for-global-climate-governance/

Liew, E. F., Ong, D. E., & Peerun, M. I. (2025). Mapping the Future of Sustainable Construction: Review on Digitalisation and Life Cycle Assessment Trends in Asia-Pacific. *Energy and Building*, *116547*. Advance online publication. DOI: 10.1016/j.enbuild.2025.116547

Lilliestam, J., Eckardt, J., & Bersalli, G. (2025). Sequencing, spending, and symbolism: Low carbon taxes primarily serve purposes other than emissions reduction. *One Earth*, *101390*. Advance online publication. DOI: 10.1016/j.oneear.2025.101390

Limna, P., & Siripipattanakul, S. (2025). Sustainable Clinic Management Based on Practices and Human Resource Management (HRM). In *Sustainable Development for Hospitals and Clinics* (pp. 65–96). IGI Global Scientific Publishing. DOI: DOI: 10.4018/979-8-3693-9800-5.ch003

Li, N., Zhang, H., Zhang, X., & Xie, X. (2025). Does Market-Based Environmental Regulatory Policy Improve Corporate Environmental Performance? Evidence from Carbon Emission Trading in China. *Sustainability*, *17*(2), 623–623. DOI: 10.3390/su17020623

Lindsey, R. (2023, August 22). *Climate Change: Global Sea Level.* Climate.Gov. https://www.climate.gov/news-features/understanding-climate/climate-change-global-sea-level

Lin, K. Y., & Wei, S. H. (2023). Advancing the industrial circular economy: The integrative role of machine learning in resource optimization. *Journal of Green Economy and Low-Carbon Development*, *2*(3), 122–136. DOI: 10.56578/jgelcd020302

Lister, J. (2018). The Policy Role of Corporate Carbon Management: Co-regulating Ecological Effectiveness. *Global Policy*, *9*(4), 538–548. DOI: 10.1111/1758-5899.12618

Liu, H., Saleem, M. M., Al-Faryan, M. A. S., Khan, I., & Zafar, M. W. (2022). Impact of Governance and Globalization on Natural Resources Volatility: The Role of Financial Development in the Middle East North Africa Countries. *Resources Policy*, *78*, 102881. DOI: 10.1016/j.resourpol.2022.102881

Liu, J., Wang, Z., Li, C., & Xu, R. (2024). Exploring pathways to digital transformation: fsQCA analysis based on the AMO framework. *PLoS One*, *19*(12), e0315249. DOI: 10.1371/journal.pone.0315249

Liu, K. W. (2024). Impact of Heterogeneous Environmental Regulation on Carbon Emissions: Firm-Level Evidence from China's Manufacturing Industry. *International Journal of Economics and Finance*, *16*(12), 44–44. DOI: 10.5539/ijef.v16n12p44

Liu, Z., Wu, C., & Xu, X. (2025). The role of artificial intelligence in sustainable development and industrial transformation. *Asia Pacific Economic and Management Review*, *2*(2), 1–6. DOI: 10.62177/apemr.v2i2.185

Li, W., Yi, P., Zhang, D., & Zhou, Y. (2020). Assessment of coordinated development between social economy and ecological environment: Case study of resource-based cities in Northeastern China. *Sustainable Cities and Society*, *59*, 102208. DOI: 10.1016/j.scs.2020.102208

Li, Y., Song, B., & Zhu, L. (2023). Integrating green governance into digital policy: Lessons from the EU circular economy. *Environmental Policy and Governance*, *33*(2), 153–169.

Li, Y., Yang, H., & Zhang, L. (2018). Green Governance: New Perspective from Open Innovation. *International Journal of Project Management*, *36*(5), 763–778. DOI: 10.3390/su10113845

Lock, I., & Seele, P. (2021). Green governance: Integrating sustainability into corporate communication and accountability. *Business Strategy and the Environment*, *30*(5), 2324–2338.

Lodha, L., Baghela, V. S., Bhuvana, J., & Bhatt, R. (2023). A blockchain-based secured system using the Internet of Medical Things (IOMT) network for e-healthcare monitoring. *Measurement. Sensors*, *30*, 100904. DOI: 10.1016/j.measen.2023.100904

Lokmic-Tomkins, Z., Davies, S., Block, L. J., Cochrane, L., Dorin, A., Von Gerich, H., & Peltonen, L. M. (2022). Assessing the carbon footprint of digital health interventions: A scoping review. *Journal of the American Medical Informatics Association : JAMIA*, *29*(12), 2128–2139. DOI: 10.1093/jamia/ocac196

Lu, Y., Zhang, M. M., Yang, M. M., & Wang, Y. (2023). Sustainable human resource management practices, employee resilience, and employee outcomes: Toward common good values. *Human Resource Management, 62*(3), 331–353. DOI: 10.1002/hrm.22153

Lyu, M., Chen, Y., & Chen, S. (2024). Prevention or ex-post forcing? The impact of dual environmental regulation on corporate environmental behavior. *Heliyon*. Advance online publication. DOI: 10.1016/j.heliyon.2024.e30850

Madanchian, M., & Taherdoost, H. (2025). Barriers and enablers of AI adoption in human resource management: A critical analysis of organizational and technological factors. *Information (Basel), 16*(1), 51. DOI: 10.3390/info16010051

Madero-Gómez, S. M., Rubio Leal, Y. L., Olivas-Luján, M., & Yusliza, M. Y. (2023). Companies Could Benefit When They Focus on Employee Wellbeing and the Environment: A Systematic Review of Sustainable Human Resource Management. *Sustainability, 15*(6), 5435. DOI: 10.3390/su15065435

Madhani, P. M. (2010). Resource based view (RBV) of competitive advantage: an overview. *Resource based view: concepts and practices, Pankaj Madhani, ed*, 3-22. https://doi.org/DOI: 10.54518/rh.2.6.2022.587-596

Mahardhani, A. J. (2023). The role of public policy in fostering technological innovation and sustainability. [ADMAN]. *Journal of Contemporary Administration and Management, 1*(2), 47–53.

Mahdia, A. (2024). The Role of Human Resource Management in Employee Well-Being and Mental Health: A Systematic Literature Review. [PRODUCTIVITY]. *Management Studies and Business Journal, 1*(3), 286–303. DOI: 10.62207/phea4z38

Mahmoud, I. H., Morello, E., Salvia, G., & Puerari, E. (2022). Greening cities, shaping cities: Pinpointing nature-based solutions in cities between shared governance and citizen participation. *Sustainability, 14*(12), 7011. Advance online publication. DOI: 10.3390/su14127011

Maleki Varnosfaderani, S., & Forouzanfar, M. (2024). The role of AI in hospitals and clinics: Transforming healthcare in the 21st century. *Bioengineering (Basel, Switzerland), 11*(4), 337. DOI: 10.3390/bioengineering11040337

Mallik, S., Mathivanan, S. K., Sangeetha, S., & Soufiene, B. O. (Eds.). (2025). *Real-world applications of AI innovation*. IGI Global Scientific Publishing., DOI: 10.4018/979-8-3693-4252-7

Mambondiyani, A. (2020, December 9). *China's billion-dollar dirty coal projects*. https://gga.org/chinas-billion-dollar-dirty-coal-projects/

Manao, A., & Hadi Senen, S. (2024). Sustainable Human Resource Management: Building an Adaptive and Inclusive Organizational Culture. *Dinasti International Journal of Education Management And Social Science*, *6*(2), 863–871. DOI: 10.38035/dijemss.v6i2.3673

Marengo, A. (2024). Navigating the nexus of AI and IoT: A comprehensive review of data analytics and privacy paradigms. *Internet of Things : Engineering Cyber Physical Human Systems*, *27*, 101318. DOI: 10.1016/j.iot.2024.101318

Marín, L., Nicolás, C., & Rubio, A. (2019). How gender, age and education influence the entrepreneur's social orientation: The moderating effect of economic development. *Sustainability*, *11*(17), 4514. DOI: 10.3390/su11174514

Marquardt, J. (2024). Promises and pitfalls of polycentric federalism: The case of solar power in India. *Global Environmental Politics*, *24*(3), 75–96. DOI: 10.1162/glep_a_01315

Martínez-Peláez, R., Ochoa-Brust, A., Rivera, S., Félix, V. G., Ostos, R., Brito, H., & Mena, L. J. (2023). Role of digital transformation for achieving sustainability: Mediated role of stakeholders, key capabilities, and technology. *Sustainability*, *15*(14), 11221. DOI: 10.3390/su151411221

Masood, R. Z. (2024). Sustainable Hrm Practices And Innovation In Healthcare Smes: Mediating Role Of Green Innovation Using Pls-Sem Analysis. *ShodhKosh: Journal of Visual and Performing Arts*, *5*(1). Advance online publication. DOI: 10.29121/shodhkosh.v5.i1.2024.3728

Massoudie, J., Afghan, H., Zamani, M., Samadi, M., & Gul, S. (2024). The Role of GHRM Practices Towards Organizational Commitment: A Mediation Analysis of Green Health and Safety Management. *Kardan Journal of Economics and Manangement Sciences*, *3*. Advance online publication. DOI: 10.31841/KJEMS.2024.168

Mathew, Dr. J. M., Fasina, M., Biju, P. J., & Hamsa, E. (2025). Role of HR in Promoting Sustainability and Corporate Social Responsibility. *International Journal of Accounting and Economics Studies*, *12*(SI-1), 371–376. DOI: 10.14419/sxztgj60

Ma, W., & Wang, M. (2023). Discussion on the relationship between environmental regulation and green technology innovation from the perspective of innovation external cooperation: Evidence from Chinese private enterprises. *Sustainability*, *15*(23), 16333.

Maza-Maza, J., Rodríguez-Caballero, E., Rodríguez-Lozano, B., & Cantón, Y. (2025). Impact of Deforestation on Ecosystem Functions of Ecuador's Tropical Dry Forest Under Different Protection Status. *Trees, Forests and People*, *22*, 101032. DOI: 10.1016/j.tfp.2025.101032

Meadows, D. H. (2008). *Thinking in systems: A primer*. Chelsea Green Publishing.

Meguro, M. (2020). State of the Netherlands v. Urgenda Foundation. *The American Journal of International Law*, *114*(4), 729. DOI: 10.1017/ajil.2020.52

Melnyk, M., Blyznyukov, A., & Cieślik, J. (2023). The impact of digital education initiatives.http://doi.org/DOI: 10.61093/sec.7(3).1-9.2023

Melville, N. P. (2010). Information systems innovation for environmental sustainability. *Management Information Systems Quarterly*, •••, 1–21. DOI: 10.2307/20721412

Memon, H., Anwar, S., & Jamali, A. (2024). The Role of Green HRM Practices and Employee Empowerment in Boosting Environmental Sustainability in Higher Educational Institutions in Sindh, Pakistan. *Progressive Research Journal of Arts & Humanities*, *6*(2), 55–67. DOI: 10.51872/prjah.vol6.iss2.357

Menton, M., Larrea, C., Latorre, S., Martinez-Alier, J., Peck, M., Temper, L., & Walter, M. (2020). Environmental Justice and the SDGs: From Synergies to Gaps and Contradictions. *Sustainability Science*, *15*, 1621–1636. https://link.springer.com/article/10.1007/s11625-020-00789-8

Mienye, I. D., Sun, Y., & Ileberi, E. (2024). Artificial intelligence and sustainable development in Africa: A comprehensive review. *Machine Learning with Applications*, *100591*. Advance online publication. DOI: 10.1016/j.mlwa.2024.100591

Mihăilescu, M. (2025). Digital Governance Assisted by Artificial Intelligence in the Energy Sector. *Advances in Computational Intelligence and Robotics*, *1*. Advance online publication. DOI: 10.4018/979-8-3373-3186-7.ch001

Miller, T., Durlik, I., Kostecka, E., Łobodzińska, A., & Matuszak, M. (2024). The emerging role of artificial intelligence in enhancing energy efficiency and reducing GHG emissions in transport systems. *Energies*, *17*(24), 6271. DOI: 10.3390/en17246271

Mishra, S. (2023). *M.C. Mehta And Anr vs. Union of India & Ors*. THE JOURNAL OF UNIQUE LAWS AND STUDENTS., DOI: 10.59126/v2i4a4

Mishra, S., Jena, L., Tripathy, H. K., & Gaber, T. (2022). Prioritized and predictive intelligence of things enabled waste management model in smart and sustainable environment. *PLoS One*, *17*(8), e0272383. DOI: 10.1371/journal.pone.0272383

Mohammadimanesh, F., Mahdianpari, M., Radman, A., Varon, D., Hemati, M., & Marjani, M. (2025). Advancements in satellite-based methane point source monitoring: A systematic review. *ISPRS Journal of Photogrammetry and Remote Sensing*, *224*, 94–112. DOI: 10.1016/j.isprsjprs.2025.03.020

Mohieldin, M., Gonzalez-Perez, M. A., & Zahran, M. (2025). AI governance, regulation, and ethical implementation. In *AI-Powered Sustainable Business. Palgrave Studies in Moral and Mindful Approaches to Leadership and Business* (pp. 81-102). Palgrave Macmillan. https://doi.org/DOI: 10.1007/978-3-031-93357-8_5

Molinaro, M., Orzes, G., & Sarkis, J. (2025). Unpacking Circular Economy Practices and Carbon Emissions Relationships: Co-benefits and Legitimacy Perspectives. *Business Strategy and the Environment*. Advance online publication. DOI: 10.1002/bse.70157

Montalvo-Falcón, J. V., Sánchez-García, E., Marco-Lajara, B., & Martínez-Falcó, J. (2023). Green human resource management and economic, social and environmental performance: Evidence from the Spanish wine industry. *Heliyon*, *9*(10), e20826. DOI: 10.1016/j.heliyon.2023.e20826

Montiel, I. (2008). Corporate social responsibility and corporate sustainability: Separate pasts, common futures. *Organization & Environment*, *21*(3), 245–269. DOI: 10.1177/1086026608321329

Moraes, T. (2025). Regulatory sandboxes for trustworthy artificial intelligence – Global and Latin American experiences. *International Review of Law Computers & Technology*, *39*(1), 55–74. DOI: 10.1080/13600869.2024.2351674

Morgan, J., & Patomäki, H. (2021). Planetary Good Governance after the Paris Agreement: The Case for a Global Greenhouse Gas Tax. *Journal of Environmental Management*, *292*, 112753. DOI: 10.1016/j.jenvman.2021.112753

Moutinho, A., Sousa, P., & Tereso, A. (2025). Applying the PRiSM™ methodology to raise awareness of the importance of using sustainable project management practices in organizations. *Systems*, *13*(2), 69. DOI: 10.3390/systems13020069

Muhammad, N., Laghari, A., Amin, D. H., Bangash, S. A., Ishfaq, U., & Nosh, N. (2025). The Green Innovation Engine: How GHRM Practices Influence Sustainable Performance in Manufacturing Industries. *Social Science Review Archives*, *3*(1), 417–435. DOI: 10.70670/sra.v3i1.321

Mukherjee, S., Dhingra, T., & Sengupta, A. (2016). Status of Electricity Act, 2003: A systematic review of literature. *Energy Policy*, *102*, 237. DOI: 10.1016/j.enpol.2016.12.001

Mutambara, A. G. O. (Ed.). (2025). *Deploying artificial intelligence to achieve the UN sustainable development goals: Enablers, drivers and strategic framework.* Springer., DOI: 10.1007/978-3-031-88423-8

Naim, A. (Ed.). (2024). *Harnessing high-performance computing and AI for environmental sustainability.* IGI Global Scientific Publishing., DOI: 10.4018/979-8-3693-1794-5

Nakra, N., Kashyap, V., & Arora, R. (2024). Interactive effects of sustainable HRM, career adaptability and career sustainability on employee well-being. *Employee Relations, 46*(8), 1706–1722. DOI: 10.1108/ER-01-2024-0021

Narula, K. (2023, June 23). *India Must Begin Paving the Way to Ending Coal.* https://www.geopoliticalmonitor.com/india-must-begin-paving-the-way-to-ending-coal

National University of Singapore (NUS). (2023). *NUS sustainability report 2023.* https://sustainability.nus.edu.sg

Nguyen, H. T., Le, D. M. D., Ho, T. T. M., & Nguyen, P. M. (2021). Enhancing sustainability in the contemporary model of CSR: A case of fast fashion industry in developing countries. *Social Responsibility Journal, 17*(4), 578–591. DOI: 10.1108/SRJ-03-2019-0108

Nguyen, T. T., Grote, U., Neubacher, F., Rahut, D. B., Do, M. H., & Paudel, G. P. (2023). Security Risks from Climate Change and Environmental Degradation: Implications for Sustainable Land Use Transformation in the Global South. *Current Opinion in Environmental Sustainability, 63*, 101322. DOI: 10.1016/j.cosust.2023.101322

Niesten, E., Jolink, A., de Sousa Jabbour, A. B. L., Chappin, M., & Lozano, R. (2017). Sustainable collaboration: The impact of governance and institutions on sustainable performance. *Journal of Cleaner Production, 155*, 1–6.

Nnorom, I. C., & Osibanjo, O. (2021). Overview of e-waste management practices and regulations in developing countries. *Waste Management & Research, 39*(5), 619–630.

Nogueira, E., Gomes, S., & Lopes, J. M. (2025). Unveiling triple bottom line's influence on business performance. *Discover Sustainability, 6*, 43. DOI: 10.1007/s43621-025-00804-x

Noor Faezah, J., Yusliza, M. Y., Chaudhary, R., Ramayah, T., & Fawehinmi, O. (2024). Green human resource management and employee ecological behaviour: The mediating role of green commitment. *Social Responsibility Journal, 20*(6), 1157–1176. DOI: 10.1108/SRJ-07-2023-0399

Noori, N. (2021). Policy transplantation for smart city initiatives. In Erasmus University of Rotterdam [Thesis]. https://repub.eur.nl/pub/135367/thesis-draftnegarversion-0044rev-est.pdf

Nor, B. A., & Yusof, Y. (2025). Environmental Degradation and Food Security in Somalia. *Discover Sustainability*, *6*, 75. https://link.springer.com/article/10.1007/s43621-024-00771-9

Núñez, R. B. C., Bandeira, P., & Santero-Sánchez, R. (2020). Social economy, gender equality at work and the 2030 agenda: Theory and evidence from Spain. *Sustainability*, *12*(12), 5192. DOI: 10.3390/su12125192

Nurgozhayeva, R. (2024). The Belt and Road Initiative and Sustainability: A Driving Force for Change or a Missed Opportunity? *Chinese Journal of Environmental Law*, *8*(2), 119. DOI: 10.1163/24686042-12340120

Nurimansjah, R. A. (2023). Dynamics of Human Resource Management: Integrating Technology, Sustainability, and Adaptability in the Modern Organizational Landscape. *Golden Ratio of Mapping Idea and Literature Format*, *3*(2), 104–123. DOI: 10.52970/grmilf.v3i2.324

Obilor, E. I. (2023). Convenience and purposive sampling techniques: Are they the same? *International Journal of Innovative Social & Science Education Research*, *11*(1), 1–7. https://seahipaj.org/journals-ci/mar-2023/IJISSER/full/IJISSER-M-1-2023.pdf

Ochoa, W. A. A., Arce, A. G. T., Neto, A. I., Munaro, M. R., Calabokis, O. P., Ballesteros-Ballesteros, V. A., Ochoa, W. A. A., Arce, A. G. T., Neto, A. I., Munaro, M. R., Calabokis, O. P., & Ballesteros-Ballesteros, V. A. (2025). Interlinking Urban Sustainability, Circular Economy and Complexity: A Systematic Literature Review. *Sustainability*, *17*(15). Advance online publication. DOI: 10.3390/su17157118

Odeyemi, O., Usman, F. O., Mhlongo, N. Z., Elufioye, O. A., & Ike, C. U. (2024). Sustainable entrepreneurship: A review of green business practices and environmental impact. *World Journal of Advanced Research and Reviews*, *21*(2), 346–358. DOI: 10.30574/wjarr.2024.21.2.0461

OECD. (2023). *AI in the climate and environmental domain: Policy opportunities and challenges*. OECD Publishing. https://www.oecd.org/en/topics/policy-issues/artificial-intelligence.html

OECD. (2023). *E-waste management and policy coherence: Bridging technology and environment. OECD Environment Working Papers, No. 194.*

OECD. (2025). Government at a Glance 2025. *Government at a Glance*. DOI: 10.1787/0efd0bcd-en

Ofosuhene, D. (2024). Adoption Of Green Human Resource Management Practices And Organisational Wellbeing In The Ghanaian Service Sector. *European Journal of Human Resource Management Studies*, *7*(2). Advance online publication. DOI: 10.46827/ejhrms.v7i2.1672

Ogbonnaya, C., & Aryee, S. (2022). HRM Practices, Employee Well-Being, and Organizational Performance. In Brough, P., Gardiner, E., & Daniels, K. (Eds.), *Handbook on Management and Employment Practices* (Vol. 3, pp. 369–391). Springer International Publishing., DOI: 10.1007/978-3-030-29010-8_17

Ojadi, J. O., Odionu, C., Onukwulu, E., & Owulade, O. (2024). Big data analytics and AI for optimizing supply chain sustainability and reducing greenhouse gas emissions in logistics and transportation. *International Journal of Multidisciplinary Research and Growth Evaluation*, *5*(1), 1536–1548. DOI: 10.54660/.IJMRGE.2024.5.1.1536-1548

Ojadi, J. O., Onukwulu, E., Odionu, C., & Owulade, O. (2023). AI-driven predictive analytics for carbon emission reduction in industrial manufacturing: A machine learning approach to sustainable production. *International Journal of Multidisciplinary Research and Growth Evaluation*, *4*(1), 948–960. DOI: 10.54660/.IJMRGE.2023.4.1.948-960

Ojo, A. O., & Raman, M. (2019). Role of Green HRM Practices in Employees' Pro-environmental IT Practices. *SciSpace - Paper*, 678–688. https://doi.org/DOI: 10.1007/978-3-030-16181-1_64

Okon, D. U. O., & Marauche, O. M. O. (2025, July 1). *The Drivers and Barriers of Corporate Carbon Accounting Adoption: A Systematic Review of Managerial, Organizational, and Policy Factors*. SciSpace - Paper. DOI: 10.35629/5252-0707717727

Oladapo, B. I., Olawumi, M. A., & Omigbodun, F. T. (2024). AI-driven circular economy of enhancing sustainability and efficiency in industrial operations. *Sustainability*, *16*(23), 10358. DOI: 10.3390/su162310358

Olatinwo, S. O., & Joubert, T. H. (2023). Resource allocation optimization in IoT-enabled water quality monitoring systems. *Sensors (Basel)*, *23*(21), 8963. DOI: 10.3390/s23218963

Olawade, D. B., Wada, O. Z., David-Olawade, A. C., Fapohunda, O., Ige, A. O., & Ling, J. (2024). Artificial intelligence potential for net zero sustainability: Current evidence and prospects. *Next Sustainability*, *4*, 100041. DOI: 10.1016/j.nxsust.2024.100041

Olawuyi, D. S., & Athwal, E. I. (2023). Energy in WTO law and policy. *Edward Elgar Publishing eBooks*, 477. DOI: 10.4337/9781800882867.00034

Omisore, A. G. (2018). Attaining Sustainable Development Goals in Sub-Saharan Africa: The Need to Address Environmental Challenges. *Environmental Development, 25*, 138–145. DOI: 10.1016/j.envdev.2017.09.002

Orenuga, A., Oyeyemi, B. B., & John, A. O. (2024). AI and sustainable supply chain practices: ESG goals in the US and Nigeria. *International Journal of Future Engineering Innovations, 1*(1), 127–137. DOI: 10.54660/IJFEI.2024.1.1.127-137

Organisation for Economic Co-operation and Development (OECD). (2023). *Digital economy outlook 2023*. OECD Publishing.

Osei, A., Zhu, N., Borgi, H., & Cobbinah, J. (2025). Empowering sustainable production: Firm governance, finance strategies, and tech-innovation in advancing circular economy for SDG 12 in the US, China and Japan. *Sustainable Development, 33*(4), 5282–5307. DOI: 10.1002/sd.3390

Osibanjo, O., & Nnorom, I. C. (2007). The challenge of electronic waste (e-waste) management in developing countries. *Waste Management & Research, 25*(6), 489–501. DOI: 10.1177/0734242X07082028

Ostrom, E. (2010). Polycentric systems for coping with collective action and global environmental change. *Global Environmental Change, 20*(4), 550–557.

Owen, R., Stilgoe, J., Macnaghten, P., Gorman, M., Fisher, E., & Guston, D. H. (2013). *Responsible innovation: Managing the responsible emergence of science and innovation in society*. John Wiley & Sons.

Oyebode, O. J. (2024, April). Management of Computer and Electronic Waste through the Fifth Industrial Revolution for the Cleaner Environment. In *2024 International Conference on Science, Engineering and Business for Driving Sustainable Development Goals (SEB4SDG)* (pp. 1-14). IEEE. DOI: DOI: 10.1109/SEB4SDG60871.2024.10629821

Paing, T. H. (2016, August 30). *Burma, India Agree to Cooperate on Defense, Energy, Infrastructure*. https://www.irrawaddy.com/news/burma/burma-india-agree-to-cooperate-on-defense-energy-infrastructure.html

Pandey, R., Bhattacharya, M., & Gupta, S. (2022). Institutional frameworks for sustainable PPP implementation: Challenges and opportunities in India. *Journal of Infrastructure Development, 14*(3), 211–229. DOI: 10.1002/wene.422

Panek, S. (2025, September 22). *The ICJ's Advisory Opinion on Climate Change: Insights and Impact.* https://www.standrewslawreview.com/post/the-icj-s-advisory-opinion-on-climate-change-insights-and-impact

Paneru, B., Mainalil, D. S., Paneru, B., & Sapkota, S. C. (2024). Advancing Sustainable Mobility: Dynamic predictive modeling of charging cycles in electric vehicles using machine learning techniques and predictive application development. *Systems and Soft Computing, 200157.* Advance online publication. DOI: 10.1016/j.sasc.2024.200157

Papagiannidis, E., Enholm, I. M., Dremel, C., Mikalef, P., & Krogstie, J. (2023). Toward AI governance: Identifying best practices and potential barriers and outcomes. *Information Systems Frontiers, 25*, 123–141. DOI: 10.1007/s10796-022-10251-y

Parent-Lamarche, A., Dextras-Gauthier, J., & Julien, A.-S. (2023). Toward a new model of human resource management practices: Construction and validation of the High Wellbeing and Performance Work System Scale. *Frontiers in Psychology, 14.* Advance online publication. DOI: 10.3389/fpsyg.2023.1151781

Pawar, P. (2025). The relationship between green hrm and employee environmental commitment. *International Journal of Research in Commerce and Management Studies, 07*(04), 481–485. DOI: 10.38193/ijrcms.2025.7436

Peng, J., Wang, H., & Wen, L. (2025). Editorial: Advancing carbon reduction and pollution control policies management: theoretical, application, and future impacts. *Frontiers in Environmental Science, 13.* Advance online publication. DOI: 10.3389/fenvs.2025.1687660

Pereira, G. V., Temple, L., & Klausner, L. D. (2024). Transferring Smart City Concepts to Smaller Urban and Rural Contexts. https://doi.org/DOI: 10.31219/osf.io/deuyc

Permadi, I. K. O., Fadah, I., & Rizki, V. L. (2024). Green Hrm and Employee Performance: How Organizational Culture Plays A Mediating Role. *Journal of Economics. Finance And Management Studies, 07*(10). Advance online publication. DOI: 10.47191/jefms/v7-i10-44

Permata, A. K., Nur W., & Hariyadi, E. (2024). Analisis Penerapan Green HRM pada Kinerja Keberlanjutan Industri Keuangan di Indonesia. *Reviu Akuntansi Dan Bisnis Indonesia, 8*(3), 371–382. DOI: 10.18196/rabin.v8i3.22777

Petcho, W., Szabo, S., Kusakabe, K., & Yukongdi, V. (2019). Farmers' perception and drivers of membership in rice production community enterprises: Evidence from the Central Region, Thailand. *Sustainability, 11*(19), 5445. DOI: 10.3390/su11195445

Petersmann, E. (2020). *International Settlement of Trade and Investment Disputes Over Chinese 'Silk Road Projects' Inside the European Union*. 45. DOI: 10.1007/978-3-030-46000-6_4

Pfajfar, G., Shoham, A., Małecka, A., & Zalaznik, M. (2022). Value of corporate social responsibility for multiple stakeholders and social impact–Relationship marketing perspective. *Journal of Business Research, 143*, 46–61. DOI: 10.1016/j.jbusres.2022.01.051

Phuangsuwan, P., Limna, P., & Siripipatthanakul, S. (2024). Ethics in the social sciences research. *Advance Knowledge for Executives, 3*(4), 1-11. https://www.researchgate.net/publication/394223864

Phuangsuwan, P., Rungsiyanont, S., & Sriboonruang, P. (2025). Sustainable Development in Hospitals for Elderly Dysphagia. In *Geriatric Dentistry. In Sustainable Development for Hospitals and Clinics* (pp. 225–252). IGI Global Scientific Publishing., DOI: 10.4018/979-8-3693-9800-5.ch008

Pinilla-De La Cruz, G. A., Rabetino, R., & Kantola, J. (2022). Unveiling the shades of partnerships for the energy transition and sustainable development: Connecting public–private partnerships and emerging hybrid schemes. *Sustainable Development, 30*(5), 1370–1386.

Pivoto, D. G., de Almeida, L. F., da Rosa Righi, R., Rodrigues, J. J., Lugli, A. B., & Alberti, A. M. (2021). Cyber-physical systems architectures for industrial internet of things applications in industry 4.0: A literature review. *Journal of Manufacturing Systems, 58*, 176–192. DOI: 10.1016/j.jmsy.2020.11.017

Policies & action. (2025, November 6). /climateactiontracker.org/countries/china/policies-action

Pongracz, E. (2020). Social economy and social entrepreneurship–The manifestation of social responsibility in the era of globalization. In *SHS Web of Conferences*. EDP Sciences. https://doi.org/DOI: 10.1051/shsconf/20207404021

Popescu, C. R., & Yu, P. (Eds.). (2024). *Intersecting environmental social governance and AI for business sustainability*. IGI Global Scientific Publishing., DOI: 10.4018/979-8-3693-1151-6

Popovski, V., & Stanchev, K. (2021). Extreme Poverty Eradication: Conceptual Evolution and Policy Changes. In Kakar, N., Popovski, V., & Robinson, N. A. (Eds.), *Fulfilling the Sustainable Development Goals on a Quest for a Sustainable World* (pp. 37–54). Routledge.

Porter, L., Rickards, L., Verlie, B., Bosomworth, K., Moloney, S., Lay, B., Latham, B., Anguelovski, I., & Pellow, D. (2020). Climate Justice in a Climate Changed World. *Planning Theory & Practice, 21*(2), 293–321. DOI: 10.1080/14649357.2020.1748959

Pourrahmani, H., Amiri, M. T., Madi, H., & Owusu, J. P. (2025). Revolutionizing carbon sequestration: Integrating IoT, AI, and blockchain technologies in the fight against climate change. *Energy Reports, 13*, 5952–5967. DOI: 10.1016/j.egyr.2025.05.042

Prange, C., Bruyaka, O., & Marmenout, K. (2018). Investigating the transformation and transition processes between dynamic capabilities: Evidence from DHL. *Organization Studies, 39*(11), 1547–1573. DOI: 10.1177/0170840617727775

Priya, M. R. (2024). The triple bottom line of green transitions: Assessing the economic, social, and environmental impacts of sustainable development goals by 2030. In Yıldırım, S., Yıldırım, D. C., Demirtaş, I., & Kandpal, V. (Eds.) *Green Transition Impacts on the Economy, Society, and Environment* (pp. 182-201). IGI Global Scientific Publishing. https://doi.org/DOI: 10.4018/979-8-3693-3985-5.ch011

Purnasari, E., Hermuningsih, S., Hidayat, R., Hadi, S., & Kumar, A. (2025). *Antecedents and Consequences of Financial Literacy: A Comprehensive Systematic Literature Review of the Past Decade*. 3–14. https://doi.org/DOI: 10.2991/978-94-6463-698-7_2

Purushothaman, R., Alamelu, R., & Sudha, M. (2025). Leveraging solutions for advancing circular economy strategies: An overview. In Pujari, P., Khan, S. A., Kumar, A., & Naim, A. (Eds.), *Fostering Economic Diversification and Sustainable Business Through Digital Intelligence* (pp. 283–304). IGI Global Scientific Publishing., DOI: 10.4018/979-8-3693-8492-3.ch014

Putnam, L. L., & Nicotera, A. M. (2010). Communicative constitution of organization is a question: Critical issues for addressing it. *Management Communication Quarterly, 24*(1), 158–165.

Qi, Y., Xie, X., & Chen, X. (2023). Effects of environmental regulation and corporate environmental commitment: Complementary or alternative? Evidence from China. *Journal of Cleaner Production*. Advance online publication. DOI: 10.1016/j.jclepro.2023.138641

Rafique, S., Tanveer, Y., Sadiq, S., & Tariq, A. (2024). Mediation of Green Perceived Organizational Support between Green Human Resource Management Practices and Employee Performance. *The Asian Bulletin of Big Data Management, 4*(3), 173–191. DOI: 10.62019/abbdm.v4i3.228

Rahman, M. A., & Saba, N. A. (2025). Driving Sustainable Business Performance through Green HRM: Examining the Mediating Roles of Green Technology Innovation and Green Organizational Culture. *Golden Ratio of Human Resource Management*, *5*(2), 501–516. DOI: 10.52970/grhrm.v5i2.1326

Rahman, N. H. A., Saad, A., & Ismail, S. (2025). Green human resource management (ghrm): A review of implementation challenges. *ADVANCED INTERNATIONAL JOURNAL OF BUSINESS, ENTREPRENEURSHIP AND SME's*, *7*(24), 315–324. DOI: 10.35631/aijbes.724021

Rajâa, O., & Mekkaoui, A. (2025). Revealing the impact of social exchange theory on financial performance: A systematic review of the mediating role of human resource performance. *Cogent Business & Management*, *12*(1), 2475983. DOI: 10.1080/23311975.2025.2475983

Rajaperumal, T. A., & Columbus, C. C. (2025). Transforming the electrical grid: The role of AI in advancing smart, sustainable, and secure energy systems. *Energy Informatics*, *8*, 51. DOI: 10.1186/s42162-024-00461-w

Ramadani, V., Agarwal, S., Caputo, A., Agrawal, V., & Dixit, J. K. (2022). Sustainable competencies of social entrepreneurship for sustainable development: Exploratory analysis from a developing economy. *Business Strategy and the Environment*, *31*(7), 3437–3453. DOI: 10.1002/bse.3093

Rana, N. P., Luthra, S., Mangla, S. K., Islam, R., & Roderick, S. (2023). Exploring the dark side of digital transformation: Environmental challenges and sustainability paradoxes. *Technological Forecasting and Social Change*, *188*, 122329.

Rauf, F., & Baolei, Q. (2025). Enhancing ESG disclosure through AI adoption: The moderating role of institutional quality. *Applied Economics*, •••, 1–19. DOI: 10.1080/00036846.2025.2547117

Robinson, D., & Igini, M. (2025, January 9). *15 Biggest Environmental Problems of 2025*. Earth.Org. https://earth.org/the-biggest-environmental-problems-of-our-lifetime/

Romprasert, S., & Trivedi, A. (2021). Sustainable economy on community enterprise and digital marketing. *ABAC Journal*, *41*(1), 62–80. http://www.assumptionjournal.au.edu/index.php/abacjournal/article/view/4195

Roscoe, S., Subramanian, N., Jabbour, C. J. C., & Chong, T. (2019). Green human resource management and the enablers of green organisational culture: Enhancing a firm's environmental performance for sustainable development. *Business Strategy and the Environment*, *28*(5), 737–749. DOI: 10.1002/BSE.2277

Ross, B., Pilz, L., Cabrera, B., Brachten, F., Neubaum, G., & Stieglitz, S. (2019). Are social bots a real threat? An agent-based model of the spiral of silence to analyse the impact of manipulative actors in social networks. *European Journal of Information Systems*, *28*(4), 394–412. DOI: 10.1080/0960085X.2018.1560920

Rosyafah, S., Hidayat, W. G. P. A., & Nitawati, E. Y. (2025). Sustainable HRM in the face of climate change: Building organizational resilience through green human capital practices. *Annals of Human Resource Management Research*, *5*(2), 113–123. DOI: 10.35912/ahrmr.v5i2.2966

Roucham, B., Lefilef, A., & Mahmoudi, H. (2023). The Forward-Looking Studies on Green Human Resources Management: A Content Analysis. *MANAGEMENT AND ECONOMICS REVIEW*, *8*(3), 259–275. DOI: 10.24818/mer/2023.10-01

Saberi, S., Kouhizadeh, M., Sarkis, J., & Shen, L. (2019). Blockchain technology and its relationships to sustainable supply chain management. *International Journal of Production Research*, *57*(7), 2117–2135. DOI: 10.1080/00207543.2018.1533261

Saeed, B. B., Afsar, B., Hafeez, S., Khan, I., Tahir, M., & Afridi, M. A. (2019). Promoting employee's proenvironmental behavior through green human resource management practices. *Corporate Social Responsibility and Environmental Management*, *26*(2), 424–438. DOI: 10.1002/CSR.1694

Sah, B. P., Hasan, M. M., Shofiullah, S., & Faysal, S. A. (2024). AI-driven IoT and blockchain integration in industry 5.0: A systematic review of supply chain transformation. *Innovatech Engineering Journal*, *1*(1), 99–116. DOI: 10.70937/itej.v1i01.12

Salamzadeh, A., Hadizadeh, M., Rastgoo, N., Rahman, M. M., & Radfard, S. (2022). Sustainability-oriented innovation foresight in international new technology based firms. *Sustainability*, *14*(20), 13501. DOI: 10.3390/su142013501

Salvia, A. L., Filho, W. L., Brandli, L. L., & Griebeler, J. S. (2019). Assessing Research Trends Related to Sustainable Development Goals: Local and Global Issues. *Journal of Cleaner Production*, *208*, 841–849. DOI: 10.1016/j.jclepro.2018.09.242

Sambhav, A. (2024). LEGAL ANALYSIS OF CARBON BORDER ADJUSTMENT MECHANISM WITHIN THE WORLD'S TRADE ORGANIZATION FRAMEWORK. SSRN *Electronic Journal*. DOI: 10.2139/ssrn.5034538

Samuels, A. B., & Singh, U. (2025). Education reimagined: South Africa's journey through the 4IR and beyond. *Transformation in Higher Education*, *10*, 482. DOI: 10.3389/frai.2024.1477044

Sánchez, E., Calderón, R., & Herrera, F. (2025). Artificial intelligence adoption in SMEs: Survey based on TOE–DOI framework, primary methodology and challenges. *Applied Sciences (Basel, Switzerland)*, *15*(12), 6465. DOI: 10.3390/app15126465

Satpathy, I., Chakravarty, D., Jain, V., & Patnaik, B. C. M. (2025). Towards Sustainable Intelligence: Enhancing Cloud Infrastructure Through Green Computing and AI-Driven Engineering. In *Advancements in Cloud-Based Intelligent Informative Engineering* (pp. 203–228). IGI Global Scientific Publishing. https://doi.org/DOI: 10.4018/979-8-3373-0781-7.ch010

Schaugg, L., Jones, N., & Qi, J. (2025, July 28). *Historic International Court of Justice Opinion Confirms States' Climate Obligations*. https://www.iisd.org/articles/deep-dive/icj-advisory-opinion-climate-change

Schuman, S., & Lin, A. (2012). China's Renewable Energy Law and its impact on renewable power in China: Progress, challenges and recommendations for improving implementation. *Energy Policy*, *51*, 89. DOI: 10.1016/j.enpol.2012.06.066

SCO Secretary-General Vladimir Norov's report at the 2020 Global Energy Interconnection Conference (Asia Energy and Power Conference). (2020, September 2). https://eng.sectsco.org/20201102/688470.html

Seng, J. K. P., Ang, K. L. M., Peter, E., & Mmonyi, A. (2022). Artificial intelligence (AI) and machine learning for multimedia and edge information processing. *Electronics (Basel)*, *11*(14), 2239. DOI: 10.3390/electronics11142239

Sengupta, U., & Sengupta, U. (2022). SDG-11 and Smart Cities: Contradictions and Overlaps between Social and Environmental Justice Research Agendas. *Frontiers in Sociology*, *7*, 995603. https://www.frontiersin.org/journals/sociology/articles/10.3389/fsoc.2022.995603/full

Sen, M., Garai, S., Dutta, S., Rakshit, D., Roy, A., Das, R., & Das, S. (2025). Harnessing artificial intelligence for circular economy in agriculture: Reducing waste and maximising efficiency. In Garwi, J., Dzingirai, M., & Masengu, R. (Eds.), *Integrating Agriculture, Green Marketing Strategies, and Artificial Intelligence* (pp. 317–362). IGI Global Scientific Publishing., DOI: 10.4018/979-8-3693-6468-0.ch011

Senthilkumar, R., Venkatakrishnan, P., & Balaji, N. (2020). Intelligent based novel embedded system based IoT enabled air pollution monitoring system. *Microprocessors and Microsystems*, *77*, 103172. DOI: 10.1016/j.micpro.2020.103172

Shafik, W. (Ed.). (2025). *Factoring technology in global sustainability: A focus on the sustainable development goals*. Springer., DOI: 10.1007/978-981-96-7299-8

Shahrulnizam, N. A. A., Hassan, M. A., Mohamad Azizie, N. A., Rashid, A. H. A., & Wahab, S. R. A. (2024). Roles of Ghrm in Sustainable Organization Practices: A Systematic Literature Review. *International Journal of Academic Research in Business & Social Sciences*, *14*(10).

Shahverdi, N., Saffari, A., & Amiri, B. (2025). A systematic review of artificial intelligence and machine learning in energy sustainability: Research topics and trends. *Energy Reports*, *13*, 5551–5578. DOI: 10.1016/j.egyr.2025.05.021

Shahzad, M. A., Du, J., Junaid, M., & Shahzad, F. (2025). From green HRM practices to green innovation performance: A mediation-moderation model. *Journal of Environmental Planning and Management*, *68*(5), 1187–1212.

Shaikh, D. M. R., Ali, M. I., Tunio, M. M. S. M., & Naeem, M. M. (2025). Green Human Resource Management and Sustainability Performance: A Systematic Review Using PRISMA Methodology. *Journal of Management & Social Science*, *2*(3), 330–344. DOI: 10.63075/80rydk30

Shamshuddin, S., Baburao, C., Shaik, H., & Prasad, Y. V. V. V. S. S. S. V. (2025). Sustainable Employee Engagement and Retention: A Review of Green HRM and Digital Integration for Modern Workplaces. In Pujari, P., Khan, S. A., Kumar, A., & Naim, A. (Eds.), (pp. 357–380). Advances in Logistics, Operations, and Management Science. IGI Global., DOI: 10.4018/979-8-3693-8492-3.ch017

Shamsudin, M. F., Hassim, A. A., & Abd Manaf, S. (2024). Mastering probability and non-probability methods for accurate research insights. *Journal of Postgraduate Current Business Research*, *9*(1), 38–53. https://www.abrn.asia/ojs/index.php/jpcbr/article/view/157

Shang, K. (2025). Balancing Corporate Social Responsibility, Environmental Policies, and Stakeholder Expectations for Sustainable Development: A Stakeholder-Centric Analysis. *Sustainable Development*. Advance online publication. DOI: 10.1002/sd.70204

Sharma, K. R. (2022). One Sun, One World, One Grid. *IPEM Law Journal*, *6*(6), 97. DOI: 10.61691/ipem_law.6.2022.97-102

Sharma, S., & Prakash, S., & Devchand. (2023). Employee Well-Being And Sustainable Hr Practices: A Pathway To Long-Term Organizational Success. *Shodh-Kosh: Journal of Visual and Performing Arts*, *4*(2), 1602–1615. DOI: 10.29121/shodhkosh.v4.i2.2023.2706

Sheldon, T. L., & Dua, R. (2020). Effectiveness of China's plug-in electric vehicle subsidy. *Energy Economics*, *88*, 104773. DOI: 10.1016/j.eneco.2020.104773

Shettigar, Dr. R., & Chakraborty, Prof. A. (2025). Sustainable Business Strategies: An HR-Centred Approach To Resilience And Sustainable Growth. *International Journal of Environmental Sciences*, •••, 551–561. DOI: 10.64252/zkhtn743

Shirzadi, A., Salvati, A., Tahan, M. H., Shahabi, H., Nodoushan, E. J., Ramezani, M., Hashim, M., & Clague, J. J. (2025). Urban flood susceptibility mapping using deep and machine learning algorithms as a management tool: A case study of Sanandaj City, Iran. *Ecological Indicators*, *178*, 113886. DOI: 10.1016/j.ecolind.2025.113886

Shi, S., & Yin, J. (2021). Global Research on Carbon Footprint: A Scientometric Review. *Environmental Impact Assessment Review*, *89*, 106571. DOI: 10.1016/j.eiar.2021.106571

Shrivastava, P., & Guimarães-Costa, N. (2017). Achieving environmental sustainability: The case for multi-layered collaboration across disciplines and players. *Technological Forecasting and Social Change*, *116*, 340–346.

Shue, H. (2014). *Climate Justice: Vulnerability and Protection*. Oxford University Press.

Singh, A. (2025). AI-driven innovations for enabling a circular economy: Optimizing resource efficiency and sustainability. In Ozen, E., Singh, A., Taneja, S., Rajaram, R., & Davim, J. P. (Eds.) *Innovating Sustainability Through Digital Circular Economy* (pp. 47-64). IGI Global Scientific Publishing. https://doi.org/DOI: 10.4018/979-8-3373-0578-3.ch003

Singh, A. R., Kumar, R. S., Bajaj, M., Khadse, C. B., & Zaitsev, I. (2024). Machine learning-based energy management and power forecasting in grid-connected microgrids with multiple distributed energy sources. *Scientific Reports*, *14*(1). Advance online publication. DOI: 10.1038/s41598-024-70336-3 PMID: 39160194

Siripipatthanakul, S. (2025). Privacy and Security Management Practices for Organizations. Open source preview, 2025, pp. 1–306

Siripipatthanakul, S., Chaiprakarn, S., Rungsiyanont, S., & Phuangsuwan, P. (2025, a). A Case Study of Sustainable Development in Hospitals in Thailand Using ChatGPT. In *Sustainable Development for Hospitals and Clinics* (pp. 1-32). IGI Global Scientific Publishing. DOI: DOI: 10.4018/979-8-3693-9800-5.ch001

Siripipatthanakul, S., Phuangsuwan, P., & Jaipong, P. (2025, e). Mobile Application

Siripipatthanakul, S., Phuangsuwan, P., & Jaipong, P., & Muthmainnha. (2025, f, September). Wi-Fi Networks and Security: A Systematic Review. In *International Conference on Artificial Intelligence and Networking* (pp. 81-92). Singapore: Springer Nature Singapore. https://doi.org/DOI: 10.1007/978-981-96-2015-9_6

Siripipatthanakul, S., Phuangsuwan, P., & Limna, P. Muthmainnha, Siripipattanakul, S., Vui, C. N., & Jaipong, P. (2025). Artificial intelligence (AI) influencing sustainable governance: Governance adopting AI. In Swaroop, A., Virdee, B., Correia, S. D., & Polkowski, Z. (Eds.) *Proceedings of Data Analytics and Management. ICDAM 2024. Lecture Notes in Networks and Systems*. Springer. https://doi.org/DOI: 10.1007/978-981-96-3372-2_38

Siripipatthanakul, S., Phuangsuwan, P., Siripipattanakul, S., & Jaipong, P. (2025, a). Sustainable Tourism Based on Challenges, Tourist Experiences, and Host Communities: A Case Study of Ayutthaya, Thailand. In *Best Practices for Human Resources Management in Tourism and Hospitality* (pp. 361-394). IGI Global Scientific Publishing.

Siripipatthanakul, S., Phuangsuwan, P., Siripipattanakul, S., & Jaipong, P. (2025, b). Sustainable Tourism Based on Challenges, Tourist Experiences, and Host Communities: A Case Study of Ayutthaya, Thailand. In Best Practices for Human Resources Management in Tourism and Hospitality (pp. 361-394). IGI Global Scientific Publishing.

Siripipatthanakul, S., Phuangsuwan, P., Siripipattanakul, S., & Jaipong, P. (2025, c). Sustainable Tourism Based on Challenges, Tourist Experiences, and Host Communities: A Case Study of Ayutthaya, Thailand. In *Best Practices for Human Resources Management in Tourism and Hospitality* (pp. 361-394). IGI Global Scientific Publishing. DOI: DOI: 10.4018/979-8-3693-8779-5.ch014

Siripipatthanakul, S., Phuangsuwan, P., Siripipattanakul, S., Limna, P., Jaipong, P., & Chaiprakarn, S. (2025, g, September). Sustainable Development Adopting Digital Intelligence Among University Students. In *International Conference on Artificial Intelligence and Networking* (pp. 211-220). Singapore: Springer Nature Singapore. https://doi.org/DOI: 10.1007/978-981-96-2015-9_14

Siripipatthanakul, S., Phuangsuwan, P., Siripipattanakul, S., Limna, P., Jaipong, P., & Chaiprakarn, S. (2025, h, September). Privacy and Information Security for Sustainable Community Among TikTok's Users. In *International Conference on Artificial Intelligence and Networking* (pp. 221-233). Singapore: Springer Nature Singapore. https://doi.org/DOI: 10.1007/978-981-96-2015-9_15

Siripipatthanakul, S., Rakpathum, W., Siripipattanakul, S., Kaewnaknaew, C., Phuangsuwan, P., & Jaipong, P. (2025, c). The Future Artificial Intelligence (AI) for the Hotel and Hospitality Industry: The Case of Hotel's Human Resource Management (HRM). In *Best Practices for Human Resources Management in Tourism and Hospitality* (pp. 395-422). IGI Global Scientific Publishing.

Siripipatthanakul, S., Rakpathum, W., Siripipattanakul, S., Kaewnaknaew, C., Phuangsuwan, P., & Jaipong, P. (2025, d). The Future Artificial Intelligence (AI) for the Hotel and Hospitality Industry: The Case of Hotel's Human Resource Management (HRM). In *Best Practices for Human Resources Management in Tourism and Hospitality* (pp. 395-422). IGI Global Scientific Publishing. DOI: DOI: 10.4018/979-8-3693-8779-5.ch015

Siripipatthanakul, S., Rungsiyanont, S., & Siripipattanakul, S. (2025, b). Sustainable Dentistry Using Artificial Intelligence (AI) for Digital Education and Business Among University Dental Hospitals. In *Sustainable Development for Hospitals and Clinics* (pp. 33-64). IGI Global Scientific Publishing. DOI: DOI: 10.4018/979-8-3693-9800-5.ch002

Siripipatthanakul, S., Rungsiyanont, S., & Siripipattanakul, S. (2025a). Sustainable Dentistry Using Artificial Intelligence (AI) for Digital Education and Business Among University Dental Hospitals. In *Sustainable Development for Hospitals and Clinics* (pp. 33-64). IGI Global Scientific Publishing.

Siripipatthanakul, S., Phuangsuwan, P., & Jaipong, P. (2025, e). Mobile Application Security: A Case Study of Bank of Ayudhya PCL.(KMA), Thailand. In *International Conference on Data-Processing and Networking* (pp. 479-493). Springer, Singapore. https://doi.org/DOI: 10.1007/978-981-96-3102-5_36

Siva, S., & Saravanakumar, A. Piyali Biswas, Shreya Verma, & Uma Shankar. (2024). Developing Sustainable HRM Practices for Retention and Satisfaction in Organizations. *Nanotechnology Perceptions*. Advance online publication. DOI: 10.62441/nano-ntp.vi.4017

Sklavos, G., Theodossiou, G., Papanikolaou, Z., Karelakis, C., & Ragazou, K. (2024). Environmental, social, and governance-based artificial intelligence governance: Digitalizing firms' leadership and human resources management. *Sustainability*, *16*(16), 7154. DOI: 10.3390/su16167154

Sobczuk, S., & Borucka, A. (2024). Recent advances for the development of sustainable transport and their importance in case of global crises: A literature review. *Applied Sciences (Basel, Switzerland)*, *14*(22), 10653. DOI: 10.3390/app142210653

Soekotjo, S., Sosidah, S., Kuswanto, H., Setyadi, A., & Pawirosumarto, S. (2024). *A Conceptual Framework for Sustainable Human Resource Management: Integrating Ecological and Inclusive Perspectives*. https://doi.org/DOI: 10.20944/preprints202412.2299.v1

Son, T. H., Weedon, Z., Yigitcanlar, T., Sanchez, T., Corchado, J. M., & Mehmood, R. (2023). Algorithmic urban planning for smart and sustainable development: Systematic review of literature. *Sustainable Cities and Society*, *94*, 104562. DOI: 10.1016/j.scs.2023.104562

SOUTH–SOUTH COOPERATION for Climate Adaptation and Sustainable Development. (2022). [UNCTAD Report]. UNITED NATIONS CONFERENCE ON TRADE AND DEVELOPMENT. https://unctad.org/system/files/official-document/tcsgdsinf2022d1_en.pdf

Sovacool, B. K., & Hess, D. J. (2017). Ordering theories: Typologies and conceptual frameworks for sociotechnical change. *Social Studies of Science*, *47*(5), 703–750. PMID: 28641502

Spandagos, C., Reaños, M. A. T., & Lynch, M. Á. (2023). Energy poverty prediction and effective targeting for just transitions with machine learning. *Energy Economics*, *128*, 107131.

Srinivas, K. (2018). Innovations and business incubation for startups for effective technology transfer. *ORYZA- An International Journal on Rice, 55*, 90. DOI: 10.5958/2249-5266.2018.00011.5

Srivastava, A., & Maity, R. (2023). Assessing the potential of AI–ML in urban climate change adaptation and sustainable development. *Sustainability*, *15*(23), 16461. DOI: 10.3390/su152316461

Srivastava, M., Aftab, J., & Tyll, L. (2025). The influence of artificial intelligence and additive manufacturing on sustainable manufacturing practices and their effect on performance. *Sustainable Futures : An Applied Journal of Technology, Environment and Society*, *10*, 100820. DOI: 10.1016/j.sftr.2025.100820

Stafford-Smith, M., Griggs, D., Gaffney, O., Ullah, I., Reyers, B., Kanie, N., Stange, B., Shrivastava, P., Leach, M., & O'Connell, D. (2017). Integration and implementation for sustainable development goals: Addressing trade-offs and building coherence. *Sustainability Science*, *12*(6), 911–919. DOI: 10.1007/s11625-016-0383-3 PMID: 30147763

Stiglitz, J. (2017, June 2). *Trump's Reneging on Paris Climate Deal Turns the US into a Rogue State*. The Guardian. https://www.theguardian.com/business/2017/jun/02/paris-climate-deal-to-trumps-rogue-america

Stoett, P., Scrich, V. M., Elliff, C. I., Andrade, M. M., Grilli, N. M., & Turra, A. (2024). Global Plastic Pollution, Sustainable Development, and Plastic Justice. *World Development*, *184*, 106756. DOI: 10.1016/j.worlddev.2024.106756

Sugiharto, B., Hiya, N., Simamora, H., Gaol, N. L. D. H. L., & Andriani, M. (2025). Strategic Human Resource Mapping and Sustainable HRM: Balancing Productivity and Employee Well-being. *Golden Ratio of Mapping Idea and Literature Format, 5*(2), 21–52. DOI: 10.52970/grmilf.v5i2.1362

Su, L. (2024). Environmental regulation and corporate green innovation: Evidence from the implementation of the total energy consumption target in China. *Journal of Business Economics*. Advance online publication. DOI: 10.1007/s11573-024-01207-6

Suljic, V. (2025). Strategic leadership in AI-driven digital transformation: Ethical governance, innovation management, and sustainable practices for global enterprises and SMEs. *SBS Journal of Applied Business Research, 13*(1), 17–35. DOI: 10.70301/JOUR/SBS-JABR/2025/13/1/2

Sultana, F. (2022). Critical Climate Justice. *The Geographical Journal, 188*(1), 118–124. https://rgs-ibg.onlinelibrary.wiley.com/doi/epdf/10.1111/geoj.12417

Supriya, S., & Padaki, S. (2016, December). Data security and privacy challenges in adopting solutions for IOT. In *2016 IEEE International Conference on Internet of Things (iThings) and IEEE Green Computing and Communications (GreenCom) and IEEE Cyber, Physical and Social Computing (CPSCom) and IEEE Smart Data (SmartData)* (pp. 410–415). IEEE. https://doi.org/DOI: 10.1109/i

Suriyankietkaew, S., Krittayaruangroj, K., & Iamsawan, N. (2022). Sustainable leadership practices and competencies of SMEs for sustainability and resilience: A community-based social enterprise study. *Sustainability, 14*(10), 5762. DOI: 10.3390/su14105762

Su, Y., Ghaderi, H., & Dia, H. (2024). The role of traffic simulation in shaping effective and sustainable innovative urban delivery interventions. *EURO Journal on Transportation and Logistics, 13*, 100130. DOI: 10.1016/j.ejtl.2024.100130

Tak, T., Agarwal, V., Mishra, S., Nilkant, D., & Ahluwalia, G. (2024). An examination of sustainable human resource management for achieving financial, social, and environmental goals. *Multidisciplinary Reviews, 6*, 2023ss090. https://doi.org/DOI: 10.31893/multirev.2023ss090

Taloba, A. I., & Rayan, A. (2025). Machine learning based on reliable and sustainable electricity supply from renewable energy sources in the agriculture sector. *Journal of Radiation Research and Applied Sciences, 18*(1), 101282. DOI: 10.1016/j.jrras.2024.101282

Tang, Y., Ye, Q., Zhou, F., & Liu, H. (2023, November 28). *A Study on the Differential Effects of Heterogeneous Environmental Regulations on the Green Transformation of Chinese Manufacturing Enterprises under 'Double Carbon' Target*. SciSpace - Paper; *Research Square*. DOI: 10.21203/rs.3.rs-3667338/v1

Tariq, M. U. (2025). Strategic place marketing in cross-border regions: Enhancing territorial cooperation and economic growth. In Sousa, B. (Ed.), *Cross-Border Regions Cooperation and Implications for Organizations* (pp. 107–136). IGI Global Scientific Publishing., DOI: 10.4018/979-8-3373-1912-4.ch004

Tawiah, V. K., Zakari, A., & Khan, I. (2021). The Environment Footprint of China-Africa Engagement: An Analysis of the Effect of China–Africa Partnership on Carbon Emissions. *The Science of the Total Environment, 756*, 143603. DOI: 10.1016/j.scitotenv.2020.143603

Terdpaopong, K., & Kraiwanit, T. (2021). The technology readiness of Thai governmental agency. *The Journal of Asian Finance, Economics and Business, 8*(12), 431–441. https://scholar.kyobobook.co.kr/article/detail/4010028724143

Tettey, D. J., Chrisben, D., Victoria, M. C., Chukwubuikem, E. P., & Abdullahi, A. (2025). Responsible AI deployment in sustainable project execution: Ensuring transparency, carbon efficiency and regulatory alignment. *International Journal of Science and Research Archive, 14*(3), 1686–1705. DOI: 10.30574/ijsra.2025.14.3.0920

The Energy Conservation (Amendment) Bill, 2022. (2022). https://prsindia.org/billtrack/the-energy-conservation-amendment-bill-2022

The Nature Conservancy. (2025). *How to Help Calculate Your Carbon Footprint*. The Nature Conservancy. https://www.nature.org/en-us/get-involved/how-to-help/carbon-footprint-calculator/

Things-GreenCom-CPSCom-SmartData. (2016). 97.Tolstolesova, L., Glukhikh, I., Yumanova, N., & Arzikulov, O. (2021). Digital transformation of public-private partnership tools. *Journal of Risk and Financial Management, 14*(3), 121. DOI: 10.3390/jrfm14030121

Tien, N. H., Anh, D. B. H., Ngoc, N. M., & Do Thi, Y. N. (2019). Sustainable social entrepreneurship in Vietnam. *International Journal of Entrepreneurship, 23*(3), 1–12. https://www.abacademies.org/articles/sustainable-social-entrepreneurship-in-vietnam-8433.html

To, M. H., & Tran, B. M. (2025). Environmental regulatory stringency and corporate reporting quality: Evidence from waste, emissions, water, and energy regulations in Vietnamese firms. *Environment and Ecology*, *16*(3), 67–81. DOI: 10.21511/ee.16(3).2025.05

Tortia, E. C., Sacchetti, S., & López-Arceiz, F. J. (2022). A Human Growth Perspective on Sustainable HRM Practices, Worker Well-Being and Organizational Performance. *Sustainability*, *14*(17), 11064. DOI: 10.3390/su141711064

Trahan, J. (2022). Vetoes and the UN Charter: The obligation to act in accordance with the 'Purposes and Principles' of the United Nations. *Journal on the Use of Force and International Law*, *9*(2), 243. DOI: 10.1080/20531702.2022.2059882

Trang, T. T. H. (2025). The Impact of Green Human Resource Management on Employee Performance in Public Sector Units: A Survey in Vietnam. *Journal of Information Systems Engineering & Management*, *10*(39s), 263–274. DOI: 10.52783/jisem.v10i39s.7149

Trends in the Investment Treaty Regime and Reform Toolbar for the Energy Transition. (2022). [UNCTAD Report]. UNCATD. https://unctad.org/system/files/official-document/diaepcbinf2023d4_en.pdf

Tripathi, S., Bachmann, N., Brunner, M., Rizk, Z., & Jodlbauer, H. (2024). Assessing the current landscape of AI and sustainability literature: Identifying key trends, addressing gaps and challenges. *Journal of Big Data*, *11*, 65. DOI: 10.1186/s40537-024-00912-x

Trist, E. L., & Bamforth, K. W. (1951). Some social and psychological consequences of the longwall method of coal-getting: An examination of the psychological situation and defences of a work group in relation to the social structure and technological content of the work system. *Human Relations*, *4*(1), 3–38. DOI: 10.1177/001872675100400101

Trombetta, M. J. (2008). Environmental Security and Climate Change: Analyzing the Discourse. *Cambridge Review of International Affairs*, *21*(4), 585–602. DOI: 10.1080/09557570802452920

Tsai, F-M., Mohamed, A.F., Bui, T.-D., Sethanan, K., & Tseng, M-L. (2025). Sustainable Development Goal 12 – Responsible Consumption and Production Challenges and Opportunities in World Regions: A Data-Driven Analysis on Circular Economy Practices. *Journal of Industrial and Production Engineering*, 1-47. DOI: 10.1080/21681015.2025.2532419

Tu, Y., Xiao, Q., Liu, Z., & Yin, S. (2024). Driving Green Development in Manufacturing Enterprises: Green Human Resource Management. *Scholars Journal of Economics, Business and Management, 11*(12), 404–412. DOI: 10.36347/sjebm.2024.v11i12.004

U.S. Department of Energy. Benes, K. J., Porterfield, J. E., Yang, C., Lawrence Livermore National Laboratory, Brownell, E., Helms, J., McGrath, J., Moughan, R., Ponce, C., Trinh, W., Wert, J., National Energy Technology Laboratory, Rose, K., National Renewable Energy Laboratory, Choi, S. L., Jain, R., Emami, P., Pacific Northwest National Laboratory, . . . Freyermuth, V. (2024). AI for Energy: Opportunities for a Modern Grid and Clean Energy Economy. https://www.energy.gov/sites/default/files/2024-04/AI%20EO%20Report%20Section%205.2g%28i%29_043024.pdf

Uakarn, C., Chaokromthong, K., & Sintao, N. (2021). Sample size estimation using Yamane and Cochran and Krejcie and Morgan and green formulas and Cohen statistical power analysis by G* Power and comparisons. *APHEIT International Journal of Interdisciplinary Social Sciences and Technology, 10*(2), 76-88. https://so04.tci-thaijo.org/index.php/ATI/article/view/254253

Ucar, A., Karakose, M., & Kırımça, N. (2024). Artificial intelligence for Predictive maintenance Applications: Key components, trustworthiness, and future trends. *Applied Sciences (Basel, Switzerland), 14*(2), 898. DOI: 10.3390/app14020898

Ukoba, K., Olatunji, K. O., Adeoye, E., Jen, T. C., & Madyira, D. M. (2024). Optimizing renewable energy systems through artificial intelligence: Review and future prospects. *Energy & Environment, 35*(7), 3833–3879. DOI: 10.1177/0958305X241256293

Ullah, A., Pandey, J., & Masengu, R. (Eds.). (2025). *Impacts of technology on operations management: Adoption, adaptation, and optimization*. IGI Global Scientific Publishing., DOI: 10.4018/979-8-3693-6205-1

Ullah, S., Awan, M. D., & Sikander Hayat Khiyal, M. (2018). Big data in cloud computing: A resource management perspective. *Scientific Programming, 2018*(1), 5418679. DOI: 10.1155/2018/5418679

UN. (2015). *The Millennium Development Goals Report 2015*. https://www.un.org/millenniumgoals/2015_MDG_Report/pdf/MDG%202015%20rev%20(July%201).pdf

UNCHS. (United Nations Commission on Human Security). (2003). *Human Security Now*. UNESCO. https://unesdoc.unesco.org/ark:/48223/pf0000130589

UNDP. (1993, January 1). *Human Development Report: People's Participation.* https://hdr.undp.org/content/human-development-report-1993

UNDP. (1994, January 1). *Human Development Report: New Dimensions of Human Security.* https://hdr.undp.org/content/human-development-report-1994

UNDP. (2021). More than just cryptocurrencies - using blockchain for climate action in agriculture | UNDP Climate Change Adaptation. (2021, April 21). https://www.adaptation-undp.org/more-just-cryptocurrencies-using-blockchain-climate-action-agriculture

UNDP. (2022, February 8). *Special Report on Human Security.* https://hdr.undp.org/content/2022-special-report-human-security

UNFCCC. (2015). *Adoption of the Paris Agreement—FCCC/CP/2015/L.9/rev.1.* https://unfccc.int/resource/docs/2015/cop21/eng/l09r01.pdf

UNGA. (2015). *Transforming Our World: The 2030 Agenda for Sustainable Development.* https://docs.un.org/en/A/RES/70/1

UN-Habitat. (2024). *World Cities Report 2024: Climate and Climate* https://unhabitat.org/sites/default/files/2024/11/wcr2024_-_full_report.pdf

United Nations Environment Programme (UNEP). (2023). Global review of the Basel Convention and e-waste enforcement. UNEP.

United Nations Environment Programme (UNEP). (2023). *Illegal e-waste trade and environmental governance.* UNEP Report Series.

United Nations Framework Convention on Climate Change (UNFCCC). (2021). *Digital innovation in climate negotiations.* UNFCCC. https://sdgs.un.org/un-system-sdg-implementation/united-nations-framework-convention-climate-change-unfcc-44210#:~:text=The%20most%20recent%20COP%2026,the%20necessary%20finance%20for%20both

United Nations Industrial Development Organization (UNIDO). (2022). *Green Industry Platform: Advancing sustainable industrialization.* UNIDO Reports.

United Nations Institute for Training and Research (UNITAR). (2024). *The Global E-Waste Monitor 2024: Facts and Figures.* https://ewastemonitor.info

Unpacking clean energy: Human rights impacts of Chinese overseas investment in transition minerals. (2023). [Unpacking clean energy: Human rights impacts of Chinese overseas investment in transition minerals]. Business & Human Rights Resource Centre. https://media.business-humanrights.org/media/documents/2023_China_TM_briefing.pdf

UNSG. (2010). *Human Security: Report of the Secretary-General*. UNGA A/64/701. https://docs.un.org/en/A/64/701

Usha, M., & Ramkumar, N. (2025). Happy Workers, Healthy Business: The Impact of Sustainable Human Resource Management and Workplace Happiness on Employee Engagement of Women Employees in Tea Plantations. *Sustainability*, *17*(3), 1047. DOI: 10.3390/su17031047

Valadas, C. (2024). Eurocities as "Collaboration Laboratories": Innovation and Policy Development within the European Union. *População e Sociedade*, (42), 90–105. DOI: 10.52224/21845263/rev42a7

van Zanten, J. A., & van Tulder, R. (2021). Towards nexus-based governance: Defining interactions between economic activities and Sustainable Development Goals (SDGs). *International Journal of Sustainable Development and World Ecology*, *28*(3), 210–226.

Vanhala, S., & Tuomi, K. (2006). HRM, Company Performance and Employee Well-being. *Management Review*, *17*(3), 241–255. DOI: 10.5771/0935-9915-2006-3-241

Veerasamy, U., Joseph, M. S., & Parayitam, S. (2024). Green human resource management practices and employee green behavior. *Journal of Environmental Planning and Management*, *67*(12), 2810–2836.

Vidyarthi, A. (2020, August 2). *Revisiting India's Position to Not Join the ICSID Convention*. https://legalblogs.wolterskluwer.com/arbitration-blog/revisiting-indias-position-to-not-join-the-icsid-convention

Viitasaari, M., Järlström, M., & Saru, E. (2022). HRM Practices in Sustainability Reports of the World's Most Sustainable Corporations. *Proceedings - Academy of Management*, *2022*(1), 15957. DOI: 10.5465/AMBPP.2022.15957abstract

Volkova, A., Hatamian, M., Anapyanova, A., & de Meer, H. (2024). Being Accountable is Smart: Navigating the Technical and Regulatory Landscape of AI-based Services for Power Grid. *arXiv*. DOI: 10.48550/ARXIV.2408.01121

Vovk, I., & Vovk, Y. (2024). Sustainable personnel management in the hospitality industry: Enhancing organizational performance through employee engagement and commitment. *Economics. Management and Sustainability*, *9*(2), 44–58. DOI: 10.14254/jems.2024.9-2.4

Wada, Y., Bierkens, M. F. P., de Roo, A., Dirmeyers, P. A., Famiglietti, J. S., Hanasaki, N., Konar, M., Liu, J., Schmied, H. M., Oki, T., Pokhrel, Y., Sivapalan, M., Troy, T. J., van Dijk, A. I. J. M., van Emmerik, T., van Huijgevoort, M. H. J., van Lanen, H. A. J., Vörösmarty, C. J., Wanders, N., & Wheater, H. (2017). Human-Water Interface in Hydrological Modelling: Current Status and Future Directions. *Hydrology and Earth System Sciences*, *21*(8), 4169–4193. https://hess.copernicus.org/articles/21/4169/2017/

Wang, M. (2025, July 28). *"Build first, break later": How China's green transformation was built to last.* https://www.weforum.org/stories/2025/07/chinas-green-transformation

Wang, F., Xie, Y., Wang, L., Shuangshuang, L., & Jin, X. (2025). Mitigating Ice Sheets and Mountain Glaciers Melt with Geoengineering. *The Science of the Total Environment*, *963*, 178450. DOI: 10.1016/j.scitotenv.2025.178450

Wang, N., & Jin, X. (2007). PPP project evaluation: A framework for sustainability assessment. *International Journal of Project Management*, *25*(2), 212–219.

Wang, Q., Li, Y., & Li, R. (2025). Integrating artificial intelligence in energy transition: A comprehensive review. *Energy Strategy Reviews*, *57*, 101600. DOI: 10.1016/j.esr.2024.101600

Wang, S., Liu, C., & Zhou, Z. (2024). Government-enterprise green collaborative governance and urban carbon emission reduction: Empirical evidence from green PPP programs. *Environmental Research*, *257*, 119335. DOI: 10.1016/j.envres.2024.119335 PMID: 38849001

Wang, W., & Wang, X. (2023). Does provincial green governance promote enterprise green investment? Based on the perspective of government vertical management. *Journal of Cleaner Production*, *396*, 136519. DOI: 10.1016/j.jclepro.2023.136519

Wang, X., Dong, X., Zhang, Z., & Wang, Y. (2024). Transportation carbon reduction technologies: A review of fundamentals, application, and performance. [English Edition]. *Journal of Traffic and Transportation Engineering*. Advance online publication. DOI: 10.1016/j.jtte.2024.11.001

Wang, Y., Ummar, R., Qureshi, T. M., Ul Haq, J., & Bonn, M. A. (2025). Employee Sustainability: How Green Practices Drive Employee Well-Being and Citizenship Behavior. *Sustainability*, *17*(3), 936. DOI: 10.3390/su17030936

Watal, J., & Caminero, L. (2019). Least-Developed Countries. *Transfer of Technology and the TRIPS Agreement*, *199*. Advance online publication. DOI: 10.1007/978-981-13-2856-5_10

WCED. (1987). *Our Common Future*. https://sustainabledevelopment.un.org/content/documents/5987our-common-future.pdf

WEF (World Economic Forum). (2024). AI and energy: Will AI help reduce emissions or increase power demand? Here's what to know. Energy Transition. World Economic Forum. https://www.weforum.org/stories/2024/07/generative-ai-energy-emissions/

Wei, J., & Shen, Y. (2025). Impact and mechanism of digital transformation on performance in manufacturing firms. *Innovation and Green Development*, *4*(1), 100205. DOI: 10.1016/j.igd.2025.100205

Wei, Y., Li, Y., Chevallier, J., & Wojewodzki, M. (2024). The efficiency of China's carbon trading schemes: A tale of seven pilot markets. *Commodities*, *3*(3), 355–375. DOI: 10.3390/commodities3030020

WHO. (2025, February 18). *Biodiversity Loss.* https://www.who.int/news-room/fact-sheets/detail/biodiversity#

Widmer, R., Oswald-Krapf, H., Sinha-Khetriwal, D., Schnellmann, M., & Böni, H. (2005). Global perspectives on e-waste. *Environmental Impact Assessment Review*, *25*(5), 436–458.

Williams, P. A., Lovelock, B., Cabarrus, T., & Harvey, M. (2019). Improving digital hospital transformation: Development of an outcomes-based infrastructure maturity assessment framework. *JMIR Medical Informatics*, *7*(1), e12465. DOI: 10.2196/12465

Winarno, A., & Silvianita, A. (2024). The role of socially responsible human resource management in improving performance: An analysis of subjective well-being as a mediating variable. *Qubahan Academic Journal*, *4*(3), 454–468. DOI: 10.48161/qaj.v4n3a878

Winkelmann, S., Guennoun, R., Moller, F., Schoormann, T., & van der Valk, H. (2024). Back to a resilient future: Digital technologies for a sustainable supply chain. *Information Systems and e-Business Management*, *22*(2), 315–350. DOI: 10.1007/s10257-024-00677-z

Wolfert, S., Ge, L., Verdouw, C., & Bogaardt, M. J. (2017). Big data in smart farming–a review. *Agricultural Systems*, *153*, 69–80. DOI: 10.1016/j.agsy.2017.01.023

Wongadisai, W., Chanchalor, S., & Murphy, E. (2020). Rural community enterprises in Thailand: A case study of participation. *World Review of Entrepreneurship, Management and Sustainable Development*, *16*(3), 276–298. DOI: 10.1504/WREMSD.2020.107255

World Bank. (2022). *Artificial intelligence for climate-smart investments.* World Bank. https://documents1.worldbank.org/curated/en/099455012062224218/pdf/P1725690cd27300c30b5f101fd728276fe7.pdf

World Bank. (2025, October 10). *Circular Economy and Pollution Management.* https://www.worldbank.org/en/topic/pollution

World Economic Forum. (2023). *Global Risks Report 2023.* WEF Publications. https://www.weforum.org

WTO's Trade and Environment Week signals new climate-trade policy momentum. (2025, July 18). https://www.c2es.org/2025/07/wtos-trade-and-environment-week-signals-new-climate-trade-policy-momentum/

Wu, Y. (2024). The Sustainable Human Resource Management (HRM): A Systematic Literature Review. In K. Szegedi (Ed.), *Sustainability Can't Wait!: VII. BBU International Sustainability Student Conference The Book of Full Paper* (pp. 468–499). Budapesti Gazdasági Egyetem. https://doi.org/DOI: 10.29180/978-615-6886-04-0_20

Wu, S., Cheng, P., & Yang, F. (2024). Study on the impact of digital transformation on green competitive advantage: The role of green innovation and government regulation. *PLoS One*, *19*(8), e0306603. DOI: 10.1371/journal.pone.0306603

Xiao, Y., & Xiao, L. (2025). The impact of artificial intelligence-driven ESG performance on sustainable development of central state-owned enterprises listed companies. *Scientific Reports*, *15*, 8548. DOI: 10.1038/s41598-025-93694-y

Xu, H., Xu, J., Wang, J., & Hou, X. (2023). Reduce production or increase efficiency? Hazardous air pollutants regulation, energy use, and the synergistic effect on industrial enterprises' carbon emission. *Energy Economics*. Advance online publication. DOI: 10.1016/j.eneco.2023.107027

Xu, Y., Loftness, V., & Severnini, E. (2021). Using machine learning to predict retrofit effects for a commercial building portfolio. *Energies*, *14*(14), 4334. DOI: 10.3390/en14144334

Yadav, A., Jindal, L., Verma, R., & Vishwakarma, K. (2024). THE ROLE OF GREEN HUMAN RESOURCE MANAGEMENT – AN ASSESSMENT. *ShodhKosh: Journal of Visual and Performing Arts*, *5*(4). Advance online publication. DOI: 10.29121/shodhkosh.v5.i4.2024.1868

Yigitcanlar, T., Mehmood, R., & Corchado, J. M. (2021). Green artificial intelligence: Towards an efficient, sustainable and equitable technology for smart cities and futures. *Sustainability*, *13*(16), 8952. DOI: 10.3390/su13168952

Yin, Q. (2023). The Impact of Green Human Resource Management on Organizational Performance. *Frontiers in Business, Economics and Management, 11*(3), 112–115. DOI: 10.54097/fbem.v11i3.13198

Yong, J. Y., Yusliza, M.-Y., Ramayah, T., Chiappetta Jabbour, C. J., Sehnem, S., & Mani, V. (2020). Pathways towards sustainability in manufacturing organizations: Empirical evidence on the role of green human resource management. *Business Strategy and the Environment, 29*(1), 212–228. DOI: 10.1002/bse.2359

You, X. (2025, January 8). *China has a new sweeping energy law. Here's what you need to know.* https://www.ciphernews.com/articles/china-has-a-new-sweeping-energy-law-heres-what-you-need-to-know/

Younas, A., & Kassim, A. A. (Eds.). (2025). *Transforming business through digital sustainability models*. IGI Global Scientific Publishing., DOI: 10.4018/979-8-3373-0608-7

Younis, N. M. M. (2023). Sustainability reports and their impact on firm value: Evidence from Saudi Arabia. *International Journal of Management and Sustainability, 12*(2), 70–83. DOI: 10.18488/11.v12i2.3275

Yu, P., Wong, S., & Prabhakar, A. (Eds.). (2025). *AI strategies for social entrepreneurship and sustainable economic development*. IGI Global Scientific Publishing., DOI: 10.4018/979-8-3693-6392-8

Yusif, S., & Hafeez-Baig, A. (2024). Impact of stakeholder engagement strategies on managerial cognitive decision-making: The context of CSP and CSR. *Social Responsibility Journal, 20*(6), 1101–1121. DOI: 10.1108/SRJ-05-2023-0295

Yu, W., Li, Z., & Hu, C. (2024). Carbon reduction and corporate sustainability: Evidence from low-carbon city pilot policy. *Heliyon*. Advance online publication. DOI: 10.1016/j.heliyon.2024.e28992

Yu, X., Fan, L., & Yu, Y. (2025). Artificial intelligence and corporate ESG performance: A mechanism analysis based on corporate efficiency and external environment. *Sustainability, 17*(9), 3819. DOI: 10.3390/su17093819

Yuxiaosi, W. (2024). *The Sustainable Human Resource Management (HRM): A Systematic Literature Review*. Budapesti Gazdasági Egyetem., https://publikaciotar.uni-bge.hu/id/eprint/2408/

Zahran, S. (2024). Investigating the Nexus between Green Supply Chain Practices and Sustainable Waste Management in Advancing Circular Economy. *Sustainability*. Advance online publication. DOI: 10.3390/su16093566

Zaid, A. (2022, July 8). *The Impact of Green Human Resource Management Practices with Sustainable and Operational Performance: A Conceptual Model.* SciSpace - Paper; Springer eBooks. https://doi.org/DOI: 10.1007/978-3-031-08090-6_37

Zaim, R. L., Lukman, S., & Syafrizal. (2023). Pengaruh Environmental Regulation terhadap Environmental Performance Melalui Mediasi Corporate Social Responsibility. *Jurnal Informatika Ekonomi Bisnis*, 331–338. DOI: 10.37034/infeb.v5i2.257

Zainuddin, S. A., Abdullah, T., Ibrahim, M. A. H., Nawi, N. C., Nasir, N. A. M., Latef, A. S. A., & Musa, Z. C. (2024). Analyzing Global Research Trends in Technology and E-Waste Management: A Bibliometric Perspective. *Technology-Driven Business Innovation: Unleashing the Digital Advantage*, *2*, 443–453. DOI: 10.1007/978-3-031-62656-2_39

Zaman, A. U., & Lehmann, S. (2011). Urban growth and waste management optimization towards 'zero waste city'. *City, Culture and Society*, *2*(4), 177–187.

Zhai, T. (2021). Environmental Challenges, Opportunities, and Policy Implications to Materialize China's Green Belt and Road Initiative. *Sustainability*, *13*(18), 10428. DOI: 10.3390/su131810428

Zhang, W., Fan, H., & Zhao, Q. (2023). Seeing green: how does digital infrastructure affect carbon-emission intensity? *Energy Economics, 127*, 107085. https://doi.org/rights and contentDOI: 10.1016/j.eneco.2023.107085Get

Zhang, G., MacCarthy, B. L., & Ivanov, D. (2022). The cloud, platforms, and digital twins—Enablers of the digital supply chain. In *The digital supply chain* (pp. 77–91). Elsevier., DOI: 10.1016/B978-0-323-91614-1.00005-8

Zhang, L., Innab, N., Shuhidan, S. M., Pan, Y., Zhang, Y., Som, H. M., & Alasbali, N. (2025). Artificial intelligence-driven internet of things-based green supply chain for carbon reduction in sustainable manufacturing. *Journal of Environmental Management, 389*, 126170. DOI: 10.1016/j.jenvman.2025.126170

Zhang, X., & Peng, L. (2023). Exploring the relationship between green human resource management (GHRM) and organizational sustainable development performance. In Qu, X. (Ed.), *Second International Conference on Sustainable Technology and Management (ICSTM 2023)* (p. 80). SPIE. https://doi.org/DOI: 10.1117/12.3004951

Zhao, L., Sun, Y., & Xiao, G. (2025). A novel machine learning-based method for identifying Industrial Symbiosis opportunities. *Journal of Cleaner Production, 513*, 145607. DOI: 10.1016/j.jclepro.2025.145607

Zhu, F., Hunjra, A. I., Roubaud, D., Zhao, S., & Grebinevych, O. (n.d.). *Assessment of Synergistic Governance of Pollution and Carbon Reduction*. SciSpace - Paper., DOI: 10.1016/j.jenvman.2025.124226

Zhuo, Z., Ye, J., Wang, Y., Chen, H., & Liang, B. (2025). Smart cities, smarter land use? Unveiling the efficiency gains from China's digital urban transformation. *Ecological Indicators*, *171*, 113151. DOI: 10.1016/j.ecolind.2025.113151

Zia, F. (2024). The Interaction between Corporate Operations and Environmental Management. *Global Regional Review*, *IX*(I), 206–217. DOI: 10.31703/grr.2024(ix-i).18

Zihan, W., & Makhbul, Z. K. M. (2024). Green Human Resource Management as a Catalyst for Sustainable Performance: Unveiling the Role of Green Innovations. *Sustainability*, *16*(4), 1453. DOI: 10.3390/su16041453

Zissler, R. (2023). *International Power Interconnections Progress and Help Countries in the Energy Crisis*. https://www.renewable-ei.org/en/activities/column/REupdate/20230119.php

Zong, Z., & Guan, Y. (2025). AI-driven intelligent data analytics and predictive analysis in Industry 4.0: Transforming knowledge, innovation, and efficiency. *Journal of the Knowledge Economy*, *16*, 864–903. DOI: 10.1007/s13132-024-02001-z

Zurlini, G., & Müller, F. (2008). Environmental Security. In Jørgensen, S. E., & Fath, B. D. (Eds.), *Encyclopedia of Ecology* (pp. 1350–1356). Academic Press., DOI: 10.1016/B978-008045405-4.00707-2

About the Contributors

Patichat Jaipong * DBA at GlobalNxt University, Malaysia * Master of Science in Management at Assumption University, Thailand * BBA in Economics at Assumption University, Thailand * Editorial Member of Advance Knowledge for Executives (ISSN 2822-0323, Online) * Assistant of Guest Editor (Special Issue: Cybersecurity & Metaverse), International Journal of Computing Sciences Research (The Philippines) * A researcher in collaboration with professors in Malaysia, Indonesia, and Thailand. *An invited reviewer for the IGI Global Scientific Publishing, USA (Scopus Indexing Books; 30 Reviews) Web of Science Researcher ID: KZT-9804-2024 Research Interests: Marketing, Management, and Strategies

***

Babita is a faculty in the Finance and Accounts area at the CHRIST (Deemed to be University). She is NET/JRF Qualified. She did her Ph.D. from the Department of Management Studies, Guru Jambeshwar University of Science and Technology, Hisar (Haryana). She received her Master degree from MDU Rohtak (Haryana), specializing in Finance and HR. She has a Bachelor of Business and Administration degree from MDU Rohtak (Haryana), specializing in General Management. Currently, she is working an Assistant Professor in Christ (Deemed to be University). Before Christ, she was associated with JC Bose University of Science and Technology, YMCA (Faridabad). She has around 2 years of experience in corporate world. Her research interest includes economic growth, corporate finance, and international trade. She has published in various international publishers like SAGE, INDERSCEINCE, BLOOMSBURRY. She has presented her research work at various National and international conferences. She has also attended more than 20 online and offline FDPs.

Sharmila A is an Assistant Professor with a background in dentistry, hospital administration, and academic research. She earned her Bachelor of Dental Surgery (BDS) from Saveetha University and subsequently completed a Fellowship in Medical Cosmetology (FMC) from ILAMED. She also holds a Postgraduate Diploma in Medico Legal Systems from Symbiosis University and an MBA in Hospital Management from Saveetha School of Management. Currently, she is pursuing her PhD in Hospital Management, further deepening her expertise in healthcare systems and operations. Her career spans over a decade with 4 years of clinical experience, 3 years in hospital administration, 1 year in human resources, and 3 years in academia. Dr. Sharmila brings a well-rounded perspective to healthcare, combining medical knowledge, legal understanding, and managerial skills to drive quality care and organizational efficiency

Gigi G.S., Associate Professor at Saveetha School of Management, SIMATS, Chennai, brings 16 years of academic experience with a strong focus on research in human resource management, particularly employee engagement in the IT sector. She is currently guiding three Ph.D. research scholars and serves as a Research Advisory Committee (RAC) member for ten others across various universities. Her scholarly work includes numerous presentations and publications at national and international conferences and in reputed journals. Her research is driven by a commitment to addressing real-world HR challenges, blending academic rigor with practical insight to advance the field meaningfully

Astri Dwi Andriani is a lecturer and the Rector of Putra Indonesia University, Indonesia. Her scholarly interests lie at the intersection of communication, digital ethics, and socio-technological transformation. With a background in communication studies and media ethics, she examines how technology-driven policies shape digital identity, privacy resilience, and sustainable communication practices. In the context of Technology-Driven Organization Policies: E-Waste Governance Framework, her contribution emphasizes the human dimension of digital transformation—highlighting communicative ethics, resilience, and intercultural awareness as foundations for sustainable technological governance. Her academic and professional pursuits reflect a commitment to integrating ethical reasoning and cultural insight into the policy frameworks that guide digital and environmental responsibility.

Utik Bidayati is the Vice Rector for Finance, Assets, and General Administration at Universitas Ahmad Dahlan (UAD). She completed her Bachelor's degree in Management at Universitas Islam Indonesia (UII), followed by a Master's degree in Management from UII, and earned her Ph.D. in Management from Universitas Brawijaya in 2023. In addition to her role at UAD, Utik has over 10 years of experience as a faculty member in the Faculty of Economics and Business, specializing in Management. She is also actively involved in research and development in the fields of management, finance, and administration, and often participates in various academic activities at the national level. As Vice Rector, Utik is responsible for managing the university's finances and general administration, with a focus on improving processes at UAD. She is also involved in various initiatives aimed at enhancing the quality of services and administrative systems within the university

Nikita Biswal is a dedicated legal scholar who earned her B.A.LL.B. (Hons.) degree from the prestigious University Law College, Utkal University, in 2024. Driven by a passion for Intellectual Property Law and its evolving dynamics, she pursued her LL.M. in Intellectual Property and Trade Law at CHRIST (Deemed to be University), Delhi NCR, completing it in 2025. She is currently pursuing her Ph.D. at CHRIST (Deemed to be University), Bengaluru, Karnataka. Her academic journey reflects a deep commitment to excellence and a strong interest in exploring the intersection of law, innovation, and creativity.

Raja Shree K. C. is a biomedical engineer with a doctorate in Healthcare Management and eight years of academic experience. She is known for her energy, enthusiasm, and multitasking abilities in academics and beyond. Her expertise lies in hospital quality management, supported by a proficiency certification as Quality Coordinator in Hospitals. She has published in Web of Science, Scopus, and UGC-listed journals and actively participates in conferences, workshops, and MOOCs to stay updated. Recipient of the Young Academician Award and Young Teacher Award, she is recognized for her dedication to teaching, research, and student development. Her professional interests also extend to organizing academic and cultural programs, contributing to a holistic learning environment

Dasmadi is the Chairman of STIE Aprin Palembang. He holds a Bachelor's degree in Economics (S.E.), a Master's degree in Management (M.M.), and a Doctorate in Management. Under his leadership, STIE Aprin has expanded its academic programs and established strategic partnerships with domestic and international institutions. Dasmadi is also actively involved in community service programs focusing on waste management and stunting prevention. As a writer, he has contributed significantly to the fields of management, economics, and human resource development. He is committed to fostering innovative education that produces graduates with both academic competence and social awareness.

Jaishri Gothania is an academician, author, researcher, and data scientist with over 9 years of experience. She holds a Ph.D. in Computer Science Engineering, an

M.Tech in Computer Science, a B.Tech in Computer Science and Engineering, and a PGD in Education. She has published 1 book, 6 book chapters, and 12 research articles; presented 6 conference papers; and holds multiple patents. Her work bridges data science and innovation, contributing significantly to academia, research, and technological advancement.

Syamsul Hadi is an esteemed academic and researcher specializing in management and organizational behavior. Currently serving as the Head of the Master of Management Program at the Faculty of Economics, Universitas Sarjanawiyata Tamansiswa (UST) Yogyakarta, he has over two decades of experience in academia. Dr. Syamsul Hadi earned his Master's degree in Management (M.M.) and his Doctorate in Economics (Dr.), with a focus on strategic management and organizational behavior. He is well-regarded for his contributions to the field through various national and international journal publications and participation in academic conferences. Aside from his academic career, Dr. Syamsul Hadi is actively involved in community empowerment initiatives aimed at human resource development and economic empowerment.

Shaikh Sahanwaz Islam is an accomplished legal academic and administrator, presently serving as Associate Professor of Law and Head of the P.G. Department of Law at Capital Law College, Bhubaneswar under Madhusudan Law University. With a distinguished teaching career of 25 years, he has contributed extensively to legal education through classroom teaching, academic leadership, research, and public engagement in Odisha and beyond. His career includes five years as Associate Professor at Uttaranchal University, Dehradun, and eighteen years as Assistant Professor at Capital Law College under Utkal University, reflecting a sustained commitment to mentoring generations of law students and strengthening institutional academic standards. Dr. Islam holds a Ph.D. in Law from Utkal University on the theme "Right to Information as a Tool of Good Governance," a subject that places him at the intersection of transparency, accountability, and democratic governance in India. His diverse academic background includes an LL.M. in Commercial Law from Utkal University and a second LL.M. with specializations in Cyber Law, Competition Law, and Intellectual Property Law from Karnataka University, equipping him with strong expertise across contemporary and technology-driven legal domains. Complementing his legal qualifications, he has obtained an MBA in Human Resource Management from Pondicherry University, a Post-Graduate Degree in Human Rights and Duties Education from Utkal University, and a Post-Graduate Degree in Sociology from Utkal University, Bhubaneswar, which together enrich his interdisciplinary perspective on law, policy, and society. An active author and researcher, Dr. Islam has written four books: "Consumer Protection Act, 1986 with

Orissa Rules" (2016), "Orissa Public Works Department Code" (2016), "Odisha Pension Rules" (2017), and "Scheduled Caste and Scheduled Tribe (Prevention of Atrocities) Act, 1989" (2017), all published by OJR Publishing Company, Cuttack. These works demonstrate his engagement with statutory interpretation, social justice, public administration, and welfare legislation, and serve as practical resources for students, practitioners, and administrators. In addition to his authored books, he has edited nine books and published 14 national and 15 international research articles, contributing to scholarly discourse on constitutional law, human rights, governance, and emerging areas of legal regulation. Dr. Islam has actively participated in academic and professional forums, having attended 64 seminars, conferences, and workshops at national and international levels. His academic visibility further extends through visiting lectures at key institutions, including delivering lectures on the Indian Penal Code at the Mahatma Gandhi Institute of Prison Administration and Correctional Homes, Bhubaneswar, and on the National Commission for Protection of Child Rights at the Navodaya Leadership Institute, Konark, under the Navodaya Vidyalaya Samiti, Ministry of Education, Government of India. He has also spoken on the Right to Information at the Judicial Academy, Cuttack, and on the new criminal law and its implementation at the Biju Patnaik Police Academy, Bhubaneswar, thereby engaging directly with judges, police officers, and public officials on critical legal reforms. Beyond law faculties, Dr. Islam has served as visiting faculty in Labour Law at both the Department of Botany and the Department of PMIR at Utkal University, underlining his interdisciplinary teaching profile and his ability to connect labour and employment law with diverse academic fields. His broad qualifications and long experience as a teacher, researcher, author, and trainer make him a respected figure in legal academia and a resource person for institutions seeking to strengthen legal awareness and governance-related capacities. Through his work in classrooms, research publications, books, and training programmes, Dr. Shaikh Sahanwaz Islam continues to contribute meaningfully to the advancement of legal education and the promotion of rights-based governance in India.

Jipson Joseph has a PhD in Theological Ethics from DVK Bangalore. Currently, he is a Research Scholar at School of Law, Christ University, Bangalore, India. His research interests are in the fields of ethics, gender studies, constitutional law, human rights, human security, environmental justice, SDGs, developmental policies, and global poverty. He has published 6 books, co-edited one book, and contributed scholarly articles to many international and national journals and edited volumes including IGI Global Scientific Publishing's edited books.

Jobin Joseph is a distinguished individual with a strong academic background and a diverse range of skills. With a foundation in philosophy, theology, English literature, and sociology, he has carved a niche for himself in the fields of education, writing, and performance. Jobin holds Master's degrees in English Literature and Sociology. Additionally, he has earned a Bachelor of Education (B.Ed.) degree, which has equipped him with the skills and knowledge to teach and inspire students. Jobin has spent eight years teaching sociology, during which he has developed a passion for sharing knowledge and shaping young minds. Apart from teaching, he is also a prolific writer who has contributed articles to various magazines on a range of secular subjects. His writing reflects his intellectual curiosity and ability to engage with diverse topics. His areas of interest include human rights, English literature, social interactions, gender justice, environmental security, climate justice, Artificial Intelligence, and Drone technology.

Kusuma Chandra Kirana is a senior lecturer and academic leader at Universitas Sarjanawiyata Tamansiswa (UST) Yogyakarta, serving in the Faculty of Economics and currently holding the position of Vice Rector II for Finance and Human Resources. She holds a Bachelor's degree in Business Education, a Master's degree in Management, and a Doctorate in Management, with a focus on human resource management, entrepreneurship, and Islamic economics. Dr. Kusuma Chandra Kirana is known for her extensive research contributions, including publications on organizational behavior, leadership, and socio-economic development. She actively collaborates in both national and international academic networks, and her work reflects a strong commitment to advancing management knowledge and strengthening institutional capacity within higher education.

Pongsakorn Limna, also known as Palm, is currently a postdoctoral researcher in the Digital Business program at the International College, Pathumthani University, Thailand. He is an active scholar with numerous publications in leading academic journals, including those indexed in Scopus and the Thai Citation Index (TCI). He also serves as a reviewer for several national and international journals and academic conferences, demonstrating his strong commitment to academic excellence and scholarly contribution. His research interests span a wide range of interdisciplinary fields, including the digital economy, digital business, business administration, management, communication, technology, artificial intelligence (AI), and consumer behavior. In 2012, Dr. Limna was awarded a prestigious scholarship jointly provided by the Office of the Basic Education Commission of Thailand and the Wisconsin Department of Public Instruction, USA, to participate in the Thailand–Wisconsin International Sunrise Program in Wausau, Wisconsin. He holds a Bachelor of Arts (B.A.) degree from the Faculty of Humanities and Social Sciences at Burapha University, Thailand, with a major in English and a minor in tourism management. He later earned an Executive Master of Business Administration (Executive MBA) from the Faculty of Business and Technology at UNITAR International University in Malaysia. He completed his Doctor of Philosophy (Ph.D.) in Digital Economy at Rangsit University, Thailand, with the support of the esteemed Prasit – Khunying Pattana Ourairat Scholarship.

Ambar Lukitaningsih is a lecturer at Universitas Sarjanawiyata Tamansiswa, Yogyakarta, Indonesia. She holds a Doctoral degree and a Master of Management (M.M.), reflecting her strong academic background in the field of management and related disciplines. As an academic, Dr. Ambar Lukitaningsih is actively involved in teaching, research, and community service, in accordance with the Tri Dharma of Higher Education. Her academic interests include management studies, organizational development, and applied management practices. She has contributed to various academic activities, including research projects, scientific publications, and seminars. Through her dedication to education and scholarship, Dr. Ambar Lukitaningsih continues to support the development of knowledge and the improvement of higher education quality at Universitas Sarjanawiyata Tamansiswa.

Abhinash Mohanta pursued a three-year LLB program at Ganjam Law College, Berhampur, under Berhampur University. He further advanced his academic journey by enrolling in the Master of Laws (LLM) program in Business Law at the P.G. Department of Law, Berhampur University, graduating in 2017. Currently, he is pursuing a Ph.D. in Law at KIIT Deemed to be University, with a research focus on critical areas such as refugee and human rights. In addition to his academic endeavors, Abhinash has authored several Scopus-indexed papers covering diverse topics, including AI, environment, migration, crime, intellectual property, and refugee studies.

Ong Lin Dar is a Senior Lecturer in the Faculty of Business and Accountancy at Universiti Malaya, bringing over 14 years of dedicated teaching and research experience to her role. She holds a Ph.D. in Human Resource Management from Universiti Malaya, building upon her earlier MBA and BBA qualifications. Her professional journey seamlessly bridges academia and industry, having served in lecturing roles at several Malaysian universities and as a Training Executive in the insurance sector. Dr. Ong's expertise centers on Human Resource Management, Organizational Behavior, Leadership, and Career Management, with a specialized interest in Malaysian employment laws and employee performance. A committed researcher and author, she has led projects on leadership and protean career attitudes as a principal investigator, contributed to publications on workplace safety and COVID-19's impact on human capital, and co-authored several practical management books. Dr. Ong actively mentors the next generation of scholars, currently supervising multiple doctorate students and serving as a Ph.D. thesis examiner.

Saizal Bin Pinjaman is a Senior Lecturer in Economics at Universiti Malaysia Sabah and Director of the Centre for Economic Development and Policy. He holds a PhD in Financial Economics from Universiti Malaysia Sabah and specialises in financial economics, governance, and macroeconomic policy. He serves as Visiting Scholar and Research Fellow at various foreign universities. He is regularly invited as a keynote speaker and guest lecturer at universities and international conferences across Indonesia, China, Australia, and the Philippines. His work combines academic research and policy engagement, with publications in Scopus and Web of Science indexed journals, principal researchers of major policy and feasibility studies, and active service on editorial boards and academic review panels.

Ashka Prince is a third-year Bachelor of Business Administration student specializing in Finance and Marketing Analytics at Christ (Deemed to be University), Delhi NCR. Her academic interests include Finance,data analytics, consumer behaviour, and the integration of gamification in educational contexts. As an active member of the university community, Ashka has also contributed to student leadership and cultural initiatives, including the Student Council, Team Lead, University Internal Commitee Member . Her research reflects a keen analytical approach and an enthusiasm for exploring the intersection of business strategy and behavioural insights.

Auliya Rosiana is a lecturer and researcher at the Faculty of Economics, Universitas Sarjanawiyata Tamansiswa (UST) Yogyakarta, specializing in management and human resource management. She holds a Bachelor's degree in English Literature (S.S.) and a Master's degree in Management (M.M.), and is a certified Certified Professional in Human Resource Management (CPHRM). Dr. Auliya Rosiana is actively involved in various academic publications and research collaborations focusing on management, digital marketing, e-learning, and organizational development. Her work has been published in national journals and indexed on academic platforms, reflecting her significant contributions to the advancement of management studies in Indonesia.

Charu Sehgal, with expertise in Human resource and Finance, hold Mphil degree and Masters of Commerce degree from Department of Commerce, Delhi School of Economics. She is silver medallist in Mcom. She is NET-JRF qualified. She has done B.com (hons) from Daulat Ram College, University of Delhi. She was awarded “Central Students Council 1966- 77 Scholarship” for obtaining highest marks in 1 st semester in M.com. She secured IInd position in “Shaheed Bhagat

Singh College, University of Delhi" in First semester. She has authored book chapters and research papers in various journals. She has more than eight years of teaching experience.

Yarnaphat Shaengchart is a lecturer at the International College, Pathumthani University. He is actively engaged in academic research and has a growing portfolio of work at the intersection of digital transformation and management. His primary research interests include information technology, the digital economy, digital business, and business management.

Sitesh Kumar Singh is an accomplished academic and Chartered Engineer with a Ph.D. in Civil Engineering and over nine years of diverse experience in education, research, consultancy, and educational administration. Currently serving as an Assistant Professor in the Department of Civil and Environmental Engineering at the National University of Science and Technology, Muscat, Sultanate of Oman. Dr. Singh holds multiple advanced degrees including M.Tech, B.Tech, PGDIPR, PGDUPDL and PGDE. His career encompasses curriculum development, research supervision, quality assurance, and strategic academic planning. He has supervised over 40 postgraduate and 2 doctoral students, contributing to academic leadership through workshops, conferences, and editorial roles, and has published over 5 books, 12 book chapters, 58 research articles & conference presentations, and 7 patents.

Supaprawat Siripipatthanakul is working as an Adjunct Professor at Manipal GlobalNxt University, Malaysia, and a Faculty Member at the Faculty of Education at Bangkok Thonburi University, Thailand. He is also a Visiting Professor at the Faculty of Economics and Business at Universitas Terbuka, Indonesia. He has published more than 90 papers and books on Google Scholar. He has a CEFR Level 5 (C1), an English Teacher certificate from AAARYA Business College in Singapore, TEFL & TESOL certificates. 9 certificates in research (Scopus & WOS), and 8 certificates of Ethics in Research. He lectures on research methodology (business/ education/ healthcare) for international universities. He has been invited to review the manuscripts of Q1 and Q2-Scopus, WOS journals/Scopus, and Springer books (217 verified reviews by WOS/ 355 verified reviews by ORCID) since 2022. He is an Editor of IGI Global Scopus Books and PLoS One, Scopus & WOS, SJR Q1 USA.

S Srinivasan is an Assistant Professor in the Department of Humanities and Social Sciences at Graphic Era (Deemed to be University), with 14 years of teaching and field expertise. He has served in the Ministry of Health and Family Welfare, as a Guest Faculty, and as a Mobilization Training Specialist with the State Government. Recently, he contributed to a tuberculosis project funded by GFATM and USAID.

He has published 82 papers, with 20 more under peer review in UGC-CARE and SCOPUS journals.

Index

A

C

D

E

G

H

I

M

N

O

P

www.ingramcontent.com/pod-product-compliance
Lightning Source LLC
LaVergne TN
LVHW062327300926
849161LV00007B/1597